TOWARDS AN OPTICAL INTERNET

IFIP - The International Federation for Information Processing

IFIP was founded in 1960 under the auspices of UNESCO, following the First World Computer Congress held in Paris the previous year. An umbrella organization for societies working in information processing, IFIP's aim is two-fold: to support information processing within its member countries and to encourage technology transfer to developing nations. As its mission statement clearly states,

IFIP's mission is to be the leading, truly international, apolitical organization which encourages and assists in the development, exploitation and application of information technology for the benefit of all people.

IFIP is a non-profitmaking organization, run almost solely by 2500 volunteers. It operates through a number of technical committees, which organize events and publications. IFIP's events range from an international congress to local seminars, but the most important are:

- The IFIP World Computer Congress, held every second year;
- open conferences;
- working conferences.

The flagship event is the IFIP World Computer Congress, at which both invited and contributed papers are presented. Contributed papers are rigorously refereed and the rejection rate is high.

As with the Congress, participation in the open conferences is open to all and papers may be invited or submitted. Again, submitted papers are stringently refereed.

The working conferences are structured differently. They are usually run by a working group and attendance is small and by invitation only. Their purpose is to create an atmosphere conducive to innovation and development. Refereeing is less rigorous and papers are subjected to extensive group discussion.

Publications arising from IFIP events vary. The papers presented at the IFIP World Computer Congress and at open conferences are published as conference proceedings, while the results of the working conferences are often published as collections of selected and edited papers.

Any national society whose primary activity is in information may apply to become a full member of IFIP, although full membership is restricted to one society per country. Full members are entitled to vote at the annual General Assembly, National societies preferring a less committed involvement may apply for associate or corresponding membership. Associate members enjoy the same benefits as full members, but without voting rights. Corresponding members are not represented in IFIP bodies. Affiliated membership is open to non-national societies, and individual and honorary membership schemes are also offered.

TOWARDS AN OPTICAL INTERNET

New Visions in Optical Network Design and Modelling

IFIP TC6 Fifth Working Conference on Optical Network Design and Modelling (ONDM 2001) February 5–7, 2001, Vienna, Austria

Edited by

Admela Jukan
Institute of Communication Networks
Vienna University of Technology
Austria

SPRINGER SCIENCE+BUSINESS MEDIA, LLC

DOI 10.1007/978-0-387-35491-0

Library of Congress Cataloging-in-Publication Data

A C.I.P. Catalogue record for this book is available from the Library of Congress.

Originally published by Kluwer Academic Publishers in 2002
MyCopy version of the original edition 2002

Printed on acid-free paper.
www.springer.com/mycopy

Contents

Part Three: Optical packet and burst switching

Part Four: Advances in optical network technologies

Preface

In these exciting times of quotidianly progressing developments in communication techniques, where more than ever in the history of a technological progress, society's reliance on communication networks for medicine, education, data transfer, commerce, and many other endeavours dominates the human's everyday life, the optical networks are certainly one of the most promising and challenging networking options. Since their commercial arrival in the nineties, they have fundamentally changed the way of dealing with traffic engineering by removing bandwidth bottlenecks and eliminating delays. Today, after the revolutionary bandwidth expansion, the networking functionality migrates more and more to the optical layer, and the need to establish fast wavelength circuits and capacity-on-demand for the higher-layer networks, in particular data networks based on Internet Protocol (IP), has become one of the central networking issues for the new century. The unifying trends toward configurable all-optical network infrastructure open up a wide range of new network engineering and design choices dealing with networks' interoperability and common platforms for control and management.

The Fifth Working Conference on Optical Network Design and Modelling, held in the Austrian capital Vienna, February 5-7, 2001, aims at presenting the most recent progress in optical communication techniques, new technologies, standardisation process, emerging markets and carriers. A short look at the Table of Contents of this book tells us, in fact, that this year's conference program reflects the current state of the art precisely. In the tradition of the previous conferences (Athens′00, Paris′99, Rome′98 and Vienna′97), which brought together university researchers, technology leaders and network operators, also this year, the conference scope spans from performance evaluation of wavelength-routed networks, over advances in optical network technologies such as optical packet and burst switching, to on-demand circuit provisioning with protection and restoration within the optical layer.

To achieve such a great success in conference actuality and attendance, as we had with ONDM'01, the first words of reward go to the authors, who came from three continents and 21 different countries to contribute their high-quality papers and presentations. Special thanks are due to the members of the Technical Committee, who, co-operatively and within a tight time schedule, have accomplished a no easy task to choose "the best among the best". Also this year, a number of invited speakers from industry and academia presented the most recent topics in the field, for which I would especially like to thank Paul Tomlinson, Piet Demeester, Roberto Sabella, Yakov Rekhter, Bala Rajagopalan, Geoff Bennett, Goff Hill, Stefano Baroni, Geert Morthier, Ronan O'Dowd, and Nico Wauters (in order of appearance) for kindly accepting the invitation to come, thus invaluably improving the technical excellence of this, and the ONDM conferences to come.

Admela Jukan, Technical Program Chair Vienna, February 2001

Technical Committee

General Chair
Harmen R. van As, Vienna University of Technology, Austria

Technical Program Chair
Admela Jukan, Vienna University of Technology, Austria

Members
Marco Ajmone Marsan, Politecnico di Torino, Torino (I)
Stefano Baroni, Corvis Corp. (USA)
Andrea Fumagalli, UT Dallas (USA)
Maurice Gagnaire, ENST Paris, France (F)
Manfred N. Huber, Siemens AG, Germany (D)
David K. Hunter, University of Strathclyde, Great Britain (GB)
Geert Morthier, Ghent University, (B)
Fabio Neri, Politecnico di Torino, Torino (I)
Roberto Sabella, Ericsson Lab Italy, Rome (I)
Bernhard Schmauss, Lucent Technologies, Nuernberg (D)
Evi Zouganeli, Telenor, Norway (N)

Local Organizing Committee

Admela Jukan, Vienna University of Technology, Austria
Zuzana Kovacovsky, Vienna University of Technology, Austria
Johanna Pfeifer, Vienna University of Technology, Austria

Part One

Performance of wavelength-routed networks

PERFORMANCE OF MULTICAST SESSIONS IN WAVELENGTH-ROUTED WDM NETWORKS*

Ahmed E. Kamal and Anwar K. Al-Yatama

Department of Computer Engineering, Kuwait University, P.O. Box 5969 Safat, Kuwait 13060

E-mail: {kamal,yatama}@eng.kuniv.edu.kw

Abstract We evaluate the blocking probabilities in optical networks which support multicast (or multipoint) service. Due to the difficulty of the problem, we restrict our attention to the case of the completely connected networks, for which the blocking probabilities act as a lower bound for other topologies. Nodes may, or may not be equipped with wavelength converters. In both cases, routing can be done either directly, or indirectly, and even in the latter case the number of hops can be either restricted to two hops, or unrestricted. We evaluate the blocking probability by first estimating the load offered to each link, and then use the iterative reduced load approximation method to refine this estimate. Using this estimate, we then obtain the probabilities of blocking and success. Numerical examples are presented, and are compared to simulation.

Keywords: Optical networks; wavelength division multiplexing; multicast service; circuit switching; blocking probability; analysis.

1. INTRODUCTION

The advent of Wavelength Division Multiplexing (WDM) [Brackett, 1990] provided a solution to the problem of speed mismatch between the fiber bandwidth and the electronics speed, which is better known as the electronic bottleneck problem [Kaminow, 1989]. As such, an aggregate bandwidth in the tens of Terabits per second range can be achieved by supporting multiple simultaneous transmissions, each utilizing a different wavelength. This can provide a means to serve those applications

*This work was supported in part by research grant EE-102 from Kuwait University Research Administration.

which individually or collectively require a large amount of bandwidth. Such applications include the class of services with multicast, or multipoint traffic which requires the delivery of data from the source to a group of destinations [Turner, 1987]. Such applications include video distribution, network news distribution, multi-party conferences, etc. Multicast service over optical networks can be implemented by either creating an independent channel from the source to each destination, which is quite expensive, or by having the destinations cooperate in relaying the traffic they receive to other destinations. In this paper we consider the second approach, and we assume that there is a provision by which a destination will relay the same traffic, after reading it, to another destination.

The purpose of this paper is to derive expressions for the call blocking probabilities in wavelength-routed optical networks using WDM and carrying multicast circuit-switched traffic. Several studies have derived the call blocking probabilities for unicast circuit-switched applications, but to the best of our knowledge, this is the first study to obtain such blocking probabilities for the multicast service.

A few studies dealt with multicast routing in optical networks, e.g., [Tridandapani and Meditch, 1995; Tridandapani and Mukherjee, 1996; Rouskas and Ammar, 1994; Modiano, 1998]. However, those studies were concerned with packet-switched communication, and were concerned with either the establishment of the routing tree, or the derivation of the packet delay. The recent study in [Sahin and Azizoglu, 1998] has also dealt with multicast service over wavelength routed networks, but from a different perspective. It found the upper bound on the number of multicast calls which can be supported on a given network using an integer linear program.

A few studies have dealt with the derivation of the blocking probabilities under unicast service and with circuit switched traffic. These include references [Barry and Humblet, 1996; Jeong and Ayanoglu, 1996; Kovacevic and Acampora, 1996; Birman, 1996] which considered similar network models. Most of those studies used Erlang's fixed point method and the reduced load approximation [Kelly, 1990; Chung et al., 1993] in order to arrive at the blocking probabilities.

The authors in [Subramaniam et al., 1996] derived the blocking probabilities when only a group of nodes are capable of wavelength conversion, and in [Subramaniam et al., 1997] they derived the blocking probabilities when the arriving traffic is non-Poisson. In [Yates et al., 1996] the blocking probabilities in all-optical networks employing limited wavelength translation were derived in terms of the link utilization, and reference [Sharma and Varvarigos, 1998] considered the same problem in a mesh network. Reference [Harai et al., 1997] obtained the blocking probabilities in all optical switching networks when alternate routing is used.

In this paper we derive the blocking probabilities under multicast service when the topology forms a completely connected graph. We have chosen this topology for two reasons. First, it is a simple topology that avoids all the routing decisions which must be made by other special topologies which will affect (and in most cases complicate) the derivation of the blocking probabilities. Second, since a link exists between each pair of nodes in the completely connected topology, and since this number of routes from a source to any group of destinations is *at least* equal to the number of routes with any other topology, the blocking probabilities derived in this study serve as lower bounds on the blocking probabilities in other topologies.

We derive the blocking probabilities with and without wavelength conversion. We also consider two cases for the maximum number of hops allowed: a single hop, which correspond to direct routing, and a two-hop dynamic indirect routing. As expected, when the limit on the number of hops is increased, the blocking probabilities improve. However, increasing such a limit requires a more complicated control mechanism, but, as will be shown in the numerical examples section, this enhances the blocking probabilities only marginally over the two hop limit case. We also use simulation in order to assess the effectiveness of the unrestricted indirect routing strategy.

The approximate approach we follow in this paper is to estimate the call arrival rate to a certain link given the number of available channels (wavelengths) on that link and on an adjacent link. This joint availability condition is necessary to capture the strong correlation that arises with multicast traffic. This estimated load is in fact the reduced load which is used to calculate the probabilities of wavelength availabilities. The procedure is iterative, and in most cases converge rapidly.

In section II we introduce the network model. In section III we derive the blocking probabilities when wavelength conversion is used, while the probabilities of blocking without wavelength conversion are obtained in section IV. Section V presents some numerical examples and discussions, and section VI concludes with some remarks.

2. NETWORK MODEL

We consider a completely connected network with N nodes such that each pair of nodes is connected by a single fibre link. Each link carries W wavelength channels. Calls are assumed to arrive at a node according to a Poisson process with rate a, and the call holding time is exponentially distributed with a mean of one time unit. Each of the arriving calls are directed to k destinations with probability r_k, and the destination nodes are randomly chosen among all the nodes, excluding the source. We assume that all possible paths are exhaustively searched, and whenever several are found, only one is chosen at random.

It is assumed that each source node is able to transmit independently to different directly connected nodes using the same, or different wavelengths. It is also assumed that each of the links used by a session must lead to one of the destinations which have not been reached before, i.e., no node which is not a destination should be visited. The rationale behind this assumption is to minimize the amount of bandwidth required to carry out that call.

In order to obtain the probability of call blocking, we require the channel occupancy distribution on the different links. The reduced load approach[1] will be employed as an approximation [Kelly, 1990; Chung et al., 1993]. We develop a model which employs a generalization of Erlang's fixed point equation. Straightforward application of Erlang's fixed point method in which the channel occupancy is modeled as a one-dimensional Markov process is not appropriate under multicasting conditions. The reason is that strong correlations exist between the channel occupancies on links originating from the same node.

We capture the correlation between adjacent links by considering the behaviour of a target link in conjunction with the behaviour of an adjacent link, i.e., two links originating from the same node. We note that a similar correlation exists between this target link and the other adjacent links. Therefore, we can use this correlation when considering the other links.

We employ a 3-dimensional Markov process, where the dimensions are:

- i: the number of available channels on the target link.
- j: the number of available channels on the adjacent link.
- k : the number of channels on the target and the adjacent link, which are used for common calls.

According to the above, k should not exceed the minimum of $W - i$ and $W - j$, where W is the total number of channels per link. Although the above process is three-dimensional, the number of states is limited to a

[1]The reduced load on a link will be that load which is carried on the link, which will be denoted by λ, and is assumed to be Poisson distributed.

moderate number since in practice the number of channels per link is small. Hence, the computational requirements imposed by this model are reasonable.

It should be noted that disjoint transitions in either of the i or j dimensions correspond to activities on the target or adjacent nodes, respectively, which does not result in any change in the third dimension, k. However, joint transitions on both dimensions correspond to using channels on both links, which also requires either increasing or decreasing k for channel use and release, respectively.

The transition rates are given by the elements of a matrix, $\mathcal{P}$. The matrix $\mathcal{P}$ depends on the routing method, and will be shown for each routing case separately. Once $\mathcal{P}$ is constructed, one can solve for the steady state probabilities, $\pi_{i,j,k}$, from which we obtain the marginal probability of having m available channels on the target link, $q(m)$.

In Appendix I we introduce the definitions which will be used in this paper, noting that all links will be treated as identical and therefore no link number or index will be used, except between the target and adjacent links. It should be noted that the terms *target* and *adjacent* links can be used interchangeably.

We will derive the probability of call blocking by first evaluating the probability of call success, Pr(success), and then using

$$\Pr(\text{blocking}) = 1 - \Pr(\text{success}).$$

Pr(success) depends on the routing strategy, and will be evaluated for each of the considered cases.

3. NETWORKS WITH WAVELENGTH CONVERSION

This section derives the session blocking probabilities under the assumption that nodes are capable of full wavelength conversion. We will consider the two cases in which routing is either direct, or indirect with a maximum of two hops.

3.1. DIRECT ROUTING

We first consider direct routing only in which the link between the source and each destination must have at least one available wavelength. Under these conditions, $\lambda^{(h)}(m,n)$ for $h > 1$ is equal to zero. Also, $\lambda^{(h)}(0,n)$ is equal to zero for all value of h.

Considering a target link, we further divide $\lambda^{(1)}(m,n)$ into two components: $\lambda_d^{(1)}(m,n)$ which is carried on the target link only, i.e., it is disjoint from the adjacent considered link; and $\lambda_s^{(1)}(m,n)$ which is carried on both the target and adjacent link, i.e., it is shared with the adjacent link. As such, these can be expressed as

$$\lambda_d^{(1)}(m,n) = \sum_{k=1}^{N-1} a_k d_k \times (1 - t_1(k))(1 - B(m))^{k-1} \quad \text{, and} \tag{1}$$

$$\lambda_s^{(1)}(m,n) = \sum_{k=2}^{N-1} a_k d_k \times t_1(k)(1 - B(m))^{k-2}, \quad \text{for } n > 0 \tag{2}$$

In equation (1) the reduced load on a certain *target* link, given that the call is destined to k stations, is the fraction of offered load that does not use the adjacent link (the probability of which is equal to $1 - t_1(k)$). The last term in the equation is the probability that the other $k-1$ destinations are reachable directly, given that the target link has m available channels. Equation (2) considers the case in which the node at the end of the adjacent link is a destination (with probability $t_1(k)$).

The elements of the transition rate matrix $\mathcal{P}$ are now given by:

$$\begin{aligned}
P_{(i,j,k),(i,j-1,k)} &= \lambda_d^{(1)}(j,i) \\
P_{(i,j,k),(i-1,j,k)} &= \lambda_d^{(1)}(i,j) \\
P_{(i,j,k),(i-1,j-1,k+1)} &= \frac{\lambda_s^{(1)}(i,j) + \lambda_s^{(1)}(j,i)}{2} \\
P_{(i,j,k),(i,j+1,k)} &= W - j - k \\
P_{(i,j,k),(i+1,j,k)} &= W - i - k \\
P_{(i,j,k),(i+1,j+1,k-1)} &= k
\end{aligned}$$

We note that the amount of traffic common to two adjacent links is not in general known. In an exact characterization, since call arrivals are independent of the link states, then $\lambda_s^{(1)}(i,j)$ and $\lambda_s^{(1)}(j,i)$ should be equal. However, since our approach is approximate they may not be exactly the same, but were found to be very close in the numerical examples. We therefore average $\lambda_s^{(1)}(j,i)$ and $\lambda_s^{(1)}(i,j)$ in order to find that fraction of traffic corresponding to the downward diagonal transition rate.

Once $q(m)$ is found, the probability of success is given by

$$\Pr(\text{success}) = \sum_{k=1}^{N-1} r_k \sum_{m=1}^{W} q(m)(1 - B(m))^{k-1} \ .$$

3.2. INDIRECT ROUTING WITH TWO LEVELS

We now allow indirect routing with a maximum of two hops. To facilitate the derivation of the reduced loads for this case, we introduce the probability $U(k)$ whose definition is given in Appendix I. This probability applies only if the target link is used on the second hop (or farther).

We can now write

$$\lambda^{(h)}(0,n) = 0, \quad \forall h, \text{ and}$$

$$\lambda^{(h)}(m,n) = 0, \quad \text{for } h > 2, \ \forall m,n.$$

The expressions for the reduced load carried on a target link, which is not shared with the adjacent link when the adjacent link is not blocked ($n > 0$) are now given by:

$$\lambda_d^{(1)}(m,n) = \sum_{k=1}^{N-1} a_k d_k (1-t_1(k)) \sum_{i=0}^{k-1} \chi(k-1,i,B(m)) H(i|k-i) \quad \text{for } n > 0, \tag{3}$$

and

$$\begin{aligned}
\lambda_d^{(2)}(m,n) &= \sum_{k=2}^{N-1} a_k e_k (N-k) P_{a\&b} \\
&\times \Bigg\{ (1-t_2(k)) \sum_{i=0}^{k-2} \chi(k-2,i,B) U(k-i-2)(1-B^{k-i-2}B(m))^i \\
&+ t_2(k) \Bigg[(1-B) \sum_{i=0}^{k-3} \chi(k-3,i,B) U(k-i-2)(1-B^{k-i-2}B(m))^i \\
&+ B \sum_{i=0}^{k-3} \chi(k-3,i,B)(1-B^{k-3-i}B(m))^i \\
&\cdot U(k-i-3)(1-U(k-i-3)) \Bigg] \Bigg\}
\end{aligned} \tag{4}$$

In deriving equation (3), we condition on having i destinations being blocked, and obtain the probability that they can be reached through the $k-i$ unblocked nodes, including the node at the end of the target link. The node at the end of the adjacent link in this case is not a destination. In equation(4), since we are concentrating on the second hop, there are $N-k$ ways of choosing the source. Also, the probability of reaching the first node directly and not reaching the node at the end of the adjacent link directly is $P_{a\&b}$. The two terms inside the curly bracket correspond to the cases in which the node at the end of the adjacent link is not a destination, and is a destination, respectively.

The expressions for the reduced load on the target link when the adjacent link is blocked are given by:

$$\lambda_d^{(1)}(m,0) = \lambda_d^{(1)}(m,1)$$

$$+\sum_{k=2}^{N-1} a_k d_k t_1(k) \sum_{i=0}^{k-2} \chi(k-2, i, B(m)) H(i+1|k-i-1), \quad (5)$$

$$\lambda_d^{(2)}(m,0) = \sum_{i=0}^{k-2} a_k e_k (N-k) P_{a\&b}$$

$$\Bigg\{ (1-t_2(k)) \sum_{i=0}^{k-2} \chi(k-2, i, B) U(k-i-2)(1-B^{k-2-i}B(m))^i$$

$$+t_2(k) \Bigg[(1-B) \sum_{i=0}^{k-3} \chi(k-3, i, B) U(k-i-2)(1-B^{k-2-i}B(m))^i$$

$$+B \sum_{i=0}^{k-3} \chi(k-3, i, B)(1-B^{k-3-i}B(m))^i \; U(k-i-3)(1-B^{k-3-i}) \Bigg] \Bigg\}$$

$$(6)$$

The above two equations are similar to equations (3) and (4), except that the node at the end of the adjacent link must be reached through other links if it is a destination. This is accounted for by the second term in equation (5), and $(1-B^{k-3-i})$ in equation (6). The load shared between the target and the adjacent links, when both m and n are greater than zero is now expressed as:

$$\lambda_s^{(1)}(m,n) = \sum_{k=2}^{N-1} a_k d_k t_1(k) \sum_{i=0}^{k-2} \chi(k-2, i, B(m)) H(i|k-i) \quad , \quad (7)$$

$$\lambda_s^{(2)}(m,n) = \sum_{k=3}^{N-1} a_k e_k (N-k) t_2(k) Pab$$

$$\cdot B(0) \sum_{i=0}^{k-3} \chi(k-3, i, B)(1-B^{k-3-i}B(m))^i U(k-i-3)^{?}(8)$$

In both of the above equations, the node at the end of the adjacent link must be a destination. However, in equation (8) that node must be unreachable on the *first* hop, and reachable through the adjacent link on the second hop. At the same time, the target link must be used to reach the node at its end. The last term in the equation accounts for these last two probabilities.

The probability $H(i|l)$ in this case is equal to $(1-B^l)^i$

The elements of the transition matrix are similar to those in the direct routing case, but take into account the fact that two hops may be used by using $\lambda^{(1)} + \lambda^{(2)}$ instead of $\lambda^{(1)}$.

Finally, after solving for the steady state probabilities, and after obtaining $q(m)$, the probability of call success can be expressed as:

$$\begin{aligned} P(success) &= \sum_{k=1}^{N-1} r_k \left\{ B \sum_{i=0}^{k-2} \chi(k-1, i, B(0)) H(i+1|k-i-1) \right. \\ &\left. + \sum_{m=1}^{W} q(m) \sum_{i=0}^{k-1} \chi(k-1, i, B(m)) H(i|k-i) \right\}. \quad (9) \end{aligned}$$

In equation (9) the first term between the curly bracket corresponds to the case in which a chosen target link is blocked, but is reachable through any of the $k-i-1$ other nodes which have been reached on the first hop. The second term corresponds to the case in which the chosen target link is not blocked, and has m available channels.

4. NETWORKS WITHOUT WAVELENGTH CONVERSION

In this section we assume that the switches do not implement wavelength conversion. The derivation of the reduced load equations in this section is more involved since one has to keep track of the number of commonly available wavelength channels on a given path.

4.1. DIRECT ROUTING

This case is similar to networks with wavelength conversion.

4.2. INDIRECT ROUTING WITH A LIMIT OF TWO LEVELS

With a limit of two hops, a call can be routed from the source to the destination either:

- on the direct link, in which case any available channel can be used on the direct link; or
- if the direct link is blocked, the call can be routed via a destination that has been reached via a direct link. This requires that the wavelength used on the first link be available on the second link.

In this section, we require a few more probabilities, which are also defined in Appendix I.

We can now obtain the reduced load on the target link. Similar to subsection III.2, we have

$$\lambda^{(h)}(0, n) = 0, \quad \forall h,$$

$$\lambda^{(h)}(m,n) = 0, \quad \text{for } h > 2, \ \forall m,n.$$

The reduced load on the target link with m available channels, when it is not shared with the adjacent link that has n, $n > 0$, channels can be derived by considering the same situations considered under wavelength conversion. However, we have to involve the probabilities of wavelength continuity. The reduced load is given by

$$\begin{aligned} \lambda_d^{(1)}(m,n) &= \sum_{k=1}^{N-1} a_k d_k (1 - t_1(k)) \sum_{i=0}^{k-1} \chi(k-1, i, B(m)) \\ &\quad \cdot \sum_{j=0}^{i} F(j|i,m) H(i-j|k-i-1) \quad \text{for } n > 0, \end{aligned} \tag{10}$$

$$\begin{aligned} \lambda_d^{(2)}(m,n) = \sum_{k=2}^{N-1} a_k e_k (N-k) P_{a\&b} \times \\ \Bigg\{ \Bigg[(1 - t_2(k)) \sum_{i=0}^{k-2} \chi(k-2, i, B) \sum_{j=1}^{i+1} G(j|i+1, m) \\ \sum_{r=0}^{k-i-2} S(i+1-j, r|k-i-2) \times \frac{1}{r+1} \Bigg] \\ + t_2(k) \Bigg[(1-B) \sum_{i=0}^{k-3} \chi(k-3, i, B) \sum_{j=1}^{i+1} G(j|i+1, m) \\ \sum_{r=0}^{k-i-2} S(i+1-j, r|k-i-2) \times \frac{1}{r+1} \\ + B \sum_{i=0}^{k-3} \chi(k-3, i, B) \Bigg(\sum_{j=1}^{i+1} G_{\overline{a}}(j|i+1, m, n) \\ \sum_{r=0}^{k-i-3} S(i+1-j, r|k-i-3) \times \frac{1}{r+1} \\ + \sum_{j=1}^{i+1} G_a(j|i+1, m, n) \sum_{s=0}^{k-i-3} \sum_{r=0}^{k-i-3} T(i+1-j, r, s|k-i-3) \\ \times \frac{1}{r+1} \times \frac{s}{s+1} \Bigg) \Bigg] \Bigg\} \end{aligned} \tag{11}$$

Equation (10) is straightforward. It is just to be noted that the last summation is the probability of reaching j out of i nodes from the end of the target link (which has m available channels), and reaching the remaining $i-j$ nodes via the $k-i-1$ nodes that have already been reached directly.

Similar to equation (8), the two terms between the curly brackets in equation (11) correspond to the cases of the node at the end of the adjacent link not being a destination, and being a destination, respectively. And the second term is divided into the probabilities of that node being already reached on the first hop, and being reached on the second hop, but through a different destinations, respectively. However, in all of these terms, we find the probability that the node at the beginning of the target link can reach j other nodes, in addition to the node at the end of this link, and we multiply it by the probability of reaching the rest of the destinations through other nodes. In doing so, we also cater for the probability that the node at the end of the target link can be reached by r other nodes, and that the target link will be used to reach it. In the case in which the node at the end of the adjacent link is a destination which will be reached on the second hop (last term which is multiplied by B for being blocked on the first hop), we distinguish between two cases: the case in which it cannot be reached on the adjacent link because of wavelength discontinuity; and the case in which it can be reached. In this last case we obtain the probability that the node at the end of the adjacent link can be reached by s other nodes, and one of those nodes will be used to reach it.

The case when the adjacent link is blocked can be considered similarly, but we take into account the fact that the node at the end of the adjacent link, if it is a destination, will be reached either on the first hop, or on the second hop but using another link. The expressions for the reduced load are given by

$$\begin{aligned}\lambda_d^{(1)}(m,0) &= \lambda_d^{(1)}(m,1)+ \\ &+ \sum_{k=1}^{N-1} a_k d_k t_1(k) \sum_{i=0}^{k-2} \chi(k-2,i,B(m)) \\ &\cdot \sum_{j=0}^{i+1} F(j|i+1,m) H(i+1-j|k-i-2)\end{aligned} \tag{12}$$

$$\begin{aligned}\lambda_d^{(2)}(m,0) = &\sum_{k=2}^{N-1} a_k e_k (N-k) P_{a\&b} \\ &\times \left\{ (1-t_2(k)) \sum_{i=0}^{k-2} \chi(k-2,i,B) \sum_{j=1}^{i+1} G(j|i+1,m) \right.\end{aligned}$$

$$\sum_{r=0}^{k-i-2} S(i+1-j,r|k-i-2) \times \frac{1}{r+1}$$

$$+t_2(k) \left[(1-B) \sum_{i=0}^{k-3} \chi(k-3,i,B) \sum_{j=1}^{i+1} G(j|i+1,m)\right.$$

$$\sum_{r=0}^{k-i-2} S(i+1-j,r|k-i-2) \times \frac{1}{r+1}$$

$$+B \sum_{i=0}^{k-3} \chi(k-3,i,B) \sum_{j=1}^{i+1} G_{\overline{a}}(j|i+1,m,0)$$

$$\left.\left.\sum_{r=0}^{k-i-3} S(i+2-j,r|k-i-3) \times \frac{1}{r+1}\right]\right\} \quad (13)$$

Equation (12) is derived in a manner similar to that used in deriving equation (10, but taking into account the fact that the adjacent link has no available channels. This requires adding a new term which caters for the case in which the node at the end of the adjacent link is a destination. Equation (13) has been derived similar to equation (11), except that the node at the end of the adjacent node cannot be reached over the adjacent link ($G_a() = 0$).

Finally, the shared load between the target and the adjacent links is

$$\begin{aligned} \lambda_s^{(1)}(m,n) &= \sum_{k=2}^{N-1} a_k d_k t_1(k) \sum_{i=0}^{k-2} \chi(k-2,i,B(m)) \\ & \quad \sum_{j=0}^{i+1} F(i,j|m) H(i-j|k-i-1) \ , \end{aligned} \quad (14)$$

$$\lambda_s^{(2)}(m,n) = \sum_{k=3}^{N-1} a_k e_k (N-k) t_2(k) P_{a\&b} B(0)$$

$$\times \left[\sum_{i=0}^{k-3} \chi(k-3,i,B) \sum_{j=1}^{i+1} G_a(j|i+1,m,n)\right.$$

$$\left.\sum_{s=0}^{k-i-3} \sum_{r=0}^{k-i-3} T(i+1-j,r,s|k-i-3) \times \frac{1}{r+1} \times \frac{1}{s+1}\right] \quad (15)$$

for $m,n > 0$. The derivation of the above equations follows earlier derivations, except that in equation (15) the node at the end of the adjacent link is reached over that link, even if that node can be reached by r other nodes.

In the Appendix II we show the derivation of $b^{(l)}(n|m)$, $F(j|i,m)$, $G(j|i,m)$, $G_{\overline{a}}(j|i,m,n)$, $G_a(j|i,m,n)$, $S(i,n|l)$, $T(i,n,p|l)$ and $H(i|l)$.

From the above,

$$P(success) = \sum_{k=1}^{N-1} r_k \left\{ B \sum_{i=0}^{k-2} \chi(k-1,i,B(0))H(i+1|k-i-1) \right. \\ \left. + \sum_{m=1}^{W} q(m) \sum_{i=0}^{k-1} \chi(k-1,i,B(m)) \sum_{j=0}^{i} F(j|i,m)H(i-j|k-i-1) \right\} \tag{16}$$

5. NUMERICAL RESULTS

The expressions for the reduced load derived in the previous section assume the knowledge of the channel availability distribution. Since these probabilities depend in turn on the knowledge of the reduced loads, through the matrix $\mathcal{P}$, an iterative solution procedure must be used, in which λ is computed and fed back to the model. The iterative procedure ends when all λ's converge, and the convergence criterion, ϵ, was chosen as 10^{-3} in all of the examples shown below. Convergence usually occurred very fast, and with a number of iterations that rarely exceeded 10.

We first validate the model by comparing the analytical results against results from simulation. In Table 1 we consider a network with six nodes, and three channels per fiber. Different load vectors are used, namely $V_1 = (1,0,0,0,0)$, $V_2 = (.2,.2,.2,.2,.2)$, $V_3 = (0,0,0,0,1)$ and $V_4 = (.5,.5,0,0,0)$. The corresponding blocking probabilities are shown in the table. Most of the analytical results shown in the table are within 10% of simulation results. Table 2 show the corresponding examples for the case of no wavelength conversion.

From the above tables we conclude two things. First, the use of indirect routing rather than direct routing enhances the performance of the system. Indirect routing with no hop limit results in a performance which is marginally better than that under indirect routing with a two hop limit. Therefore, it is not really justifiable to use a significantly more complex control function in order to achieve that marginal enhancement. Second, the use of wavelength conversion reduces the probability of call blocking as compared to the absence of wavelength conversion.

In Figure 1 we study the effect of the number of channels per fiber on the call blocking probability for all routing strategies. We assume

Table 1 Blocking probabilities for a network with wavelength conversion; N = 6 and W = 3

Network parameters			P(Blocking):model		P(Blocking):simulation		
Load per node	Load vector	Load per ch.	direct routing	indirect with 2 hops	direct routing	indirect with 2 hops	indirect with no limit
1	V_1	.06667	.00109	.00109	.00103	.00103	.00103
1	V_2	.2	.0371	.0069	.0343	.0079	.0078
1	V_3	.3333	.0623	.0619	.0625	.0625	.0624
1	V_4	.1	.0049	.0017	.0049	.0017	.0017
2	V_1	.1333	.007156	.007156	.007268	.007268	.007268
2	V_2	.4	.1378	.0423	.1320	.0514	.048
2	V_3	.6667	.21056	.2105	.21169	.2106	.2106
2	V_4	.2	.0279	.01129	.0279	.0114	.0114

Table 2 Blocking probabilities for a network without wavelength conversion; N = 6 and W = 3

Network parameters			P(Blocking): model		P(Blocking):sim.		
Load per node	Load vector	Load per ch.	direct routing	indirect with 2 hops	direct routing	indirect with 2 hops	indirect with no limit
1	V_1	.06667	.00109	.00109	.00107	.00107	.00107
1	V_2	.2	.0371	.00777	.0343	.01016	.00875
1	V_3	.3333	.0623	.06263	.0625	.06248	.06244
1	V_4	.1	.0049	.001775	.0049	.0019	.0019
2	V_1	.1333	.007156	.007156	.007268	.007067	.007067
2	V_2	.4	.1378	.055355	.1320	.06653	.05817
2	V_3	.6667	.21056	.21052	.21169	.21052	.21051
2	V_4	.2	.0279	.01196	.0279	.01328	.01328

that the network has 10 nodes, with each node offering a total load of 2 Erlangs. Since in most networks, unicast applications are expected to dominate, we assume that the probability of a unicast call is 0.5, while the probability of a call with k destinations for $k = 2, 3, \ldots, 9$ is equal to 0.0625. As expected, increasing the number of channels per fiber significantly reduces the call blocking probability. Adding one more channel per fiber reduces the call blocking probability by almost one order of magnitude for all cases. For example, under wavelength conversion, and with indirect routing with a limit of two hops, increasing the number of fibers per channel from 6 to 7 reduces the call blocking probability from 4.9×10^{-5} to 5×10^{-6}. We notice also that direct routing results in the worst performance, while the other indirect routing strategies result in almost the same performance. This confirm our earlier conclusion that the complexity of indirect routing with a number of hops beyond two is not justifiable.

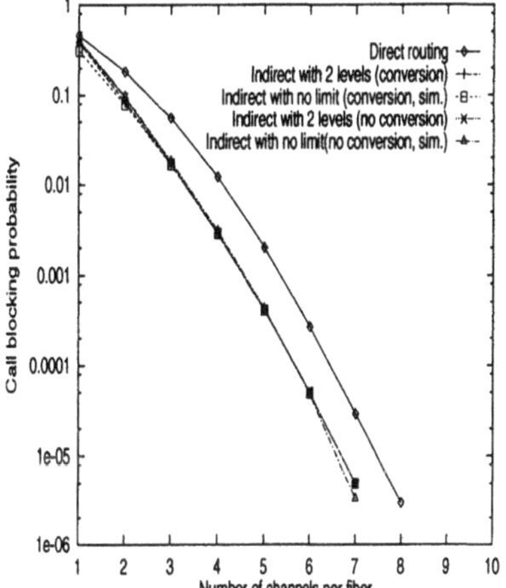

Figure 1 Blocking probabilities versus the number of wavelength channels per fiber, when the node offered load is 2 Erlangs

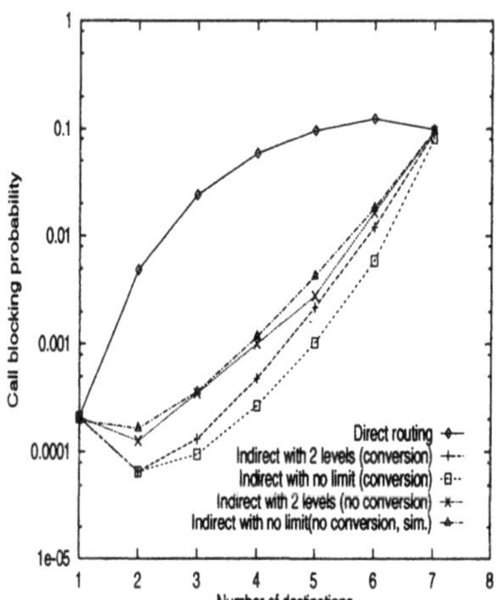

Figure 2 Blocking probabilities versus the number of wavelength channels per fiber, when the node offered load is 2 Erlangs

In the last example, we study the effect of the number of destinations per call. We consider a network with 8 nodes, and 4 channels per fiber. Seven experiments are performed in which the number of destinations is fixed at k, where $k = 1, 2, \ldots, 7$. Similar to the above, we assume a fixed offered load per node that is equal to 2 Erlangs. The results are shown in Figure 2. First, it should be noticed that in both the unicast and the broadcast cases, all routing strategies yield exactly the same performance. This is true for the broadcast case since all links leaving any of the sources will be either all blocked, or all unblocked. It is also noticed that with direct routing, increasing the number of destinations increases the blocking probability. However, the broadcast case, $k = 7$, results in a slightly better performance because of the above mentioned correlation. With the indirect routing strategies, increasing the number of destinations from the unicast case improves the performance first, before it starts to degrade. This can be explained since with two destinations, a source has three alternative routes of reaching all the destinations compared to just one route for the case of one destination. This reduces the probability of call blocking. After that, the increased loading on the links degrades the performance.

6. CONCLUSIONS

In this paper we derived expressions for the call blocking probabilities in optical networks under multicast service in a completely connected network topology. We considered the case in which full wavelength conversion is allowed, and the case in which no wavelength conversion is used. For both cases, we considered the direct routing method, and the indirect routing method in which the maximum number of hops is restricted to two hops. The case in which there is no restriction on the number of hops is very involved, and the blocking probabilities were obtained using simulation only. The solution method was based on the reduced load approach, and was iterative in nature.

Based on the numerical examples, it was found that although indirect routing improved the blocking probabilities, the enhancement achieved by using more than two hops is marginal compared to the enhancement achievable with two hops compared to direct routing. Therefore, it is advantageous to use indirect routing with multicast service, but restricting the maximum number of hops to two usually yields a sufficiently good performance. It was also shown that increasing the number of destinations does not necessarily result in a degraded performance. On the contrary, since increasing the number of destinations, when employing indirect routing, increases the number alternative routes from the source to all the destinations, a better performance can be achieved.

Appendix: I

This appendix gives the definitions and notations used in this paper.

N number of nodes in the system.

W number of wavelength channels per link.

a offered load per node.

r_k probability that a call is destined to k destinations.

a_k offered load per node with k destinations $= ar_k$.

d_k probability that the node at the end of the link is chosen as a destination, given that the call is destined to k nodes out of $N-1$ nodes $= \frac{k}{N-1}$.

e_k probability that the two nodes connected by a given link. are chosen as destinations, given that the call is destined to k nodes out of $N-1$ nodes $= \frac{k(k-1)}{(N-1)(N-2)}$.

$t_i(k)$ probability that a given node is chosen as a destination for a call with k destinations, given i known destinations $= \frac{k-i}{N-i-1}$.

$\lambda^{(h)}(m,n)$ load carried on the link[2], given that the target and the adjacent links have m and n available channels, respectively, and the load is experiencing its hth hop.

$\pi_{m,n,l}$ the steady state probability that the target and the adjacent links have m and n available channels, respectively, and that l channels are being used for common calls.

$q(m)$ probability that the link has m idle wavelengths.

$B(m)$ probability that a link is blocked, given that the adjacent target link has m available channels $= \frac{\sum_{k=0}^{W-m} \pi_{0,m,k}}{q(m)}$.

B the unconditional link blocking probability $= q(0) = \sum_{m=0}^{W} B(m)q(m)$.

$P_{a\&b}$ probability that a link is not blocked, while its adjacent link is blocked $= \sum_{m=1}^{W} B(m) \cdot q(m)$.

$P_{b\&b}$ probability that the target and the adjacent links are both blocked $= B(0)q(0)$.

$\chi(i,j,p)$ probability of choosing j out of i events according to the binomial distribution with parameter $p = \binom{i}{j} p^j (1-p)^{i-j}$.

$H(i|l)$ probability of reaching i out of i destinations, given that l nodes have been reached.

[2]The load carried on a link is the reduced load offered to the link.

$U(k)$ probability of using the target link to reach the node at its end, given that k other links are available for that purpose $= \sum_{i=0}^{k} \binom{k}{i} (1-B)^i B^{k-i} \times \frac{1}{i+1} = \frac{1-B^{k+1}}{(k+1)(1-B)}$.

$c(n|m, f, W)$ probability of n common available channels between two links, given m available channels on the first, f available channels on the second, and a total of W channels per link $=$ $\binom{m}{n}\binom{W-m}{f-n}/\binom{W}{f}$.

$b^{(l)}(n|m)$ probability that a link with m available channels has n available channels in common with l other links originating from the end of this particular link.

$F(j|i, m)$ P(reaching j destinations out of i destinations from the end of the first link, given m wavelengths on the link).

$G(j|i, m)$ P(reaching j out of i, including the node at the end of a particular second link, given m wavelengths on that particular link, and the first link is not blocked).

$G_{\overline{a}}(j|i, m, n)$ P(reaching j out of i nodes from the node at the end of the first link, and **not reaching** the node at the end of the adjacent link, given that m and n channels are available on the target and adjacent links, respectively).

$G_a(j|i, m, n)$ P(reaching j out of i nodes from the node at the end of the first link, and **reaching** the node at the end of the adjacent link, given that m and n channels are available on the target and adjacent links, respectively). P(reaching i out of i over the second hop, via l nodes

$T(i, n, p|l)$ P(reaching i out of i nodes over the second hop, via ^{l} nodes that are reached directly, and n of the l nodes can also reach the destination at the end of the target link, while p of the l nodes will reach the destination at the end of the adjacent node).

$S(i, n|l)$ P(reaching i out of i nodes over the second hop, via l nodes that are reached directly, and n out of the l nodes can also reach the destination at the end of the target link) $= \sum_{p=0}^{l} T(i, n, p|l)$.

Appendix: II

In this appendix we derive the auxiliary probabilities $b^{(l)}(n|m)$, $F(j|i, m)$, $G(j|i, m)$, $G_{\overline{a}}(j|i, m, n)$, $G_a(j|i, m, n)$, $S(i, n|l)$, $T(i, n, p|l)$ and $H(i|l)$

which are required in the derivation of the blocking probability under multicast service when no wavelength conversion is used, and when a maximum of two hops is imposed.

We first define $\eta_{i,j}$ and $\delta_{i,j}$ in exactly the same way e_i and d_i are defined, except that we condition on having a total of j possible destinations. Therefore, we have

$$\eta_{i,j} = \frac{i}{j}$$

$$\delta_{i,j} = \frac{i(i-1)}{j(j-1)}$$

The probability $b^{(l)}(n|m)$ can be obtained recursively as follows:

$$\begin{aligned} b^{(1)}(n_1|m) &= \sum_{f_1=n_1}^{W-m+n_1} c(n_1|m, f_1, W)q(f_1) \\ b^{(l)}(n_l|m) &= \sum_{n_{l-1}=n_l}^{m} \sum_{f_l=n_l}^{W-n_{l-1}+n_l} c(n_l|n_{l-1}, f_l, W)b^{(l-1)}(n_{l-1}|m)q(f_l) \\ & \qquad \text{for } l > 1 \end{aligned}$$

$F(j|i,m)$ can now be computed using the following equation:

$$F(j|i,m) = \binom{i}{j} \sum_{n=1}^{m} b^{(j)}(n|m)F(0|i-j,n) \tag{17}$$

with the initial condition

$$F(0|i,m) = \left[\sum_{f=0}^{W-m} c(0|m,f,W)q(f)\right]^i = \left[b^{(1)}(0|m)\right]^i$$

To compute $G(j|i,m)$ it is required that out of the i destinations which must be reached on the second hop, j will be reached, including the one at the end of the target link which has m available wavelength channels. $G(j|i,m)$ can therefore be expressed as

$$G(j|i,m) = \sum_{f=1}^{W} \sum_{n=1}^{\min(m,f)} c(n|m,f,W)\frac{q(f)}{1-B}F(j-1|i-1,n) \tag{18}$$

In equation (18), $c(n|m,f,W)$ is the probability that, given that the considered second link has m available channels, and the direct link has

f channels, they both have n common available channels. $F(j-1|i-1,n)$ indicates the probability of reaching $j-1$ destinations out of the remaining $i-1$ destinations given those available n channels.

Once $G(j|i,m)$ has been calculated, one can proceed to calculate

$$G_{\overline{a}}(j|i,m,n) = c(0|m,n,W)G(j|i,m), \quad \text{and}$$

$$G_a(j|i,m,n) = \sum_{r=1}^{\min(m,n)} c(r|m,n,W)G(j|i,r)$$

Finally,

$$S(i,n|l) = \sum_{j=0}^{i}\left(\sum_{f=1}^{W} F(j|i+1,f)\frac{q(f)}{1-B}\right)\cdot\frac{i+1-j}{i+1}S(i-j,n|l-1)$$
$$+\sum_{j=0}^{i}\left(\sum_{f=1}^{W} F(j+1|i+1,f)\frac{q(f)}{1-B}\right)\cdot\frac{j+1}{i+1}S(i-j,n-1|l-1) \tag{19}$$

with $S(0,0|0)=1$, $S(i,n|l)=0$ for $n>l$, $S(i,n|0)=0$ for $i>0,\ n\geq 0$, $S(0,0|l)=1$ for $l\geq 0$, $S(0,n|l)=0$ for $l,n>0$ and

$$S(i,0|l) = \sum_{j=0}^{i}\left(\sum_{f=1}^{W} F(j|i+1,f)\frac{q(f)}{1-B}\right)\cdot\frac{i+1-j}{i+1}S(i-j,0|l-1)\ . \tag{20}$$

We can also obtain

$$T(i,n,p|l) =$$
$$\sum_{j=0}^{i}(\sum_{f=1}^{W} F(j|i+2,f)\frac{q(f)}{1-B})(1-\eta_{j,i+2}-2\delta_{j,i+2})T(i-j,n,p|l-1)$$
$$+\sum_{j=0}^{i}(\sum_{f=1}^{W} F(j+1|i+2,f)\frac{q(f)}{1-B})g_{j+1,i+2}\sum_{k=0}^{1} T(i-j,n-k,p|l-1)$$
$$+\sum_{j=0}^{i}(\sum_{f=1}^{W} F(j+2|i+2,f)\frac{q(f)}{1-B})\eta_{j+2,i+2}T(i-j,n-1,p-1|l-1) \tag{21}$$

with the following initial conditions:

$$\begin{aligned}
T(i,n,p|0) &= 0 && \text{for } i \text{ or } n \text{ or } p>0\\
T(i,n,p|l) &= 0 && \text{for } n \text{ or } p>l\\
T(0,0,0|l) &= 1 && \text{for } l\geq 0\\
T(0,n,p|l) &= 0 && \text{for } n \text{ or } p>0
\end{aligned}$$

Similarly, but with less involvement we can compute the following:

$$H(i|l) = \sum_{j=0}^{i} \left(\sum_{f=1}^{W} F(j|i,f) \frac{q(f)}{1-B} \right) H(i-j|l-1) \qquad (22)$$

with $H(0|l) = 1$ for $l \geq 0$, and $H(i|0) = 0$ for $i > 0$.

References

Barry, R. A. and Humblet, P. A. (1996). Models of blocking probability in all-optical networks with and without wavelength changers. *IEEE Journal on Selected Areas in Communications*, 14(5):858–867.

Birman, A. (1996). Computing approximate blocking probabilities for a class of all-optical networks. *IEEE Journal on Selected Areas in Communications*, 14(5):852–857.

Brackett, C. A. (1990). Dense wavelength division multiplexing networks: Principles and applications. *IEEE Journal on Selected Areas in Communications*, 8(6):948–964.

Chung, S., Kashper, A., and Ross, K. W. (1993). Computing approximate blocking probabilities for large loss networks with state-dependent routing. *IEEE/ACM Transactions on Networking*, 1(1):105–115.

Harai, H., Murata, M., and Miyahara, H. (1997). Performance of alternate routing methods in all-optical switching networks. In *Proceedings of the IEEE INFOCOM*, pages 516–524.

Jeong, G. and Ayanoglu, E. (1996). Comparison of wavelength-interchanging and wavelength-selective cross-connects in multiwavelength all-optical networks. In *Proceedings of the IEEE INFOCOM*, pages 156–163.

Kaminow, I. P. (1989). Photonic multiple-access networks: Topologies. *AT&T Tech. Jour.*, 68(2):61–71.

Kelly, F. P. (1990). Routing and capacity allocation in networks with trunk reservation. *Math. of Oper. Res.*, 15:771–792.

Kovacevic, M. and Acampora, A. (1996). Benefits of wavelength translation in all-optical clear-channel networks. *IEEE Journal on Selected Areas in Communications*, 14(5):868–880.

Modiano, E. (1998). Unscheduled multicasts in wdm broadcast-and-select networks. In *Proceedings of the IEEE INFOCOM*, pages 86–93.

Rouskas, G. N. and Ammar, M. H. (1994). Multi-destination communication over single-hop lightwave WDM networks. In *Proceedings of the IEEE INFOCOM*, pages 1520–1527, Toronto, Canada.

Sahin, G. and Azizoglu, M. (1998). Multicast routing and wavelength assignment in wide area networks. In *SPIE Conf. on All-Optical Networking*.

Sharma, V. and Varvarigos, E. A. (1998). Limited wavelength translation in all-optical wdm mesh networks. In *Proceedings of the IEEE INFOCOM.*

Subramaniam, S., Azizoglu, M., and Somani, A. (1996). Connectivity and sparse wavelength conversion in wavelength-routing networks. In *Proceedings of the IEEE INFOCOM*, pages 148–155.

Subramaniam, S., Somani, A., Azizoglu, M., and Barry, R. (1997). A performance model for wavelength conversion with non-poisson traffic. In *Proceedings of the IEEE INFOCOM*, pages 499–506.

Tridandapani, S. and Meditch, J. S. (1995). Supporting multipoint connections in multi-hop wdm optical networks. *Journal of High-Speed Networks*, 4(2):169–188.

Tridandapani, S. B. and Mukherjee, B. (1996). Multicast traffic in multihop lightwave networks: Performance analysis and an argument for channel sharing. In *Proceedings of the IEEE INFOCOM*, pages 345–352, San Francisco, CA.

Turner, J. S. (1987). The challenge of multipoint connection. In *Proc. 5th ITC Seminar: Traffic Engineering for ISDN Design*, Cernobbio, Italy.

Yates, J. et al. (1996). Limited-range wavelength translation in all-optical networks. In *Proceedings of the IEEE INFOCOM*, pages 954–961.

ILP FORMULATION OF GROOMING OVER WAVELENGTH ROUTING WITH PROTECTION

Tibor Cinkler
High-Speed Networks Laboratory, Department of Telecommunications & Telematics
Budapest University of Technology & Economics, Budapest, Hungary
cinkler@ttt-atm.ttt.bme.hu

Abstract An efficient and general graph-theoretic model (the Wavelength-Graph (WG)) has been proposed which enables Integer Linear Programming (ILP) formulation of static Routing and Wavelength Assignment (RWA) problems in Multihop Wavelength Routing (WR) Dense Wavelength Division Multiplexing (DWDM) Networks.

The topology of the physical layer, the type of each node (e.g., OADM, OXC or EXC), the number of available wavelengths per link and the capacity of each wavelength-channel are assumed given with the aggregated traffic demand of each node-pair. The output of the optimisation is the system of wavelength-paths, lightpaths and semi-lightpaths.

The objective of the optimisation is to reduce resource usage at upper (electrical) layers, subject to constrained amount of capacity of each wavelength and limited number of wavelengths.

1. INTRODUCTION

WDM has been introduced to increase the transmission capacity of existing optical links. Instead of using one several transmitter and receiver pairs were used over the same fibre at different wavelengths forming independent channels and over-bridging the speed limitations of electronics. It has been soon recognised that the switching decision can be made according to the incoming wavelength without any processing of the data stream. In single-hop DWDM based All-Optical Networks (AON) a wavelength is assigned to a connection in such a way that each connection (wavelength) is handled (switched) in the optical domain without any electrical conversion during the transmission [1]. This will be referred to as a *wavelength-path*. Wavelength (WL) reuse is allowed in parts of the network where that WL was not used. This WDM AON would require a huge number of different WLs. The technology sets the limit nowadays to around 160 different WLs per fibre in the 1550 nm window with 25 GHz (about 0.2 nm) spacing in the flat operating gain band (1530-1560 nm) of the present Erbium-Doped Fibre Amplifiers (EDFA) according to the recently completed ITU-T Recommendation G.692. For this reason WL conversions are needed. The most expensive way is to make optical WL conversion which ensures transparency of the network. Then, the end-to-end channel will be referred to as a *lightpath*. Simpler and cheaper method is to do first opto-electrical conversion, electrical space-switching and then electro-optical conversion. The end-to-end connection will then use a *semi-lightpath*. This is the idea for realising so called opaque networks, where systems using different sets of WLs are to be interconnected. The optimisation method proposed in this paper can be applied to all, wavelength-paths, lightpaths and semi-lightpaths.

It has been shown in [2] that the required number of different wavelengths per fibre for networks with and without WL conversion capability is about the same. In general, for networks of practical size, the number of available wavelengths is lower by a few orders of magnitude than the number of connections to be established. The only solution here is to join some of the connections to fit into the available wavelength-links. This is referred to as traffic grooming). This can be done at electrical layer only since re-multiplexing, i.e., time division multiplexing is required. For this reason, taking not only the optical, but both, optical and electrical layer into account when configuring the system is demanded.

Many excellent papers deal with design, configuration and optimisation of WDM Networks. See, e.g., [2 – 8]. The widely accepted approach is to decompose the problem to the following sub-problems in given order. First, determine the virtual topology (route the light-paths); second, assign a wavelength to each light-path (WA); and third, route the traffic over the light-paths.

In [3] a heuristic (greedy) algorithm is proposed for WA followed by routing over established light-paths. In [4] is defined a bound for carried traffic in all-optical networks and shown that the proposed heuristic Routing and Wavelength Assignment (RWA) algorithm gives results very close to this bound. A performance study has been carried out. In [5] mathematical formulation of the design problem is given along with heuristics for solving the sub-problems relaxing some of the constraints one-by-one. In [2] there is also given a heuristic algorithm, and is showed that networks with and without wavelength converters require about the same number of wavelengths. A hybrid solution is also proposed where wavelengths are electrically regenerated in some specific nodes. In [6] multi-commodity flow model with randomised rounding is applied followed by graph colouring algorithms. In [7] an algorithm is given for light-path routing by transforming the network to a special structure called wavelength graph (Although we will use the same term the structure covered by it differs significantly). In this graph costs are assigned to edges and shortest path algorithms are run. The optimality of the algorithm is also proved. In [8] design principles of Optical Networks are explored. The design is formulated as an optimisation problem with two objectives to be solved by heuristics. The (weighted) average delay trough the network is minimised, while the total carryable traffic over the network is maximised. In [9] the extended layered graph is used which is a bit similar to our model, but not flexible enough. In [10] the optical path routing strategies for WDM networks are investigated. The performance of networks with and without wavelength converters is evaluated. In [11] is described a method for planning WDM layer for carrying ATM traffic over it with the survivability constraint using Tabu Search. In [12] a reconfigurable OADM is demonstrated. In [13] hitless reconfiguration of WDM networks is investigated. [14] gives a performance evaluation and comparison of WDM networks with and without wavelength interchange capability. [15] proposes an algorithm for simultaneous Routing and Wavelength Assignment on a Path Graph. Channel capacities are not taken into account. In [16] an exact linear programming formulation is presented and the "closest" virtual topology is chosen for reconfiguration. [17] investigates the performance of partial reconfiguration on a SDM/WDM architecture. [18] proposes a model and algorithm for finding the globally optimal WL assignment with high probability using generally applicable heuristics for global optimisation. In [19] Mixed Integer Linear Programming (MILP) formulation of the static RWA problem is given for the case without WL changers as a minimax problem. The MILP formulation is then relaxed to LP and solved, followed by rounding.

Our subject is to configure the light-path system optimally without separating the network-layers. This improves the quality of results, but on the other hand the complexity of the problem grows.

As the optimisation result we decrease the traffic to be processed and carried in the electrical domain over-bridging the speed limits of electronics. Since a considerable part of the load of electrical, e.g., SDH, ATM or MPLS switches is undertaken by the optical switches much larger networks with higher loads can be realised by the current technology offering better granularity and using optimally any limited number of WLs.

In Section 2 we present the model of the network with different node-types. In Section 3 we formulate the problem first informally, and then formally as an Integer Linear Program (ILP) for both, directed and undirected graph models. In Section 4 th protection alternatives are given, for protection in both, upper electrical layer and lower wavelength layer. Section 5 adds some comments on the problem, while in Section 6 solution alternatives are proposed.

2. THE WAVELENGTH GRAPH (WG)

The task was to provide a general model for configuration of WDM networks with different types of nodes and arbitrary topologies. Although the most popular topology is ring or interconnected rings, the model must be able to handle any specific or mesh topology. The nodes can also be quite different: Optical Add-and-Drop Multiplexers (OADM), Optical Cross-Connects (OXC) with full or limited (optical or opto-electrical) WL conversion or even an Opto-Electrical Cross Connect (OEXC). The protection strategies can also be quite different. All these aspects are taken into account in the proposed model. First the link model is described followed by models of different nodes. In this section we assume that all traffic demands are bidirectional and symmetrical. In this case the network can be modeled by an undirected graph. The model can be simply generalised for un-symmetrical demands, by using directed graphs. In later case the model is more complex and for this reason the algorithms will run slower.

2.1. MODEL OF LINKS

A network consists of nodes, and links connecting the nodes. This can be modeled by a graph: a node is a vertex and a link is an edge. Having multiple WLs we will represent a WL of a link as an edge in the graph of wavelengths according to Figure 1 for the network proposed in [20]. To prioritise filling up WLs one-by-one we can assign slightly different weights to different channels of one link. For example, edges representing WL1, WL2 and WL3 will have weights 101, 102 and 103 respectively.

2.2. MODEL OF NODES

A node is modeled by a subgraph. The subgraph-nodes are the certain WLs at the switch-ports, while the weighted edges represent the costs of transitions, terminations, conversions, etc. There are different types of nodes. Models of nodes differ for these. Here will be shown some examples. In similar manner a model can be derived for any additional

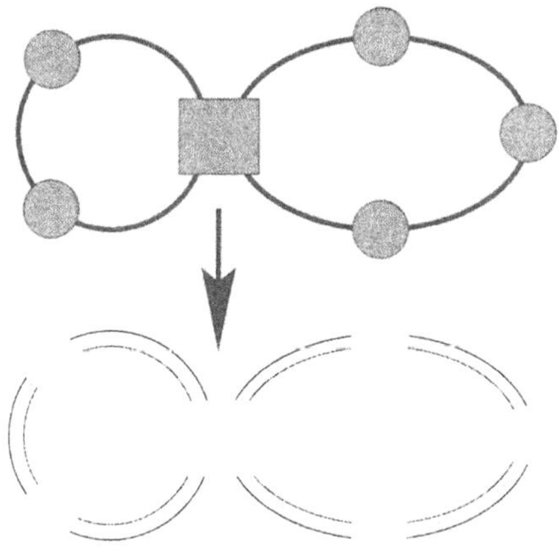

Figure 1 Modeling edges.

node-type. The models proposed here are similar to those described in [21], but those were used for setting up connections one-by-one using shortest path algorithms, while here is the emphasis on global simultaneous configuration requiring special node-models.

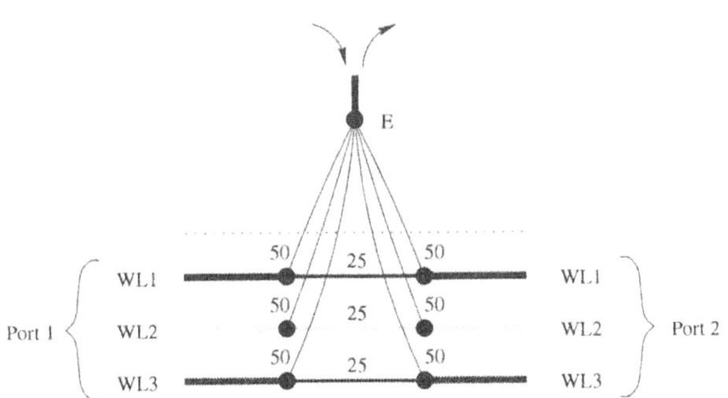

Figure 2 Model of OADM Nodes.

Optical Add-and-Drop Multiplexer: OADM. The OADM Nodes have in general two bi-directional ports (4 fibres). Their function is either to transmit a WL channel or to terminate it and usually they do not allow WL-conversion. These devices are capable to add or drop wavelength channels.

The weights assigned to edges representing termination (e.g., 50) are higher than weights of transition (e.g., 25), because transition is preferred to termination. According to the proposed model (Figure 2) the traffic streams can enter or exit the OADM crossing vertex E or can be even re-multiplexed.

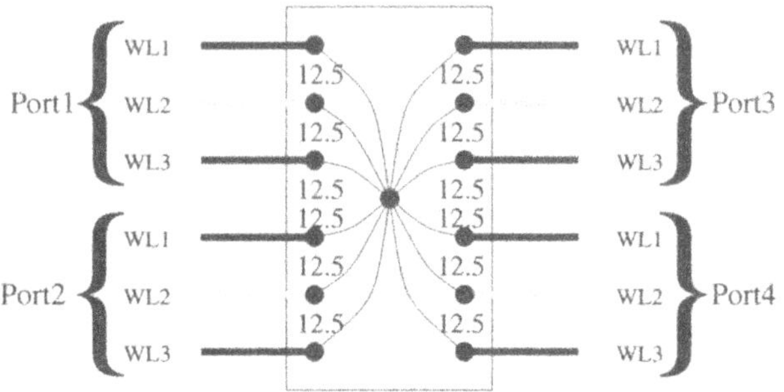

Figure 3 Model of OEXC Nodes.

Optical Cross-Connect with Electrical Core: OEXC. In the model shown in Figure 3 each pair of nodes should be connected by an edge. All edges should have equal weights. Instead of connecting all pairs using nxn edges we use n edges and one node. This simplifies the model. Each incoming channel is converted to electrical domain switched by a space-switch and again converted to the optical domain to arbitrary WL. Each termination, transition or WL change of a light-path has the same cost (e.g., 25). Therefore all edges have the same weight (e.g., 25/2).

Optical Cross-Connect with All-Optical Core: OXC. An optical Cross-Connect has more than two ports, e.g., four bi-directional ports according to Figure 1. In an OXC a light-path can make transition to any output port which supports that WL, and that WL is not yet used. This OXC type (without WL change capability) will be referred to as *simple* OXC (see Figure 4). In this case one incoming channel can exit at any of the remaining output ports where that WL is supported and not yet used.

In some OXC devices WL translation (change) is also supported. This node will be called OXC. Its model is showed in Figure 5. For this node any incoming channel can exit at one of 11 remaining channels. Now there are 3 possibilities for light-path transition since there are channels

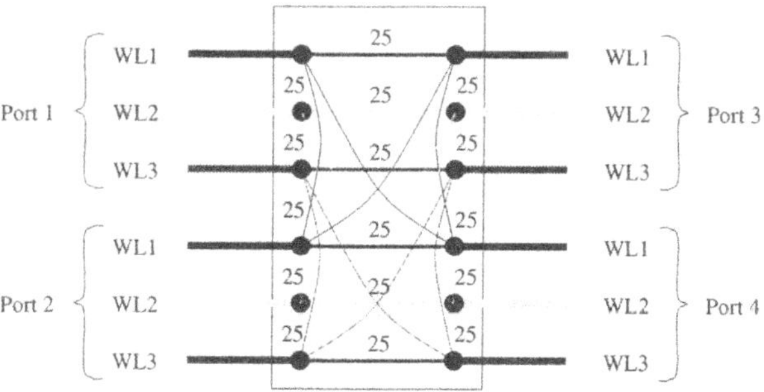

Figure 4 Model of a Simple Optical Cross-Connect Node (without WL conversion)

of the same wavelength on all ports in our example. For WL change there are 8 possibilities. It has higher cost (e.g., 2x50=100) than the WL transition. WL change is modeled by conducting the traffic stream through node E.

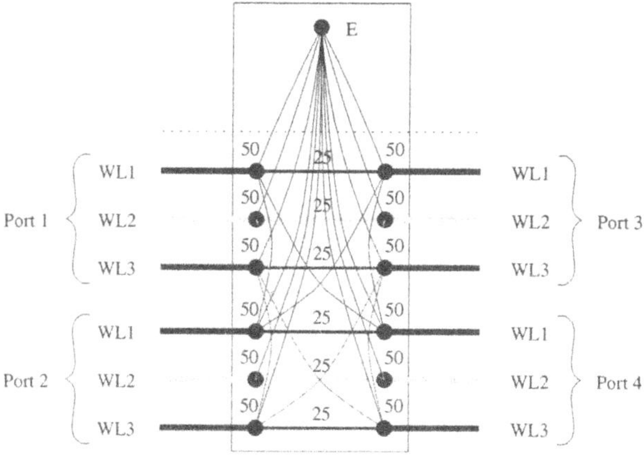

Figure 5 Model of an Optical Cross-Connect Node (with WL conversion)

In some cases the traffic stream termination is also among the functions of an OXC. In that case the model does not need any change. The only difference will be that there will be some traffic offered to that OXC node which can be modeled by offering traffic to node E and considering it as an end-node. In this case traffic-stream re-multiplexing capability is also required.

Modeling opto-electro-optical conversions, multiplexing and re-multiplexing. If we want to differentiate the simple wavelength-change from the electrical signal re-multiplexing a more complex model is needed. An example has been shown in Figure 6 for an OADM node for simplicity reasons, which can be extended to any other node-type. As can be seen node E has been substituted by a fully connected sub-graph. In this case assigning costs to internal edges the costs of wavelength-change and signal re-multiplexing can be differentiated.

All-Optical WL conversion is not supported by all OXCs. Therefore the optical signal is terminated and passed to the electrical layer where space switching or space switching with time switching (re-multiplexing) is done and then the resulting electrical signal passed back to the optical layer.

In cross connects three levels of cross-connecting and switching are to be differentiated:

- WL transition - This is done by the optical layer without any processing. This is the preferred and cheapest function. The signal can bypass the electrical layer using lightpaths of the optical layer.
- WL translation - It can be carried out by optical WL shifters, or by opto-electrical conversion, space-switching and electro-optical conversion. It is more expensive than the previous function, but still cheaper than the next one. Here is the switching very simple and no traffic stream processing is needed.
- multiplexing and re-multiplexing - In a larger WDM transport network there are considerably less available WLs per fibre than it would be needed for full interconnection of the end nodes by single-hop lightpaths. For this reason some of the traffic streams have to be multiplexed along a lightpath, i.e., in some cases the lightpath termination is not a traffic stream termination. In these cases time-division re-multiplexing is needed. In this case the part of the equipment performing the Time Division Multiplexing is able to add traffic to that node or to drop it. In this paper we will not take this capability into account.

3. PROBLEM FORMULATION

It is algorithmically very complex to obtain globally optimal solution for the global simultaneous routing and wavelength assignment problem. (The problem can be expressed as well as configuration of the lightpath system of a WDM network.) This problem very likely belongs to the class

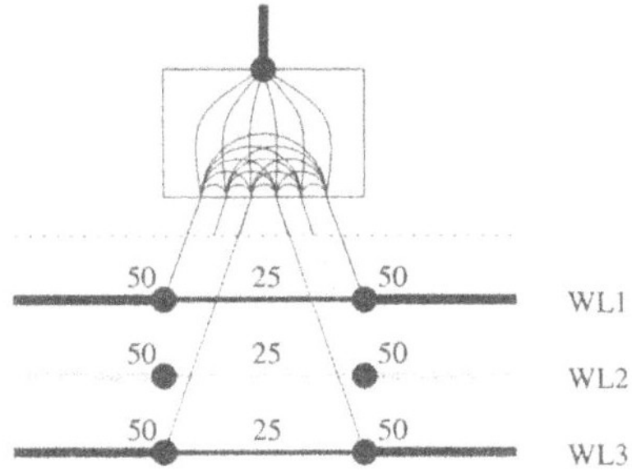

Figure 6 Modeling opto-electro-optical conversions and re-multiplexing: the complex model

of NP-hard problems [22], because its sub-problem, the Static Lightpath Establishment (SLE) has been shown to be NP-hard [23].

The task was to find a shortest path in the obtained graph (which is built up of the link and node models) between all pairs of nodes simultaneously. There are alternatives for choosing the objective of the optimisation, e.g.,:

1. Minimise the total number of used WLs per fibre.
2. Decrease the total amount of used resources at the optical layer.
3. Decrease the total amount of used resources and processing at the electrical layer.
4. Decrease the total amount of lost traffic. It is also possible to introduce a scale-up factor for all traffic demands. For example scaling up all traffic demands by 10% the network should still work properly.
5. Minimise the number of WL conversions in total and for each path.

Our objective function will optimise 3. and 5. simultaneously, as will be discussed.

3.1. ILP FORMULATIONS FOR DIRECTED GRAPHS

The above described problem can be formulated as an Integer Linear Program using the proposed model. For this purpose we will first formulate the problem for the directed graph model. This formulation has slight similarities with Minimal Cost Multicommodity Flow (MCMCF) problem formulation [24].

Let D(V,A,C) be a directed graph (digraph) where V is the set of vertices, $A = \{(i,j) \in A, i,j \in V\}$ is the set of arcs (directed edges) and $C = \{c_{ij} \in R^+, (i,j) \in A\}$ is the cost of using arc (i,j). Set $O = \{o(s^o, t^o, b^o) : s^o, t^o \in V; b^o \in R^+\}$ is the set of all demands characterised by its source s^o sink t^o and bandwidth requirement b^o. $V_E \subset V$ is the set of vertices representing electrical nodes capable of performing grooming, i.e., time division multiplexing. These nodes are denoted by 'E' in Figures 2 and 5. $V_i^+ = \{k : (i,k) \in A\}$ and $V_i^- = \{k : (k,i) \in A\}$ denote all arcs entering and leaving node i respectively.

The number of the wavelength channels is assumed to be equal for all fibers, and each link consists of two fibers carrying information in opposite directions. Without loss of generality for simplicity reasons the capacity B of each wavelength channel is assumed to be equal as well.

Variable $x_{ij}^o \in \{0,1\}$ denotes flow of commodity o in arc (i,j), while variable $y_{ij} \in \{0,1\}$ indicates, whether lighpath spanning between i and j is used (1) or not (0). For illustration see Figure 7, where the used internal light-links are denoted by solid lines.

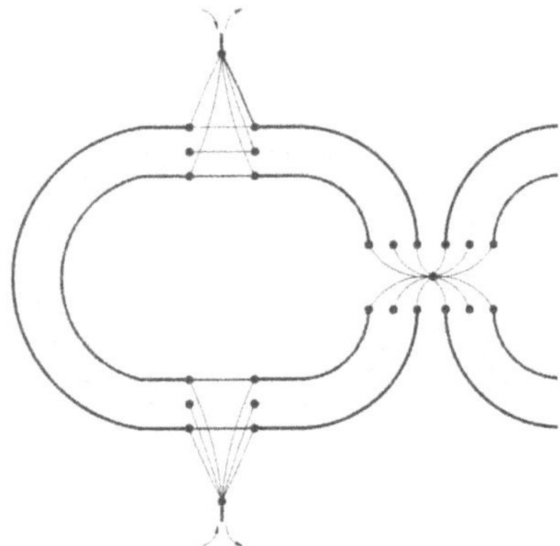

Figure 7 The graph model of a part of the network shown in Figure 1 with 2 OADM and 1 OEXC node.

Objective:

$$\text{minimise} \quad (\alpha C_E + (1-\alpha) C_O) \tag{1}$$

$$\text{where} \quad 0 \leq \alpha \leq 1 \tag{2}$$

$$\text{while} \quad C_E = \sum_{(i,j)\in A, i\in V_E \vee j \in V_E} \sum_{o\in O} c_{ij} b^o x_{ij}^o \tag{3}$$

$$\text{and} \quad C_O = \sum_{(i,j)\in A, i\notin V_E \wedge j \notin V_E} \sum_{o\in O} c_{ij} y_{ij}^o \tag{4}$$

Subject to constraints:

$$\sum_{\forall o \in O} x_{ij}^o b^o \leq B \qquad \forall (i,j) \in A \tag{5}$$

$$x_{ij}^o \leq y_{ij} \qquad \forall (i,j) \in A, \quad \forall o \in O \tag{6}$$

$$y_{ij} \leq \sum_{\forall o \in O} x_{ij}^o \qquad \forall (i,j) \in A \tag{7}$$

$$\sum_{\forall j \in V_i^+} x_{ji}^o - \sum_{\forall k \in V_i^-} x_{ik}^o = \begin{cases} -1 & \text{if } i = s^o \\ 0 & \text{if } i \in V \setminus \{s^o, t^o\} \\ 1 & \text{if } i = t^o \end{cases}$$
$$\text{for all nodes } i \in V \text{ and commodities } o \tag{8}$$

$$y_{ji} = \sum_{\forall k \in V_i^-} y_{ik} \qquad \forall i \in V \setminus V_E, \forall j \in V_i^+ \tag{9}$$

Variables:

$$x_w^o \in \{0,1\} \qquad \forall (i,j) \in A, \quad \forall o \in O \tag{10}$$

$$y_{ij} \in \{0,1\} \qquad \forall (i,j) \in A \tag{11}$$

The objective is to minimise the number of hops for each traffic demand weighted by the required capacity of that traffic stream and by the cost of using those light-links subject to the following constraints. Constraint (5) states that the amount of traffic using a light-link may not exceed the capacity B of that light-link.

Constraint (8) ensures that traffic-streams are to be terminated at end-nodes and the traffic flows must be conserved at each non-end node.

Constraints (6) and (7) guarantee that traffic streams may use available light-paths only, and a light-path will be established only if it is needed for carrying a traffic flow. Ninth constraint expresses that a light-path can not branch.

The last two constraints (10) and (11) mean that both variables can take values 0 or 1 only.

The optimisation will result in a single-hop configuration (Wavelength-paths) whenever possible or in a multihop configuration with as few WL translations and re-multiplexing as possible, i.e., the largest possible part of the load of the electrical layer will be overtaken by the optical layer.

If the aim was to decrease the number of used WLs in total, the value of α should be strictly less than 1, possibly as close to 0 as possible. Then constraint (7) can be avoided.

3.2. ILP FORMULATIONS FOR UNDIRECTED GRAPHS

To decrease the number of both, variables and constraints in this Sub-section the undirected graph model will be used.

Let U(V,E,C) be the undirected graph (ugraph) where V is the set of vertices, $E = \{(i,j) \in E, i,j \in V\}$ where $(i,j) = (j,i)$ is the set of (undirected) edges and $C = \{c_{ij} \in R^+, (i,j) \in E\}$ is the cost of using edge (i,j). Set $O = \{o(s^o, t^o, b^o) : s^o, t^o \in V; b^o \in R^+\}$ is the set of all demands characterised by its two ends s^o and t^o and bandwidth requirement b^o. $V_E \subset V$ is the set of vertices representing electrical nodes capable of performing grooming, i.e., time division multiplexing. V_i is the set of neighbour-nodes of node i (nodes adjacent to i) defined as $V_i = \{j : (i,j) \in E\}$, remember, that $(i,j) = (j,i)$.

As for the directed case the number of the wavelength channels is assumed to be equal for all fibers, and each link consists of two fibers carrying information in opposite directions. Without loss of generality for simplicity reasons the capacity B of each wavelength channel is assumed to be equal as well.

Variable $x^o_{ij} \in \{0,1\}$ denotes flow of commodity o in edge (i,j), while variable $y_{ij} \in \{0,1\}$ indicates, whether lighpath spanning between i and j is used (1) or not (0).

Objective:

$$\text{minimise} \quad (\alpha C_E + (1-\alpha) C_O) \tag{12}$$

$$\text{where} \quad 0 \leq \alpha \leq 1 \tag{13}$$

$$\text{while} \quad C_E = \sum_{(i,j)\in E, i\in V_E \vee j \in V_E} \sum_{o\in O} c_{ij} b^o x^o_{ij} \tag{14}$$

$$\text{and} \quad C_O = \sum_{(i,j)\in E, i\notin V_E \wedge j \notin V_E} \sum_{o\in O} c_{ij} y^o_{ij} \tag{15}$$

Subject to constraints:

$$\sum_{\forall o \in O} x^o_{ij} b^o \leq B \qquad \forall (i,j) \in E \tag{16}$$

$$x^o_{ij} \leq y_{ij} \qquad \forall (i,j) \in E, \quad \forall o \in O \tag{17}$$

$$y_{ij} \leq \sum_{\forall o \in O} x^o_{ij} \qquad \forall (i,j) \in E \tag{18}$$

$$\sum_{\forall j \in V_i} x^o_{ij} = 1 \qquad \forall i \in V_E, \quad \forall o \in O \tag{19}$$

$$\sum_{\forall j \in V_i} x_{ij}^o = 2z_{xi}^o \qquad \forall i \in V \setminus V_E, \quad \forall o \in O \tag{20}$$

$$\sum_{\forall j \in V_i} y_{ij} = 2z_{yi} \qquad \forall i \in V \setminus V_E \tag{21}$$

$$\tag{22}$$

Variables:

$$x_w^o \in \{0,1\} \qquad \forall (i,j) \in E, \quad \forall o \in O \tag{23}$$

$$y_{ij} \in \{0,1\} \qquad \forall (i,j) \in E \tag{24}$$

$$z_{xi}^o \in \{0,1\} \qquad \forall i \in V \setminus V_E, \quad \forall o \in O \tag{25}$$

$$z_{yi} \in \{0,1\} \qquad \forall i \in V \setminus V_E \tag{26}$$

where the variable z has been used for emulating the condition "equal to 0 or 2"

As in the case for directed graphs the objective is to minimise the number of hops for each traffic demand weighted by the required capacity of that traffic stream and by the cost of using those light-links subject to the following constraints. Constraint (16) states that the amount of traffic using a light-link may not exceed the capacity B of that light-link.

Constraints (19) and (20) ensure that traffic-streams are to be terminated at end-nodes and the traffic flows must be conserved at each non-end node.

Constraints (17) and (18) guarantee that traffic streams may use available light-paths only, and a light-path will be established only if it is needed for carrying a traffic flow. 21th constraint expresses that a light-path can not branch.

The last four constraints (23) - (26) mean that all variables can take values either 0 or 1.

The problem formulated here is the same as it was for the directed graph, with only difference, that it is less complex, i.e., needs less variables and constraints.

If the aim was to decrease the number of used WLs in total, the value of α should be strictly less than 1, possibly as close to 0 as possible. Then constraint (18) can be avoided.

4. PROTECTION

Here we assume dedicated (i.e., either 1+1 or 1:1) protection. The aim is to protect all demands either at the electrical or at the optical layer by using link- or node-disjoint working and protection path for each demand. ILP formulation of these cases follows.

4.1. ILP FORMULATION OF PROTECTION AT UPPER LAYERS

Protection at upper layers means that the end-to-end connections at the uppermost layer have to be protected by configuring a protection path at the uppermost layer for each working path which is physically disjoint of the working one. That is the two paths may not use a common cable or node at the same geographical location.

Link Disjoint Case. To formulate the Link Disjoint case we have to introduce sets of edges (i,j) which share a common optical cable or even a duct. Let L be the set of all physical links l. Then $L = \{l : l \subset V, \forall(i,j) \in l$ belong to the same physical link $l\}$. Here we will continue using directed graphs, however, as shown in Section 3 we can reformulate it for directed graphs whenever needed.

We need a set of new variables. Instead of x^o_{ij} we will use $x1^o_{ij}$ and $x2^o_{ij}$ for indicating working and protection flow of demand o over edge (i,j). In this way we have doubled the number of variables, and slightly increased the number of constraints as well.

Now the formulation is as follows:

Objective:

$$\text{minimise} \quad (\alpha C_E + (1-\alpha)C_O) \tag{27}$$

$$\text{where} \quad 0 \leq \alpha \leq 1 \tag{28}$$

$$\text{while} \quad C_E = \sum_{(i,j)\in A, i\in V_E \vee j\in V_E} \sum_{o\in O} c_{ij} b^o (x1^o_{ij} + x2^o_{ij}) \tag{29}$$

$$\text{and} \quad C_O = \sum_{(i,j)\in A, i\notin V_E \wedge j\notin V_E} \sum_{o\in O} c_{ij} y^o_{ij} \tag{30}$$

Subject to constraints:

$$\sum_{\forall o\in O} (x1^o_{ij} + x2^o_{ij}) b^o \leq B \qquad \forall(i,j)\in A \tag{31}$$

$$x1^o_{ij} + x2^o_{ij} \leq y_{ij} \qquad \forall(i,j)\in A, \quad \forall o \in O \tag{32}$$

$$y_{ij} \leq \sum_{\forall o\in O} (x1^o_{ij} + x2^o_{ij}) \qquad \forall(i,j)\in A \tag{33}$$

$$\sum_{\forall(i,j)\in l} (x1^o_{ij} + x2^o_{ij}) \leq 1 \qquad \forall l \in L, \quad \forall o \in O \tag{34}$$

$$\sum_{\forall j \in V_i^+} x1^o_{ji} - \sum_{\forall k \in V_i^-} x1^o_{ik} = \begin{cases} -1 & \text{if } i = s^o \\ 0 & \text{if } i \in V \setminus \{s^o, t^o\} \\ 1 & \text{if } i = t^o \end{cases}$$

for all nodes $i \in V$ and commodities o (35)

$$\sum_{\forall j \in V_i^+} x2^o_{ji} - \sum_{\forall k \in V_i^-} x2^o_{ik} = \begin{cases} -1 & \text{if } i = s^o \\ 0 & \text{if } i \in V \setminus \{s^o, t^o\} \\ 1 & \text{if } i = t^o \end{cases}$$

for all nodes $i \in V$ and commodities o (36)

$$y_{ji} = \sum_{\forall k \in V_i^-} y_{ik} \qquad \forall i \in V \setminus V_E, \forall j \in V_i^+ \tag{37}$$

Variables:

$$x1^o_w \in \{0,1\} \qquad \forall (i,j) \in A, \quad \forall o \in O \tag{38}$$

$$x2^o_w \in \{0,1\} \qquad \forall (i,j) \in A, \quad \forall o \in O \tag{39}$$

$$y_{ij} \in \{0,1\} \qquad \forall (i,j) \in A \tag{40}$$

In this formulation constraint (34) is that one which ensures diversity (i.e., link disjointness) of working and protection paths. Note, that we have implicitly assumed that arcs between the same pair of nodes of different direction (i,j) and (j,i) belong to the same set l. This constraint allows to simply substitute x^o_{ij} by $(x^o_{ij} + x^o_{ij})$ in all constraints, except for (35) and (36).

Node Disjoint Case. If we want to make the protection and working paths not only link- but node-disjoint as well then instead of (34) we should write any of the following constraints:

$$\sum_{\forall i \in n \setminus V_E^o} \sum_{\forall j \in A_i^+} (x1^o_{ij} + x2^o_{ij}) \leq 1 \qquad \forall n \in N, \quad \forall o \in O \tag{41}$$

$$\sum_{\forall i \in n \setminus V_E^o} \sum_{\forall j \in A_i^-} (x1^o_{ij} + x2^o_{ij}) \leq 1 \qquad \forall n \in N, \quad \forall o \in O \tag{42}$$

In this case constraints (35) and (36) can be written as one constraint, i.e., instead of them we should write

$$\sum_{\forall j \in V_i^+} (x1^o_{ji} + x2^o_{ji}) - \sum_{\forall k \in V_i^-} (x1^o_{ik} + x2^o_{ik}) = \begin{cases} -1 & \text{if } i = s^o \\ 0 & \text{if } i \in V \setminus \{s^o, t^o\} \\ 1 & \text{if } i = t^o \end{cases}$$

for all nodes $i \in V$ and commodities o(43)

4.2. ILP FORMULATION OF PROTECTION AT THE DWDM LAYER

Protection at the DWDM layer means that the end-to-end connections at the uppermost layer have to be protected by configuring a protection wavelength path for each working wavelength path which is physically disjoint of the working one. That is the two paths may not use a common cable or a node at the same geographical location.

Link Disjoint Case. In this case we should substitute all variables y_{ij} by $y1_{ij}$ and $y2_{ij}$ and rewrite the objective functions and constraints in way to ensure, that for each working wavelength-path there should be a disjoint protection one.

Objective:

$$\text{minimise} \quad (\alpha C_E + (1-\alpha)C_O) \tag{44}$$

$$\text{where} \quad 0 \leq \alpha \leq 1 \tag{45}$$

$$\text{while} \quad C_E = \sum_{(i,j)\in A, i\in V_E \vee j\in V_E} \sum_{o\in O} c_{ij} b^o x_{ij}^o \tag{46}$$

$$\text{and} \quad C_O = \sum_{(i,j)\in A, i\notin V_E \wedge j\notin V_E} \sum_{o\in O} c_{ij}(y1_{ij}^o + y2_{ij}^o) \tag{47}$$

Subject to constraints:

$$\sum_{\forall o\in O} x_{ij}^o b^o \leq B \qquad \forall (i,j)\in A \tag{48}$$

$$x_{ij}^o \leq y_{ij} \qquad \forall (i,j)\in A, \quad \forall o \in O \tag{49}$$

$$y_{ij} \leq \sum_{\forall o\in O} x_{ij}^o \qquad \forall (i,j)\in A \tag{50}$$

$$\sum_{\forall j\in V_i^+} x_{ji}^o - \sum_{\forall k\in V_i^-} x_{ik}^o = \begin{cases} -1 & \text{if } i = s^o \\ 0 & \text{if } i \in V \setminus \{s^o, t^o\} \\ 1 & \text{if } i = t^o \end{cases}$$

$$\text{for all nodes } i \in V \text{ and commodities } o \tag{51}$$

$$y1_{ji} = \sum_{\forall k\in V_i^-} y1_{ik} \qquad \forall i \in V \setminus V_E, \forall j \in V_i^+ \tag{52}$$

$$y2_{ji} = \sum_{\forall k\in V_i^-} y2_{ik} \qquad \forall i \in V \setminus V_E, \forall j \in V_i^+ \tag{53}$$

$$y2_{ji} = \sum_{\forall k \in V_i^-} y1_{ik} \qquad \forall i \in V_E, \forall j \in V_i^+ \tag{54}$$

$$\sum_{\forall (i,j) \in l} (y1_{ij} + y2_{ij}) \leq 1 \qquad \forall l \in L \tag{55}$$

Variables:

$$x_w^o \in \{0,1\} \qquad \forall (i,j) \in A, \quad \forall o \in O \tag{56}$$

$$y1_{ij} \in \{0,1\} \qquad \forall (i,j) \in A \tag{57}$$

$$y2_{ij} \in \{0,1\} \qquad \forall (i,j) \in A \tag{58}$$

Here constraint (52) state that each working wavelength-path is continuous between its electrical terminations, constraint (53) states the same for protection wavelength-paths, while constraint (54) states, that where a working wavelength-path is terminated, there must start a protection wavelength-path. Constraint (55) guarantees that the working wavelength-path and protection wavelength-path must be disjoint.

Node Disjoint Case. To obtain protection path which is node-disjoint of the working one we need to substitute constraint (55) by the following one.

$$\sum_{\forall i \in n \setminus V_E} \sum_{\forall j \in A_i^+} (y1_{ij} + y2_{ij}) \leq 1 \qquad \forall n \in N \tag{59}$$

5. COMMENTS ON THE PROBLEM

In paper [2] the authors present different ways of formulating the problem for both single-hop and multihop lightpaths referred to as Wavelength-Path (WP) and Virtual Wavelength-Path (VWP) respectively. Among other methods they also use ILP. The drawback of the VWP formulation using ILP (and also of other methods applied to VWP) is that they implicitly assume pure electrical nodes, where not only electrical space-switching but also time-switching, i.e., re-multiplexing has to be performed. This approach degrades the Wavelength Routing network to a network employing WDM links and therefore it requires electrical (e.g., ATM) switches of large capacities increasing the costs and deteriorating the performance. The advantage of the method proposed in this paper is that it can differentiate various nodes with flexible functionality.

In paper [7] also a graph-model has been used but vertices are arranged in a matrix like grid, and edges representing lightpaths afterwards. Although the method is advantageous because of the applicability of fast shortest path algorithms, it does not take into account different node-types except the electrical switches. Furthermore, the method is used for routing of lightpaths only, not for configuration, i.e., simultaneous routing of all traffic demands.

The bin-packing problem, where we want to pack objects of different size into bins (of equal size) in optimal way is NP-hard. Its generalisation for the multidimensional case is called vector-packing. If the wavelength-channel capacities over all links are considered to be "bins" which form a vector indexed by links and we want to "pack" them optimally by traffic streams of different demands (different "objects"), we would have the same problem. However, our problem is even more complex, since there is "interaction" between "bins", because loading one bin will induce loading one of the neighbouring bins. However, these relations change dynamically, since if two neighbouring bins are filled by an object, it will avoid filling some other bins. For the above reasons this method is expected to be also NP-hard.

6. SOLUTION ALTERNATIVES

The presented network model allows many methods for solving the above formulated problem. The alternatives include, but are not limited to:

- ILP Solvers
- Relaxing ILP to LP and then rounding the variables.
- Generally applicable heuristics for global optimisation directly applied for 0-1 programming (e.g., Simulated Annealing (SA), Genetic Algorithm (GA), Threshold Accepting (TA), Tabu Search (TS), Go with the Winners (GW), etc.).
- Generally applicable heuristics for global optimisation (SA, GA, TA, TS, GW,...) applied through specially matched models.
- Heuristic Algorithms based on decomposition.

Solving the problem by any available ILP solver (e.g., LP_ SOLVE or CPLEX) will be possible for very small networks only. The reason is that the number of both, variables and constraints will grow by increasing the size of the network and this will result in exponential growth of the alternatives to be investigated by ILP software. Although ILP solvers give global optimum, they do not solve problems of larger scale. There are alternatives.

It is possible to relax integrality condition, i.e., to use continuous variables bounded by the closed interval $[0, 1]$ inserted of binary ones.

In this case the result can be obtained in shorter time even for problems of large scale, however, rounding of variables can be needed.

Another alternative is to apply directly to ILP formulation randomised methods, e.g., Simulated Annealing, Genetic Algorithm or Tabu Search - all using binary encoding. All constraints are evaluated, and if violated a penalty term is added to the objective. The penalty term for each constraint is a function of the penalty violation. Although different functions were tried out, ranging from linear to those which punish large deviations from the constraint more strictly, the method does not give expected results. Fine tuning of coefficients in the linear combination of the objective and penalty terms was needed. However, it has happened sometimes, that some traffic demands were not satisfied (x, y and z variables were 0), but non of the constraints was violated.

Based on the above formulation even more sophisticated heuristic approximation methods can be used. The idea is to exclude a part of the state-space which is not of interest, instead of using penalty term as a means of obeying constraints.

The decomposition according to the node-pairs appeared to be very promising as well. This is approximation only, but a very fast one. In this case the demands are routed one-by-one.

7. CONCLUSION

In this paper the general model for DWDM networks is proposed which enables the ILP formulation of the static RWA problem including traffic grooming and protection at either electrical or WDM layer. ILP formulation is proposed for both, directed and undirected graph models.

In static case this approach allows configuration of combined MPLS-MPλS (e.g., GMPLS) networks. This approach can be used for dynamic RWA as well, furthermore it can be combined by routing techniques like OSPF, OMP-OSPF or PNNI.

The advantage of our approach is that it optimises jointly the electrical layer and the underlying wavelength or fiber system, it includes various protection mechanisms, and it supports traffic grooming whenever the number of available wavelengths is not sufficient, but certain wavelength channels have some spare capacity.

8. ACKNOWLEDGEMENTS

This work has been done in the research co-operation framework between Ericsson and the High-Speed Networks Laboratory (HSN*Lab*) at the Department of Telecommunications and Telematics, Budapest University of Technology and Economics. The authors are grateful to Miklós Boda (Ericsson) and Tamás Henk (HSN*Lab*) for their support.

References

[1] P.E. Green, "Optical Networking Update", *IEEE Journal on Selected Areas in Communications*, vol. 14, no. 5, pp. 764-779, June 1996

[2] N. Wauters, P. Demeester, "Design of the Optical Path Layer in Multiwavelength Cross-Connected Networks", *IEEE Journal on Selected Areas in Communications*, vol. 14, no. 5, pp. 881-892, June 1996

[3] Z. Zhang, A.S. Acampora, "A Heuristic Wavelength Assignment Algorithm for Multihop WDM Networks with Wavelength Routing and Wavelength Re-Use" *Transactions on IEEE Networking*, vol. 3, no. 3, pp. 281-288, June 1995

[4] R. Ramaswami, K.N. Sivarajan, "Routing and Wavelength Assignment in All-Optical Networks", *IEEE Transactions on Networking*, vol. 3, no. 5, pp. 489-500, Oct. 1995

[5] R. Ramaswami, K.N. Sivarajan, "Design of Logical Topologies for Wavelength-Routed Optical Networks", *IEEE Journal on Selected Areas in Communications*, vol. 14, no. 5, pp. 840-851, June 1996

[6] D. Banerjee, B. Mukherjee, "A Practical Approach for Routing and Wavelength Assignment in Large Wavelength-Routed Optical Networks", *IEEE Journal on Selected Areas in Communications*, vol. 14, no. 5, pp. 903-908, June 1996

[7] I. Chlamtac, A. Faragó, T. Zhang, "Lightpath (Wavelength) Routing in Large WDM Networks" *IEEE Journal on Selected Areas in Communications*, vol. 14, no. 5, pp. 909-913, June 1996

[8] B. Mukherjee, D. Banerjee, S. Ramamurthy, A. Mukherjee, "Some Principles for Designing a Wide-Area WDM Optical Network", *IEEE Transactions on Networking*, vol. 4, no. 5, pp. 684-696, Oct. 1996

[9] H. Harai, M. Murata, H. Miyahara, "Performance of All-Optical Networks with Limited-Range Wavelength Conversion" *ICC'97 - IEEE International Conference on Communications*, Montréal, June 1997

[10] M. Garnot, M. Sotom, F. Masetti, "Routing Strategies for Optical Paths in WDM Networks", *ICC'97 - IEEE International Conference on Communications*, Montréal, June 1997

[11] J. Armitage, O. Crochat, J.-Y. Le Boudec, "Design of a Survivable WDM Photonic Network", *INFOCOM 97*, Kobe, April 1997

[12] L. Berthelon et.al., "Towards photonic networking: experimental demonstration of a reconfigurable survivable WDM ring network", *GLOBCOM*, London, November 1996

[13] K. Bala, G. Ellinas, M. Post, "Towarwds Hitless Reconfiguration in WDM Optical Networks for ATM Transport", *GLOBCOM*, London, November 1996

[14] E. Karasan, E. Ayanoglu, "Effects of Wavelength Routing and Selection Algorithms on Wavelength Conversion Gain in WDM Networks", *GLOBCOM*, London, November 1996

[15] S. Banerjee, C. Chen, "Design of Wavelength-Routed Optical Networks for Circuit Switched Traffic", *GLOBCOM*, London, November 1996

[16] D. Banerjee, B. Mukherjee, "Wavelength-Routed Optical Networks: Linear Formulation, Resource Budgeting Tradeoffs and a Reconfiguration Study", *INFOCOM 97*, Kobe, April 1997

[17] J.-F. P. Labourdette, "Performance Impact of Partial Reconfiguration on Multihop Lightwave Networks", *IEEE Transactions on Networking*, vol. 5, no. 3, pp. 351-358, June 1997

[18] T. Cinkler, "Heuristic Algorithms for Configuration of the ATM-Layer over Optical Networks" *ICC'97 - IEEE International Conference on Communications*, Montréal, June 1997

[19] R.M. Krishnaswamy, K.N. Sivarajan, "Design of Logical Topologies: a Linear Formulation for Wavelength Routed Optical Networks with no Wavelength Changers", *IEEE INFOCOM'98*, March 1998

[20] S. Johansson, "Transport Network Involving a Reconfigurable WDM Network Layer - A European Demonstration", *IEEE Journal on Lightwave Technology*, vol.14, no.6, pp. 1341-1348, June 1996

[21] T. Cinkler, R.S. Castro, S. Johansson, "Configuration and Reconfiguration of WDM Networks", *NOC'98 - Conference on Networks and Optical Communications*, Manchester, June 1998

[22] M.R. Garey, D.S. Johnson, "Computers and Intractability – A Guide to the Theory of NP-completeness", W.H. Freeman and Co., New York, 1979

[23] I. Chlamtac, A. Ganz, G. Karmi, "Lightpath Communications: An Approach to High Bandwidth Optical WANs", *IEEE Transactions on Communications*, vol. 40, no.7, pp. 1171-1182, July 1992

[24] R.K. Ahuja, T.L. Magnanti, J.B. Orlin, "Network Flows, Theory, Algorithms and Applications", *Prentice-Hall*, 1993

MAPPING OF ARBITRARY TRAFFIC DEMAND AND NETWORK TOPOLOGY ON A MESH OF RINGS NETWORK

Christian Mauz
Communication Technology Laboratory
ETH Zurich
Switzerland
mauz@nari.ee.ethz.ch

Abstract An efficient algorithm is presented to map an arbitrary traffic pattern and network topology on a mesh of rings network. The network is described by a two-layer model. The first layer defines the ring configuration and the second the connection allocation. Rings are constructed systematically from faces. It is assumed that a ring provides a full-meshed network with a certain capacity between every node pair. Simulated annealing (SA) is used to optimize the mapping to best fit the traffic demands. For that, an objective function with certain parameters is introduced. It is possible to incorporate boundary conditions, e.g. the maximum number of nodes in a ring in this function. The algorithm is tested on the basis of the COST 239 Pan-European case study as an example.

1. INTRODUCTION

Wavelength-division multiplexing is used to an increasing extent for point-to-point connections in wide area networks to satisfy the demand for transport capacity. SDH (Synchronous Digital Hierarchy) or SONET (Synchronous Optical Network) represent the classic transport protocol.

Two approaches can be applied to design a communication network. The first one is connection-based i.e. the routing will be computed for every connection. The second one is structure-based: the network will be built up from blocks like rings or meshes. Several architectures based on regular topologies for all-optical networks e.g. the Colored Section Ring [Ham96] or meshed topologies e.g. the Gridconnect [Jag99], have been proposed and investigated. These concepts assume a certain topology and a homogeneous traffic matrix. The mapping of the real-world topologies and the traffic demand on such simple networks is a challenging task due to the many degrees of freedom. H. Jäger proposed a mapping for the Gridconnect which has the disadvantage of high computational complexity [Jag96].

The aim of this work is to propose an efficient semi-heuristic algorithm with rather low complexity for the mapping of arbitrary topology and traffic pattern on a mesh of rings. It is possible to set in advance a certain fiber topology as well as to compute an optimum one. To justify this approach, an implementation of the algorithm using the `C++` library `LEDA` [Meh96] has been made and the "Pan-European Network" of the COST 239 project has been used as a test-scenario.

To clarify the presentation, not a certain realization of a ring is considered. The ring network is treated as a black box which provide a full-meshed network with a certain transport capacity between the node pairs. For the same reason, protection has been neglected even though this is the main reason to introduce rings.

The structure of the paper is as following. First a two layer graph model is introduced. The ring layer and a systematic way to construct rings out of smaller pieces is presented. Then, the resulting effective graph and the routing thereon is explained. For the optimization "Simulated Annealing" (SA) is used. The results for the Pan-European Network are analyzed. The short summary at the end includes an outlook for further work.

2. MULTI-LAYER GRAPH MODEL

For the description of the network, a two-layer graph model is used (see Fig. 1). The lower layer contains the real network nodes and the real fiber topology. The actual configuration is a set of rings: each ring consists of a set of nodes and edges of the network.

The upper layer is the connection layer. It describes the effective, logical topology which results from the set of rings of the lower layer. Here, for every connection request of the traffic matrix, a path through this logical network has to be found.

The general aim is to minimize the ring configuration with respect to a given objective function so that all traffic demands are fulfilled and the boundary conditions, e.g. the capacity per link, are not violated.

In this work, only the ring identification and the routing in the logical network is investigated. It is assumed, that for the realization of one of these rings, an efficient solution is known (e.g. the Colored Section Ring or an SDH ring). Therefore, for example, the mapping of the resulting channels to distinct wavelengths is not treated. This could be easily

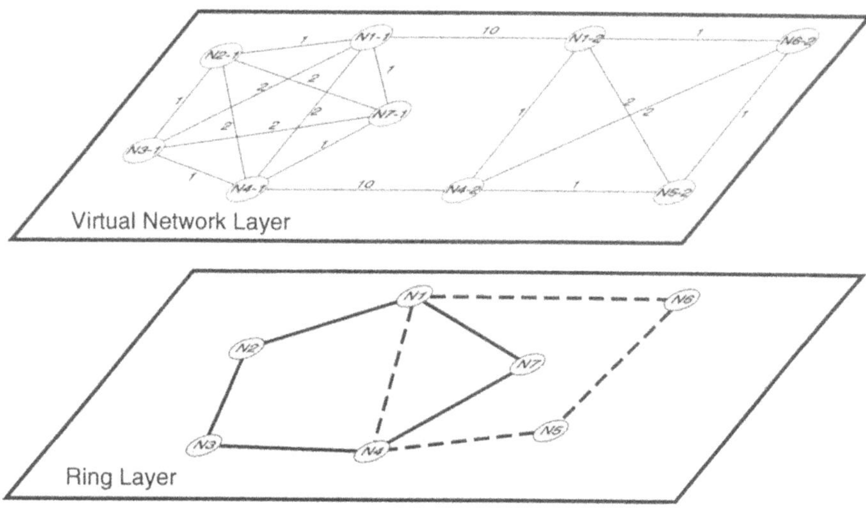

Figure 1 Two layer graph model. The lower layer shows a two ring configuration (solid and dashed edges). In the virtual network layer, the resulting effective graph is depicted. The edges are labeled with the distances. N1-1 and N1-2 are the two virtual copies of N1.

integrated in the algorithm by adding a third layer and using one of the known algorithms for wavelength allocation e.g. "First-Fit" [Har98].

3. THE RING-LAYER

A map is a graph for which the reversal information for every edge is defined. An embedding of a graph is a drawing, where no edge is drawn across a node, images of distinct edges do not cross, and multiple edges between two nodes are embedded the same. A map for which an embedding in the plane exists, is a planar map. A famous theorem of Kuratowski states [Kur30], that every map can be decomposed into a planar map and Kuratowski sub-graphs K_5 (complete graph of degree 5) and $K_{3,3}$ (complete bipartite graph with 2×3 nodes). For the two sub-graphs, a realization by an all-optical network is straight-forward. In addition, due to the high connectivity of these graphs, they will normally not be found in transport networks. Therefore it is sufficient to assume a planar map in the following. Details for graph theory may be found in [Whi73].

In graph theory, there exists the combinatorial and geometrical concept of a face. A face is a list of edges. To construct a face, one starts at

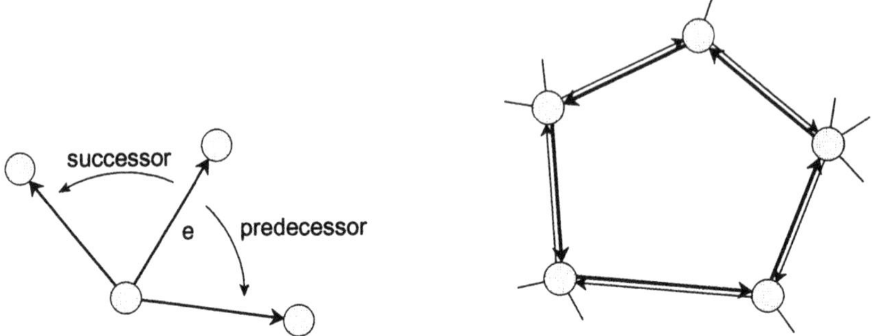

Figure 2 On the left side the definition of the successor and predecessor of the edge e is displayed. On the right side, the constructed face and its edges are marked with thick lines.

a node, follows the edges, and takes at every node the preceding outgoing edge to the next node until one returns to the starting node (see Fig. 2). It is possible to decompose every planar map into faces with complexity linear to the number of edges in the graph.

These faces are the building blocks for rings and form a basis for the space of cycles in a graph. A face is the smallest possible ring in a network. By the addition of a neighbor face to a ring, the ring may be enlarged. On the other hand, to shrink the ring, one of the faces which share an edge with the border of the ring has to be removed. It is worth to note, that this operation is self-inverse: addition of two times the same face to a ring does not change the ring. Some cautions has to be taken for both operations. In some special cases (see an example in Fig. 3), the result is no longer a ring. By using the face concept, it is possible to construct rings in a systematic way with low complexity. Every ring is described by a set of faces. The removal of the last face of a ring destroys the ring. On the other side, the addition of one face to an empty set creates a new ring.

One configuration of the ring layer stands for a set of rings or in other words a mesh of rings with distributed interfaces.

For a given node configuration and edge topology, all faces may be computed. On the other hand, it is possible to use a triangulation algorithm of the plane for the generation of a triangular mesh when only the nodes and their positions are given to gĕt a starting topology. At the end, some of these links may not be used.

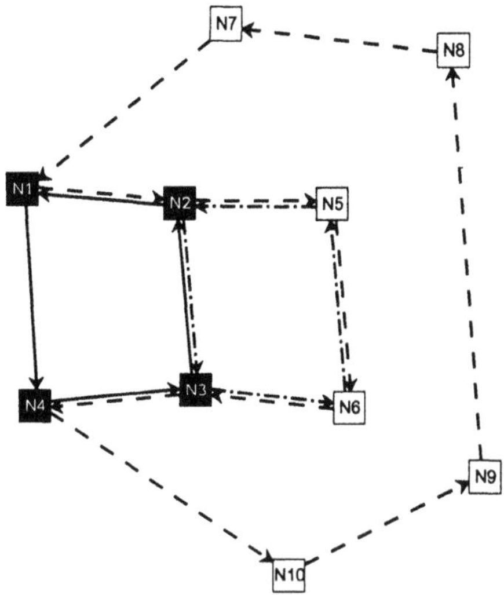

Figure 3 Problem which occurs when enlarging a ring (solid line and black nodes) by a face. The dashed face can't be added to the ring because the result would not be a ring, only the addition of the dot-dashed face is possible.

4. THE VIRTUAL NETWORK LAYER

The ring configuration provides an effective network topology for the connections. Every ring in the lower layer is translated into a complete graph of the virtual network layer. In a complete graph a direct link exists between every node. Nodes which belong to several rings, create multiple copies in the effective graph, where all copies are connected by a bridging link.

For the simulation it is assumed that every ring link has a capacity of four connections. This is e.g. the case for a traffic matrix with granularity of STM-16 (2.5 Gbit/s) and a ring technology which offers STM-64 (10 Gbit/s) channels. The connections could be multiplexed using SDH/SONET technology. The bridging links are not limited in the number of connections they can carry.

For the routing, it is necessary to assign a distance to every link. Here, the inter-ring distance is the number of nodes which are passed in the real network topology and the distance for the bridging links is choosen as $d_{bridge} = 10$ in the examples. This takes into account that changing the ring results in a demultiplexing and a multiplexing operation, which is rather expensive. In general, the value of d_{bridge} depends on the ratio between the cost for bandwith and the cost for multiplexing or demultiplexing, and may even be a function of the network size.

In Fig. 1 an example for a ring-configuration and the resulting effective graph in the higher layer is shown.

For every connection request of the traffic matrix, a connection has to be assigned on the effective graph. Since there is the abstraction of the concrete realization of one ring, this is trivial for connections which can be handled within a direct channel of the ring. For the remaining ones (in the following referred to as "hop connections"), a minimization of the path-length is aspired. Possible paths, e.g. the ten shortest paths, through the network for every node pair are computed with one of the standard algorithms [Had99]. In case that the start or the end node belong to several rings, all virtual copies of it are treated in the same way.

The hop connection requests are assigned in random order. For every request, starting with the shortest possible path, it is checked if enough capacity remains on all links. If this is the case, a solution has been found, otherwise the next longer possible path is examined (in the simulation at least $n = 10$ paths are investigated). When there is no possible routing found among the computed possible paths, the connection is added to a special list of not-fulfilled connections. It would be possible to route this traffic manually in the standard connection-oriented style.

Due to the fact that the granularity of the ring connections is higher than that of the traffic matrix, also the grooming of the different connections can be solved.

For simplicity, protection against link or node failure has not been investigated. Though it is straightforward to take a combination of working and protection paths to provide $1 + 1$ protection, instead of handling only one path between the node pairs. This is solely important for the connections passing through several rings because normally, the ring management will provide inter-ring protection.

5. THE OPTIMIZATION PROCESS

Due to the high degrees of freedom in network optimization problems, it is impossible to compute all configurations to find the optimum solution. Therefore one has to rely on approximation methods. One very well-known method which has proved in this class of problems is "Simulated Annealing". Any other known optimization method like tabu-search or genetic optimization, which are not trapped in local minima, could be used as well.

As a starting configuration, all faces of the graph are computed and assigned as a ring to the set of rings. The translation in an effective network graph and the routing on this network is done as described above.

In every step, a neighbor configuration is derived from the actual configuration. One of the following operations is chosen at random to derive the neighbor configuration:

- adding one face to one of the rings,
- creating a new ring with a random face,
- deleting a face of a ring (when there are no faces left in the set of faces of the ring, the ring is destroyed),
- change of the order in which the hop connections are assigned.

For the optimization process the evaluation of the configuration is of crucial importance. Here the sum of the length of all connections is used. For every connection which could not be routed, a penalty of p_{np} is added

$$C_{\mathrm{cfg}} = \sum_{p \in R_p} d(p) + \sum_{np \in R_{\mathrm{np}}} p_{np}, \tag{1}$$

where C_{cfg} denotes the cost of the configuration, R_{p} and R_{np} are the sets of routed paths and the node-pairs for which no connection has been found, respectively. For the example, $p_{np} = 1000$ has been choosen. The distance for the path p is devoted as $d(p)$. It is simple to incorporate any other design criteria, e.g. the maximum diameter or the maximum number of nodes for a ring.

The objective function for the actual configuration and the neighbor configuration are compared. When the cost for the neighbor configuration C_{ncfg} is lower than for the actual configuration C_{cfg}, it is accepted in any case. Otherwise the worse neighbor configuration is accepted for

$$r < \exp\left(-\frac{C_{\mathrm{ncfg}} - C_{\mathrm{cfg}}}{T}\right) \tag{2}$$

with $r \in [0,1]$ a random number with uniform distribution. This prevents to be trapped in a local minima of the objective function. T is a state variable of the optimization process which is referred to as the temperature. At the beginning, T is chosen such that more or less every configuration is accepted. It is lowered in the following steps for that at the end, only better configuration with lower cost are accepted. Here, for T an exponential cooling has been chosen.

One process of SA is used for changes of the routing and topology. To improve the results, a two-step way may be favorable where one SA process optimizes the ring-configuration and a second SA tries to minimize the resulting virtual network.

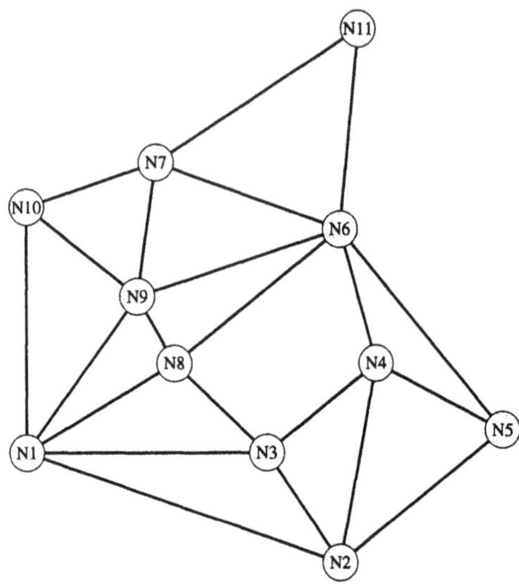

Figure 4 Fiber topology according to the COST 239 case study for a Pan-European Network.

6. A CASE STUDY FOR THE "PAN-EUROPEAN NETWORK" OF THE COST 239 PROJECT

In the framework of the European project COST 239, several case studies for a Pan-European network have been performed [Bat00]. In Fig. 4 the given fiber topology and in Tab. 1 the corresponding traffic matrix are depicted. For the example, a granularity of 2.5 Gbit/s per connection has been assumed. The presented algorithm has been implemented in `C++` using the library for discrete math and graphs `LEDA` and tested using the COST 239 scenario as a test example. All parameters of the objective function (distances, capacities, penalties) has been choosen as descripted before. In total 348 connections have to be established.

A typical example for the resulting ring configuration is

```
Ring #0:   (N8 N9 N6 N4 N3 N1)
Ring #1:   (N3 N8 N6 N4 N2)
Ring #2:   (N2 N1 N8 N9 N6 N5)
Ring #3:   (N6 N7 N11)
Ring #4:   (N10 N7 N11 N6 N8 N3 N1)
Ring #5:   (N3 N8 N9 N7 N11 N6 N5 N4 N2 N1)
Ring #6:   (N1 N10 N7 N11 N6 N8)
Ring #7:   (N3 N8 N9 N6 N5 N2 N4)
Ring #8:   (N7 N6 N4 N5 N2 N1 N9 N10)
```

	N1	N2	N3	N4	N5	N6	N7	N8	N9	N10	N11
N1	0	12.5	15	2.5	5	27.5	12.5	2.5	17.5	25	2.5
N2	12.5	0	15	2.5	7.5	22.5	5	2.5	5	7.5	2.5
N3	15	15	0	2.5	7.5	27.5	7.5	2.5	7.5	7.5	2.5
N4	2.5	2.5	2.5	0	2.5	5	2.5	2.5	2.5	2.5	2.5
N5	5	7.5	7.5	2.5	0	22.5	2.5	2.5	2.5	5	2.5
N6	27.5	22.5	27.5	5	22.5	0	20	5	15	20	7.5
N7	12.5	5	7.5	2.5	2.5	20	0	2.5	10	12.5	2.5
N8	2.5	2.5	2.5	2.5	2.5	5	2.5	0	2.5	2.5	2.5
N9	15	5	15	2.5	2.5	15	10	2.5	0	10	2.5
N10	25	7.5	7.5	2.5	5	20	12.5	2.5	10	0	2.5
N11	2.5	2.5	2.5	2.5	2.5	7.5	2.5	2.5	2.5	2.5	0

Table 1 Traffic matrix for the COST 239 case study for a Pan-European Network in units of Gbit/s.

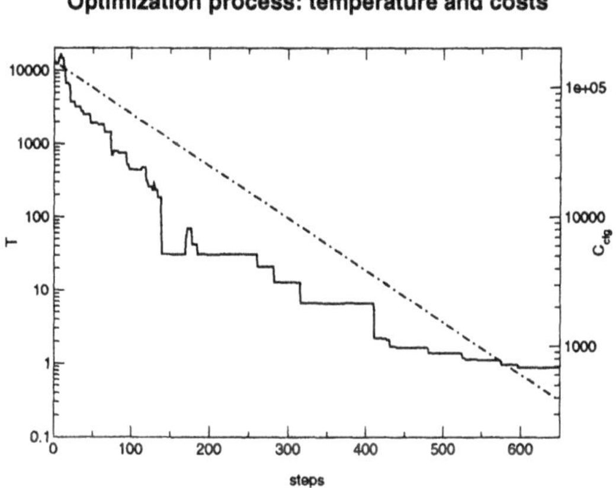

Figure 5 Trend of temperature and cost function during the optimization process without restriction on ring size. Dashed line and right axis belongs to the temperature T, solid line and left axis belongs to the configuration costs C_{cfg}.

In Fig. 5 the corresponding course of the cooling and cost is depicted. At the beginning 131 connection requests can not be established. After 410 steps, for every demand a connection can be allocated. The final configuration with 9 rings has a total cost of 683 and has evolved after 647 steps. Despite the fact that the implementation is not optimized (e.g. in every step a complete routing and not only the necessary changes due to a change in topology or other possible routings for every node pair are calculated even if the specific connection can be handled by a ring), the running time on a Pentium 500 computer is below one hour.

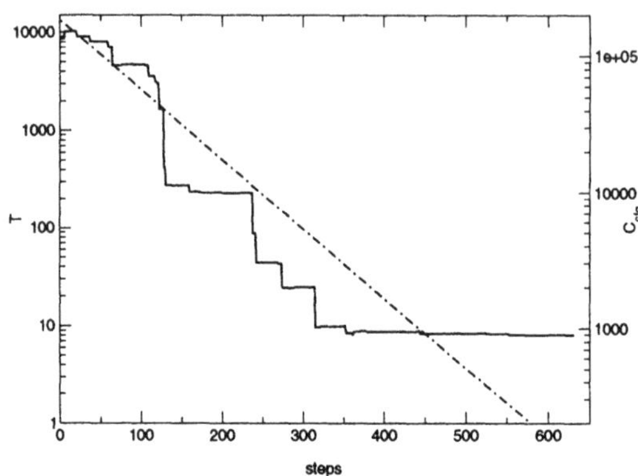

Figure 6 Trend of temperature and cost function during the optimization process with restriction on ring size.

All connections are handled by rings and no hop connections are necessary. It is worth to note, that no restriction on the ring size has been imposed. For the example this results in rather large rings with up to 10 nodes which are difficult to realize. Therefore, in a second run a limit in the ring size of 8 nodes is set by adding a penalty of 1000 to larger rings. The result for the changed objective is

```
Ring #0:   (N7 N11 N6 N9 N10)
Ring #1:   (N4 N3 N1 N8 N9 N6 N5)
Ring #2:   (N9 N7 N6 N4 N3 N8 N1)
Ring #3:   (N8 N3 N4 N2 N1 N10 N9 N6)
Ring #4:   (N8 N6 N5 N2 N3 N1 N10 N9)
```

In Fig. 6 the corresponding trend for cost function and temperature is depicted. Now, in total 5 rings are necessary to fulfill all traffic demands. The final configuration with a cost of 898 has evolved after 630 steps. Thus, 30 hop connections are necessary due to the smaller size of the rings. Therefore, the cost of the solution is slightly higher than the previous one without that boundary condition.

7. CONCLUSIONS AND OUTLOOK

A two-layer graph model with a ring layer and an effective network layer has been presented. Using faces as building blocks allows a systematic and efficient description of the ring configuration. The allocation of the connection for the actual traffic demands takes place on the resulting effective network. SA is used to optimize the ring configuration with respect to an objective function. Boundary conditions can be incorporated in this function. Therefore, it is possible to map an arbitrary network topology and traffic demand on a mesh of rings network. The algorithm has been tested with the COST 239 Pan European Case study. Only 300-700 steps are necessary to find a good realization for this network.

There are several possibilities to improve the efficiency of the implementation, especially with respect to the routing in the effective network layer. Also, an interleaved optimization of the ring configuration and the routing on top of it with two separate SA would be favorable.

A specific realization of the rings has not been treated until now. The method can be easily extended to incorporate a third layer for the wavelength paths. This would enable, e.g. an optimization with respect to the number of necessary wavelengths or the treatment of the wavelength allocation problem.

Acknowledgments

The author is grateful to Dr. J. Kemper for the carefully reading of the manuscript and his various suggestions and to the reviewers helping to clarify the presentation. The author would also like to thank Prof. P. Leuthold, Director of the Communication Technology Laboratory, ETH Zürich, for his support. This work has been performed in the framework of the COST 266 action which was partly founded by the Swiss BBW (Bundesamt für Bildung und Wissenschaft).

References

[Ham96] A. Hamel *et al*, "Increased Capacity in an MS Protection Ring using WDM Technique and OADM: The Coloured Section Ring", IEE Electr. Lett., Vol. 32, p. 234-235, Feb. 1996

[Jag99] H.A. Jäger, "The Modular WDM-Gridconnect as a Passive Routing Structure with Distributed Interfacing Capabilities", Photon. Netw. Comm., Vol. 1, No. 1, p. 23-34, June 1999

[Jag96] H.A. Jäger and T. Gipser, "Mapping of Virtual Regular Mesh Network Topologies onto Arbitrary Geographical Node Distributions", Proc. of Conf. on Netw. and Opt. Comm. (NOC), ISBN 90 5199 274, Vol. 3, p. 64-71, June 1996

[Har98] H. Harai, M. Murata und H. Miyahara, "Performance Analysis of Wavelength Assignment Policies in All-Optical Networks with Limited-Range Wavelength Conversion", IEEE J. on Selected Areas in Comm., Vol. 16, No. 7, p. 1051-60, Sept. 1998

[Kur30] C. Kuratowski, "Sur le problème the courbes guaches en topologie", Fundamenta Mathematicae, Vol. 15, p. 271-83, 1930

[Meh96] K. Mehlhorn and S. Näher, "LEDA, A Platform for Combinatorial and Geometric Computing", Cambridge University Press, 1996, `http://www.mpi-sb.mpg.de/LEDA/leda.html`

[Whi73] A.T. White, "Graphs, Groups, and Surfaces", North Holland, 1984

[Had99] E. Hadjiconstantinou and N. Christofides, "An efficient implementation of an algorithm for finding K shortest simple paths", Networks, vol. 34, no. 2, p.88-101, Sept. 1999

[Bat00] P. Batchelor *et al*, "Study on the implementation of optical transparent transport networks in the European environment-results of the research project COST 239", Photon. Netw. Comm., Vol. 2, No. 1, p.15-32, Jan.-March 2000

A DESIGN METHOD OF LOGICAL TOPOLOGY FOR IP OVER WDM NETWORKS WITH STABLE ROUTING

Junichi Katou
Department of Informatics and Mathematical Science
Graduate School of Engineering Science, Osaka University
1–3 Machikaneyama, Toyonaka, Osaka 560–8531, Japan
j-katou@ics.es.osaka-u.ac.jp

Shin'ichi Arakawa
Graduate School of Economics, Osaka University
1–7, Machikaneyama, Toyonaka, Osaka 560-0043, Japan
arakawa@ics.es.osaka-u.ac.jp

Masayuki Murata
Cybermedia Center, Osaka University
1–30, Machikaneyama, Toyonaka, Osaka 560-0043, Japan
murata@cmc.osaka-u.ac.jp

Abstract

An IP (Internet Protocol) over WDM network is expected to be an infrastructure for the next–generation Internet by directly carrying IP packets on the WDM–based network. Among several architectures for IP over WDM networks, one promising way is to overlay a logical topology consisting of lightpaths over the physical WDM network, so that IP packets are carried using the lightpaths. The conventional design methods of the logical topology have been focusing on maximizing throughput of the traffic. However, when the WDM network is applied to IP, the end-to-end paths provided by the logical topology of the WDM network is not suitable to IP routing since IP has its own metrics for route selection. We therefore propose a new heuristic algorithm to design a logical topology by considering the delay between nodes as an objective metric. We use a non-bifurcated flow deviation method to obtain a set of routes that IP packets are expected to traverse. Our proposal is then compared with conventional methods in terms of the average packet delays and throughput. It is shown that our method becomes effective when the number of wavelengths is a limited resource.

Keywords: IP over WDM, logical topology, flow deviation, route stability

1. INTRODUCTION

IP (Internet Protocol) over WDM networks where IP packets are directly carried over the WDM network is expected to offer an infrastructure for the next generation Internet. A currently available product for IP over WDM networks is only for providing the large bandwidth on point-to-point links. That is, each wavelength on the fiber is treated as a physical link between the conventional IP routers. In this way, the link capacity is certainly increased by the number of wavelengths multiplexed on the fiber, but it is insufficient to resolve the network bottleneck against an explosion of traffic demands since it only results in that the bottleneck is shifted to an electronic router.

One promising way to alleviate the bottleneck is to configure the wavelength paths over the WDM physical network and to carry IP packets utilizing the wavelength paths. Here, the physical network means an actual network consisting of the optical nodes and optical–fiber links connecting nodes. Each node has optical switches directly connecting an input wavelength to an output wavelength, by which no electronic processing is necessary at the node. Then, the wavelength path can be set up directly between two nodes via one or more optical switches. Hereafter, we will call the wavelength path directly connecting two nodes as a *lightpath.*

By utilizing the logical topology consisting of lightpaths, the physical structure of the WDM networks is embedded, and the logical topology is viewed as a underlying network by IP. In such a network, if the lightpaths are placed between every two end nodes, then no electronic processing is necessary within the network. However, too many wavelengths are necessary to establish such a network [Ramaswami and Sivarajan, 1995]. By limiting the number of lightpaths, on the other hand, we need less wavelengths though a routing capability should be provided at nodes, which will be described in more detail in the next section. In this approach, lightpaths are first established by using the available wavelengths as much as possible. If the direct lightpath cannot be set up between two nodes, two or more lightpaths are used for packets to reach the destination.

Many researchers have discussed the design methods of the logical topology. See, e.g., [R.Dutta and G.N.Rouskas, 2000] and references therein. For example, the authors in [Mukherjee et al., 1996] formulate a design method of the logical topology as an optimization problem, and show that the problem is NP-hard. In [Ramaswami and Sivarajan, 1996], the authors combine the logical topology design problem and routing problem so as to maximize the network throughput. Since the combined problem is computationally hard to solve, they split it into two subproblems, and solve those two subproblems independently. The routing problem is formulated as a linear programming problem by imposing the delay constraint for each node pair. Several heuristics are also proposed to relax the computational burden.

We should note here that MPLS (Multi-Protocol Label Switching) is now being developed by IETF [R.Callon et al., 2000, Jamoussi, 2000, R.Callon et al., 1999], and is considered to be applied to IP over WDM networks [D. O. Awduche and Coltun, 2000]. Among several options of MPLS, the route the packet traverse may be determined explicitly (explicit routing). In such a network, the lightpath should be prepared among every end node pairs within the MPLS domain, which requires too many wavelengths as described in the above. To alleviate the problem, we split the lightpath within the network. In this approach, it may take two or more lightpaths within the IP over WDM network for the packets to be forwarded. Then, the IP routing capability becomes necessary within the network. See Section 2 for more detail.

In our network, packet route is determined by the routing protocol provided by the IP layer, and the WDM network only provides (logical) paths between nodes. Then, in designing the logical topology, routes of the lightpaths should be determined by considering the nature of the IP routing protocol. That is, we place lightpaths such that the IP packet experiences smaller delays on its end-to-end path as much as possible. For this purpose, we try to reduce the number of (electronic) nodes in addition to small propagation delays between two end nodes.

A routing stability of IP is another important issue in designing IP over WDM networks. Most of conventional researches assume the amount of traffic between nodes are given and fixed. In building IP networks, however, the issue on routing stability should also be considered. In our experiments, we compare the delays of first and second shortest end-to-end paths, and if packet delays experienced by those two paths are different, we will conclude that the logical topology is "robust" against the traffic fluctuation. Actually, we will show through numerical examples that our proposed method is robust against the routing stability.

The paper is organized as follows. In Section 2, we describe our architecture model of optical node. In Section 3, we propose the logical topology design method considering the route stability. A flow deviation method, one of methods for flow assignment on the logical topology, is shown in Section 4. In Section 5, we compare and evaluate our proposed algorithm with the conventional algorithm. Finally, Section 6 concludes this paper.

2. ARCHITECTURAL MODEL OF NODES

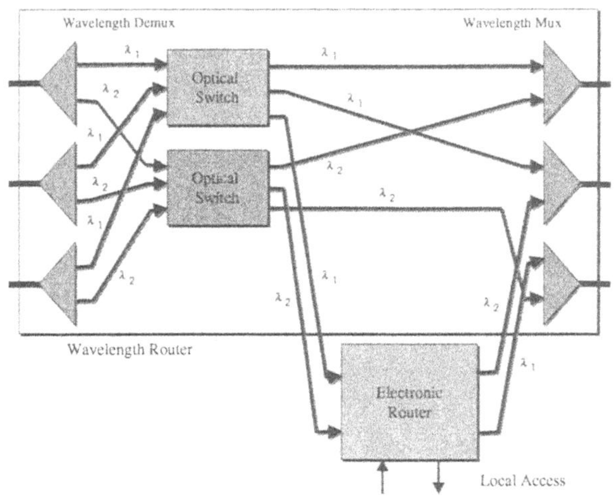

(a) Node Architecture

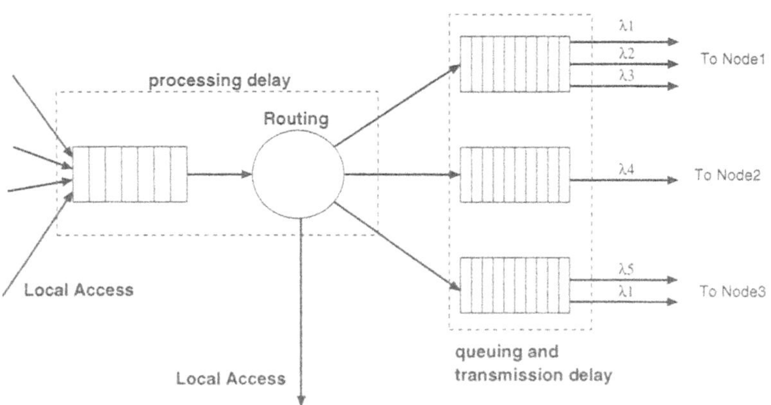

(b) Model of Electronic Router

Figure 1 Node Architecture Model

Figure 1 shows our architecture model of the optical node. Every node is equipped with an optical switch and an electronic router. The optical switch consists of three main blocks; input section, a non-blocking switch, and output section. At input section, optical signals are demultiplexed into W fixed wavelengths, $\lambda_1, \ldots, \lambda_w$. Each wavelength is switched into an appropriate output port at non-blocking switch without wavelength changes. Finally, wavelengths on the fiber are again multiplexed, and go to the next node. Note that a lightpath is placed by configuring the non-blocking switch along the path, so that the traffic on a particular wavelength from the input port to the output port is forwarded without any electronic processing.

As having been described in the previous section, all end node pairs are not always provided a one-hop lightpath in our target system, by which the use of wavelengths can be reduced. If the lightpath is terminated at the node, then IP packets on that lightpath is converted to electronic signals and forwarded to the electronic router. The electronic router processes the packet forwarding, just same as the conventional routers. If the packet should be further forwarded to other nodes, it is put on the adequate lightpath.

In our study, an electronic router is modeled as shown in figure 1(b). IP packets, which come from an optical switch or local access, are first buffered for processing, and then these packets are processed on FIFO (First In First Out) basis. In the case where the packets are forwarded onto the network, those are queued on the appropriate output port buffer. In this paper, we assume that multiple lightpaths between an adjacent node pair share the same buffer.

We last note that the other structures of optical nodes can also be considered, but the above–mentioned node architecture is preferable since there is no need to modify the IP routing mechanism.

3. LOGICAL TOPOLOGY DESIGN ALGORITHM

In [R.Ramaswami and Sivarajan, 1995], the authors propose a heuristic algorithm called MLDA (Minimum delay Logical topology Design Algorithm) to establish a logical topology. MLDA works as follows. First, it places the lightpath between two nodes if there exists a fiber directly connecting those nodes. Then, attempts are made to place lightpaths between nodes in the order of descending traffic demand. Finally, if there still exist non-utilized wavelengths, lightpaths are placed randomly utilizing those wavelengths as much as possible. We again note that a lot of conventional methods including MLDA are focusing mainly on maximizing throughput of traffic and those are not adequate to design a logical topology suitable to carry IP traffic.

We therefore introduce a new logical topology design algorithm called SHLDA (Shortest-Hop Logical topology Design Algorithm) to resolve the above-mentioned problems. As described before, we assume that a routing function is performed only at the IP layer. Then, the logical topology should be designed by incorporating the nature of route selection adopted in the IP routing protocol. It is natural that a shorter path would be selected by the IP routing protocol for forwarding packets. (We note here that by *a short path*, we mean that the number of lightpaths between an end node pair is small.) That is, hop counts of lightpapths (i.e., the number of lightpaths that the packet traverses) should be reduced as much as possible, which is our primary objective in the proposed algorithm. Once the lightpath is split between some two end node pair, and a series of lightpaths is necessary to reach the destination, the processing delay at the electronic router must be considered. It will be incorporated in the final determination of the lightpaths setup, which will be described in the next subsection.

MLDA uses traffic demand between node pairs to set up the next lightpath in the algorithm. On the contrary, we use the performance metric F_{ij} for node pair ij, which is determined by the following equation,

$$F_{ij} = \gamma_{ij} \times h_{ij}, \tag{1}$$

where γ_{ij} is the traffic demand from node i to j, and h_{ij} is the hop-count of the minimum hop route between node i and node j on the physical topology. Here, the hop-count of the lightpath refers to the number of physical links that the lightpath traverses. Note that F_{ij} is equal to γ_{ij} in MLDA, i.e., MLDA does not consider the hop-count of the lightpath, and only uses the propagation delay in determining the shortest route of the lightpath.

On the other hand, SHLDA first uses the hop-count as the metric to calculate the shortest route in configuring the lightpath. It enables us to establish the lightpath consisting of the minimum number of electronic routers that IP packets traverse. The detailed description of the SHLDA algorithm is as follows.

Step 1: Calculate the metric F_{ij} for each node pair ij from the traffic matrix $Q = q_{ij}$. In initially determining F_{ij}, h_{ij} is simply set as the hop-count of the shortest physical path.

Step 2: Place the lightpath between two nodes if there exists a fiber.

Step 3: Select the node pair $i'j'$, where i' and j' are indices giving $\max_{ij}(F_{ij})$. If $F_{i'j'} = 0$, go to Step 5. Otherwise, go to Step 4.

Step 4: Find the minimum-hop route between the node pair $i'j'$, and check the availability of wavelengths in order to configure the lightpath. If so, use the wavelength with lowest index to establish the lightpath. Then set $F_{i'j'} = 0$ and go back to Step 3. If there is no available wavelength, set $F_{i'j'} = 0$ and go back to Step 3.

Step 5: If there still exist non-utilized wavelengths, lightpaths are configured randomly as much as possible utilizing those wavelengths as in MLDA.

By the above algorithm, we obtain the logical topology, but it is insufficient in building IP over WDM networks. We consider the routing stability by next applying the flow deviation method.

4. FLOW DEVIATION METHOD

In this section, we first introduce the flow deviation method [L.Fratta et al., 1973] in Subsection 4.1. Then, it is applied to our case in Subsection 4.2.

4.1. DESCRIPTION OF FLOW DEVIATION METHOD

In this subsection, we summarize the flow deviation method [L.Fratta et al., 1973]. It incrementally changes the flow assignment along the feasible and descent direction. It works as follows. Given an objective function T, the flow deviation method sets l_{ij} as the partial derivative with respect to λ_{ij}, where λ_{ij} is the flow rate of lightpath(s) between nodes i and j. Then, the new flow assignment is solved by using the shortest path algorithm in terms of the l_{ij}. By incrementally changing from the old flow assignment to the new one, the optimal flow assignment is explored. The detailed description of the flow deviation method is as follows.

Step 1: Prepare a starting feasible flow assignment f^0. Let $n = 0$.

Step 2: Set $g \leftarrow f^n$. Assume that flow assignment f^n is represented as $\{x_{11}, \dots, x_{pq}, \dots, x_{NN}\}$.

Step 3: Calculate $l_{ij} = \frac{\partial T}{\partial \lambda_{ij}}$. Then, set the new flow assignment $R(g)$ to $\{x'_{11}, \dots, x'_{pq}, \dots, x'_{NN}\}$ by solving the shortest path algorithm using the metric l_{ij}.

Step 4: For each node pair ij, do following.

Step 4.1: Let v be the flow assignment by deviating flow ij from g toward $R(g)$. That is, the resulting flow assignment, v, is set to $\{x_{11}, \dots, x'_{ij}, \dots, x_{NN}\}$.

Step 4.2: Check whether v is feasible. Feasible v means that the processing capability of IP routers and/or the capacity of a lightpath does not exceed its limit. If not, then the deviation at Step 4.1 is rejected, and go back to Step 4.

Step 4.3: Check whether the v is descent. If $T(v) < T(g)$, g is allowed to deviating toward v. Then, $g \leftarrow v$. Then go back to Step 4. If $T(g) \leq T(v)$, the deviation from the g toward $R(g)$ is rejected, and go back to Step 4.

Step 5: If $g = f^n$, then stop iteration. Note that $g = f^n$ means there is no improvement of performance by deviating the flow. Otherwise, set $n \leftarrow n + 1$, and go back to Step 2.

4.2. DERIVATION OF THE METRIC l_{ij}

In this subsection, we will show the metric l_{ij} in our case. We will use the following notations.

N: the number of nodes in the network

P_{ij}: the propagation delay of lightpath ij

C: the capacity of each wavelength

μ: processing capability at an electronic router, which is assumed to be identical among all routers.

We also introduce the following variables.

a_{ij}^{sd}: when the packets are routed from node s to node d via the lightpath ij, the value is set to be 1. Otherwise, 0.

δ_i: the sum of all traffic switched by the IP electronic router at node i except the traffic flow λ_{ij}.

The objective function T in our case is given as the average of T_{sd}'s (the delay between node s and d), i.e.,

$$T = \frac{1}{N(N-1)} \sum_{s=1}^{N} \sum_{d=1}^{N} T_{sd} \tag{2}$$

As shown in figure 1, the delay experienced at a node consists of processing delay and transmission delay. Henceforth, the delay between nodes s and d consists of the propagation delay, processing delay and transmission delay. That is, we have

$$T_{sd} = \left[\sum_{ij} a_{ij}^{sd} P_{ij} \right] + \left[\sum_{ij} a_{ij}^{sd} Q_{ij} \right] + \left[\sum_{ij} (a_{ij}^{sd} R_i) + R_d \right] \tag{3}$$

where Q_{ij} is the transmission delay of the packets on lightpath ij, and R_i is the processing delay at the electronic router of node i. In this paper, Q_{ij} is determined by a $M/M/k_{ij}$ (where k_{ij} shows the number of lightpaths between node pair ij) queuing system and R_i by a $M/M/1$ queueing system. Recall that we allow a multiple number of lightpaths between the node pair, and those lightpaths share same buffer (see Section 2). Q_{ij} and R_i are derived as follows.

$$Q_{ij} = \frac{X_l}{l \cdot C - \lambda_{ij}} + \frac{1}{C} \tag{4}$$

$$R_i = \frac{1}{\mu - (\lambda_{ij} + \delta_i)} \tag{5}$$

where

$$X_l = \frac{p_0\,(l\rho)^l}{(1-\rho)\,l!} \tag{6}$$

$$\rho = \frac{\lambda_{ij}}{k_{ij} \cdot C} \tag{7}$$

$$p_0 = \left\{ \sum_{x=0}^{k_{ij}-1} \frac{(k_{ij}\rho)^x}{x!} + \frac{(k_{ij}\rho)^{k_{ij}}}{k_{ij}!(k_{ij}-\rho)} \right\}^{-1} \tag{8}$$

Three kinds of packets arrive at the electronic router at node i; packets destined for node i, packets arriving at node i from local access, and packets changing the lightpath at node i. Thus, δ_i is given by the following equation.

$$\delta_i = \left[\sum_j \gamma_{ji} + \sum_j \gamma_{ij} + \sum_{sd, s \neq i, d \neq i} \sum_j a_{ij}^{sd} \gamma_{sd} \right] - \lambda_{ij} \tag{9}$$

Note that λ_{ij} is the flow rate of lightpath(s) between nodes i and j. That is, we have

$$\lambda_{ij} = \sum_{sd} a_{ij}^{sd} \gamma_{sd} \tag{10}$$

Using equations (4) and (5), we finally obtain l_{ij} as

$$l_{ij} = \frac{\partial T}{\partial \lambda_{ij}} = \frac{1}{N\,(N-1)} \sum_{s=1}^{N} \sum_{d=1}^{N} a_{ij}^{sd} \alpha_{sd},$$

where

$$\alpha_{sd} = \frac{X_{k_{ij}}}{(l \cdot C - \lambda_{ij})^2} + \frac{1}{(\mu - (\lambda_{ij} + \delta_i))^2} \tag{11}$$

5. NUMERICAL EVALUATION AND DISCUSSIONS

In this section, we evaluate our SHLDA by comparing with MLDA. In addition to MLDA, we also consider WLA (WDM Link Approach), where a WDM technology is only utilized for point-to-point link between adjacent IP routers.

5.1. NETWORK MODEL

As a network model, we consider 14–node NSFNET shown in figure 2. A traffic matrix given in [Ramaswami and Sivarajan, 1996] is used in numerical evaluation. Since the traffic matrix is given by a relative value, we introduce *traffic scale* α, and actual traffic demands between nodes are given by a product of the traffic matrix and α. We also assume the value of the given traffic matrix is represented in Gbps. We set the capacity of each wavelength to 10 Gbps. The packet processing capability of the electronic router, μ, is represented in pps (packet per second) assuming that the mean packet size is 1,000 bits long.

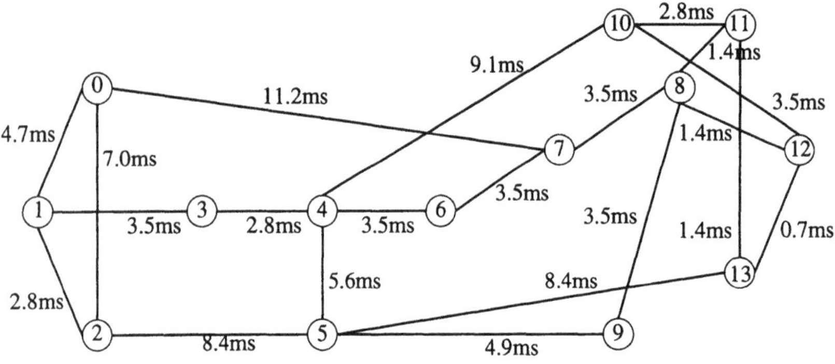

Figure 2 NSFNET

5.2. NUMERICAL RESULTS AND DISCUSSIONS

Figure 3 compares average delays obtained by three algorithms. The horizontal axis shows the traffic scale α. In obtaining the figure, we set the number of wavelengths, W, to eight and the packet processing capacity of IP router, μ, to 40 Mpps. In the figure, when the traffic scale α is small, we cannot observe significant differences among three algorithms, SHLDA, MLDA and WLA, and delays are suddenly increased as α becomes large in three algorithms. We also notice that our SHLDA is

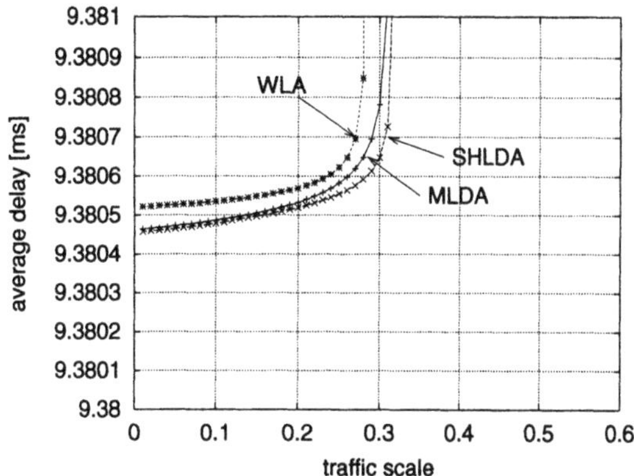

Figure 3 Average delay of each algorithm : $W = 8$, $\mu = 40$ Mpps

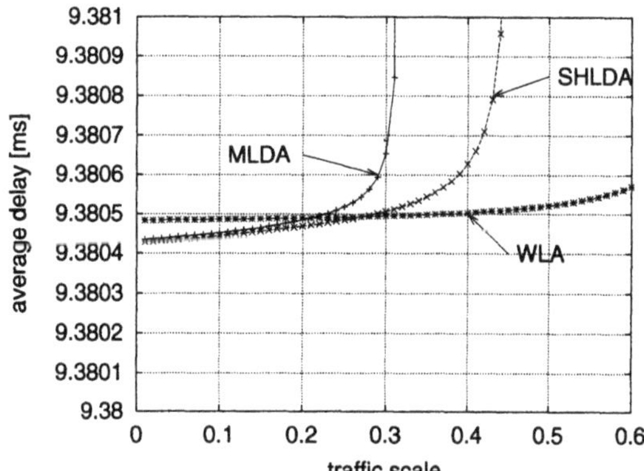

Figure 4 Average delay of each algorithm : $W = 8$, $\mu = 100$ Mpps

slightly better than MLDA in terms of the maximum throughput which is a saturation point of the delays.

We next show the effect of increasing the packet forwarding capability of IP routers. Figure 4 shows the results by changing μ to 100 Mpps. For other parameters, same values in obtaining figure 3 are used. By comparing these two figures, we can observe that the maximum throughputs in SHLDA is increased as the IP router has enough capacity. However, improvement in the maximum throughput cannot be seen when we apply MLDA. To explain this, let us look at the nodal delays in more detail. Figures 5 and 6 show the processing delay and transmission delay dependent on α. As expected, the effect of increasing the capability of the IP routers can be observed in these figures. As the processing delay is reduced with high capability of the IP router, the transmission delay beccomes the bottleneck of the network. Then, our SHLDA becomes superior to MLDA.

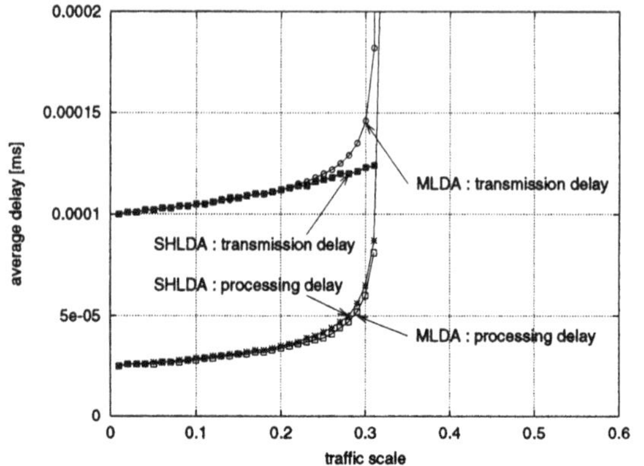

Figure 5 Average delay on node : $W = 8$, $\mu = 40$ Mpps

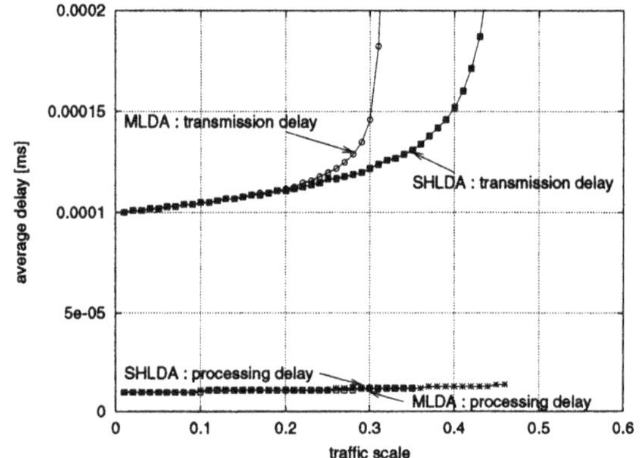

Figure 6 Average delay on node : $W = 8$, $\mu = 100$ Mpps

We next set the number of wavelengths W to twelve and μ to 40 Mpps. Results are shown in figure 7. By comparing figures 3 and 7, it is apparent that SHLDA exhibits the largest improvement on the maximum throughput. To see this more clearly, figure 8 presents components of delays in figure 7. By comparing figures 5 and 8, we can see SHLDA can provide much improvements on the maximum throughput. In Figure 5 and 8, we can see SHLDA does not show any difference in the transmission delays although the number of available wavelengths is increased from eight to twelve. On the other hand, the transmission delay by MLDA is decreased. Its reason can be explained as follows. SHLDA places lightpaths in a descending order of the product of the hop-count and traffic demand. As a result, a lightpath placed by SHLDA tends to utilize more links than the one by MLDA. Thus, MLDA can establish more lightpaths than SHLDA as the number of available wavelengths

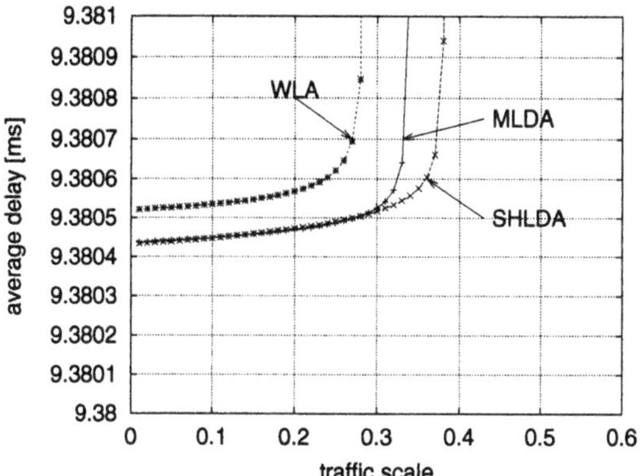

Figure 7 Average delay : $W = 12$, $\mu = 40$ Mpps

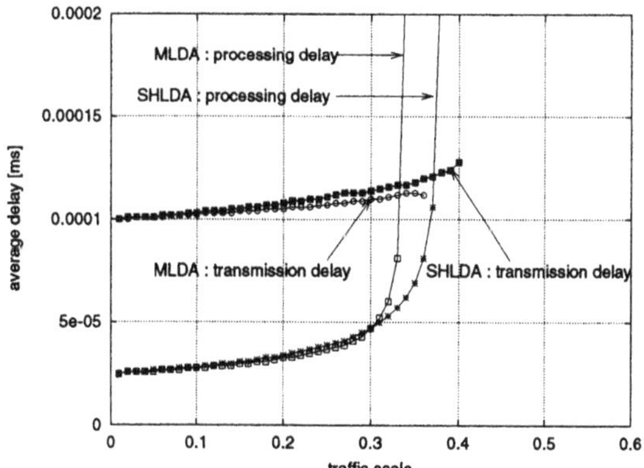

Figure 8 Delay on node : $W = 12$, $\mu = 40$ Mpps

increases. It leads to decrease the transmission delay in MLDA as the number of wavelengths becomes large.

The processing delay at the IP router is reduced as the number of available wavelengths increases. Its effect is larger in SHLDA. As mentioned before, the lightpaths placed by SHLDA tend to utilize more physical links. It results in more reduction of electric processing in SHLDA than MLDA as the number of available wavelengths increases.

We last show the average delay obtained by our SHLDA by increasing the number of wavelengths. The results are plotted in figure 9 where we set $W = 20$ and $\mu = 40$ Mpps. In this figure, SHLDA still shows higher throughput than MLDA, but the difference comparatively becomes smaller than the previous cases (figures 3 and 7). The reason is that by increasing the number of wavelengths, the logical topologies obtained by SHLDA or MLDA become close to a fully meshed network. Then, the advantage of SHLDA becomes smaller since SHLDA tries to

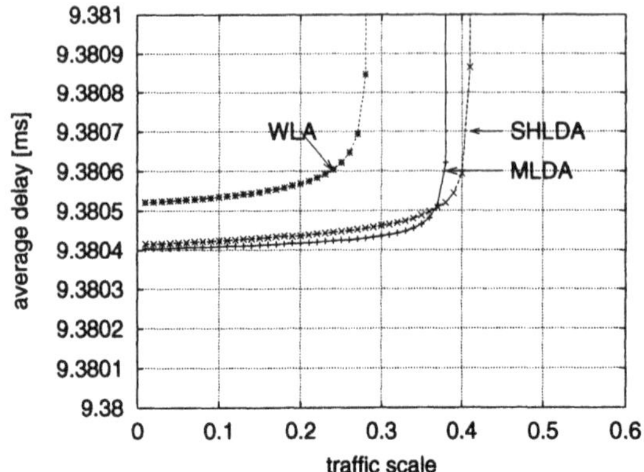

Figure 9 Average delay : $W = 20$, $\mu = 40$ Mpps

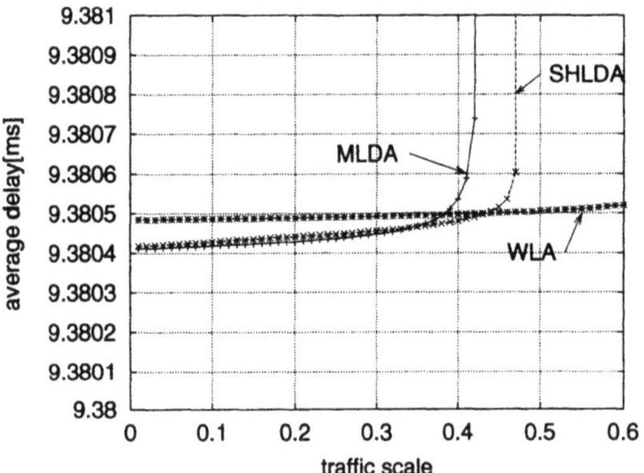

Figure 10 Average delay : $W = 12$, $\mu = 100$ Mpps

reduce the traffic load on the IP router. We also show the case that μ is 100 Mpps. The result is plotted in figure 10 where we set $W = 12$. By comparing figure 7 and 10, we can also observe the effectiveness of our SHLDA as well as in figure 3 and 4.

Lastly, we summarize the characteristics of WLA by observing figures 3, 4 and 7. By comparing figures 3 and 7, the improvement of the maximum throughput in WLA is very limited. This is because the processing delay at the electronic router is the primary bottleneck of the network, and henceforth, the effect of increasing the number of wavelengths cannot be seen. As one can easily imagine, the results of WLA is greatly improved as the capability of IP router becomes large (see figures 3, 4 and 10). Only in that case, WLA is not a bad approach for IP over WDM networks.

5.3. INVESTIGATION ON ROUTING STABILITY

We finally discuss our logical topology design algorithm from a viewpoint of the stability of IP routing. In IP networks, it is significant to avoid or at least to reduce unnecessary changing of the route, which is caused by dynamically changing traffic demand. To evaluate this, we examine the packet delays of first and second shortest end-to-end paths (lightpaths) prepared by our logical topology design algorithm. If those two values are close, the route of IP packets may frequently change against the traffic fluctuation.

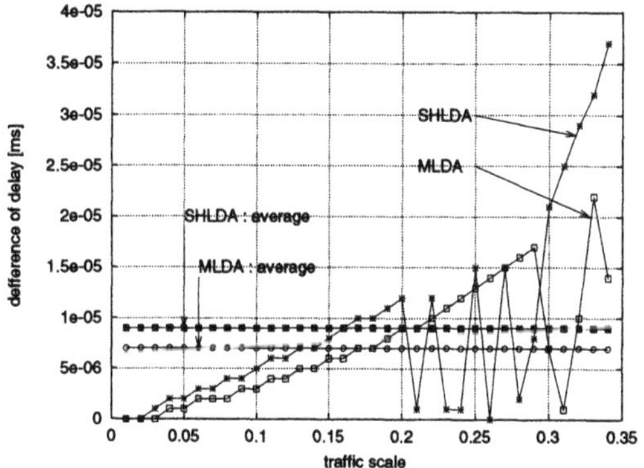

Figure 11 Route Stability : $W = 8$, $\mu = 40$ Mpps

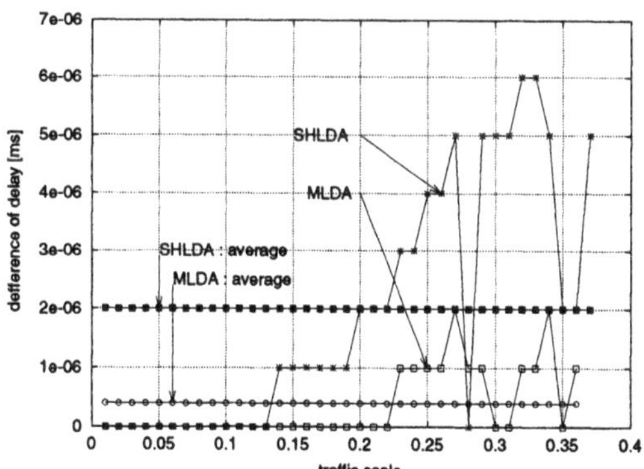

Figure 12 Route Stability : $W = 12$, $\mu = 40$ Mpps

Let us introduce the metric d_{sd} which defines the difference of delays of the first and second shortest routes between source node s and destination node d. From all possible combinations of source and destination node pairs, we choose the smallest one as d_{min}, i.e., $d_{min} = min_{sd}\{d_{sd}\}$. We consider that the design algorithm providing the larger d_{min} gives a higher routing stability. In figures 11 and 12, we plot d_{min} obtained from SHLDA/MLDA as a function of α, where the number of wavelengths W is set to be eight and twelve, respectively. The processing capacity of the IP router, μ, is identically set to be 40 Mpps in both figures. The average of d_{min} is also shown in the figures. We can observe that when $W = 8$ (figure 11), SHLDA is not very good especially when the traffic scale becomes large. However, it gives higher stability than MLDA when the number of wavelength to be twelve (figure 12) because of the design principle of our SHLDA.

The problem found in both of MLDA and SHLDA is that at several values of α, d_{min} takes very small value. It is mainly because SHLDA as well as MLDA is a "one–way algorithm". That is, there are no step back operation in the algorithms; if the nodal delay is high, it tends to lead to the situation that the delay of the first shortest route becomes close to the delay of the second shortest one, since the nodal delay becomes domiant of the delay in such a region. We believe the situation can be avoided by reassembling the lightpaths to reduce the nodal delay, but it is one of our future research topics.

6. CONCLUSION

In this paper, we have proposed a new heuristic algorithm, SHLDA, to design a logical topology by considering the delay between nodes as an objective metric. Then, we have compared our proposed algorithm with conventional methods in terms of the average packet delay and throughput. The results have shown that SHLDA becomes effective when the number of wavelengths are small and the processing capacity of IP router is large. Furthermore, we have evaluated our proposed algorithm from a viewpoint of the routing stability. It is shown that SHLDA can obtain a stable network than MLDA.

However, in several values of the traffic scale, it has also been shown that SHLDA leads to the network having routes that can cause the routing instability than MLDA. To alleviate this problem, we need to reconfigure lightpaths in order to increase the routing stability. This is one of our future research topics.

References

[D. O. Awduche and Coltun, 2000] D. O. Awduche, Y. Rekhter, J. D. and Coltun, R. (2000). Multi-protocol lambda switching: Combining MPLS traffic engineering control with optical crossconnects. *draft-awduche-mpls-te-optical-02.txt.*

[Jamoussi, 2000] Jamoussi, B. (2000). Constraint-based LSP setup using LDP. *draft-ietf-mpls-cr-ldp-04.txt.*

[L.Fratta et al., 1973] L.Fratta, M.Gerla, and L.Kleinrock (1973). The flow deviation method: An approach to store-and-forward communication network design. *Networks*, 3:97–133.

[Mukherjee et al., 1996] Mukherjee, B., Banerjee, D., S.Ramamurthy, and Mukherjee, A. (1996). Some principles for designing a wide-area WDM optical network. *IEEE/ACM Transactions on Networking*, 4(5):684–695.

[Ramaswami and Sivarajan, 1995] Ramaswami, R. and Sivarajan, K. N. (1995). Routing and wavelength assignment in all–optical networks. *IEEE/ACM Transactions on Networking*, 3:489–500.

[Ramaswami and Sivarajan, 1996] Ramaswami, R. and Sivarajan, K. N. (1996). Design of logical topologies for wavelength-routed optical networks. *IEEE Journal on Selected Areas in Communications*, 14:840–851.

[R.Callon et al., 2000] R.Callon, A.Viswanathan, and E.Rosen (2000). Multiprotocol label switching architecture. *draft-ieft-mpls-arch-07.txt.*

[R.Callon et al., 1999] R.Callon, G.Swallow, N.Feldman, A.Viswanathan, P.Doolan, and A.Fredette (1999). A framework for multiprotocol label switching. *draft-ietf-mpls-framework-05.txt.*

[R.Dutta and G.N.Rouskas, 2000] R.Dutta and G.N.Rouskas (2000). A survey of virtual topology design algorithms for wavelength routed optical networks. *Optical Network Magazine*, 1(1):73–89.

[R.Ramaswami and Sivarajan, 1995] R.Ramaswami and Sivarajan, K. N. (1995). Design of logical topologies for wavelength-routed all-optical networks. In *IEEE Infocom*, pages 1316–1325.

Influence of Chord Length on the Blocking Performance of Wavelength Routed Chordal Ring Networks

MÁRIO M. FREIRE[1] AND HENRIQUE J. A. DA SILVA[2]
[1]*Department of Computer Science, University of Beira Interior*
Rua Marquês d'Ávila e Bolama, 6200-001 Covilhã, Portugal
Phone: +351 275 319732; Fax: +351 275 319888; E-mail: mario@noe.ubi.pt
[2]*Department of Electrical Engineering, University of Coimbra - Pole II*
3030-290 Coimbra, Portugal
Phone: +351 39 796248; Fax: +351 39 796293; E-mail: hjas@dee.uc.pt

Key words: Wavelength routed optical networks, chordal rings, blocking performance.

Abstract: In this paper, we present an assessment of the blocking performance of wavelength routed optical networks with chordal ring topology. The performance is focused on chord lengths of 3, $N/4$, and $\sqrt{N}+3$, being N the number of nodes. It is shown that, in a 100-node chordal ring network, the use of the chord length of $N/4$ instead of 3 clearly reduces the blocking probability. A reduction of the blocking probability is also observed when the chord length of $\sqrt{N}+3$ is used, instead of $N/4$ or $w=3$. The increase of the node degree from 3 (chordal ring with 100 nodes and a chord length of $\sqrt{N}+3$) to 4 (mesh torus with 100 nodes) has also revealed an important feature: very small blocking gains were observed due to the increase of the node degree from 3 to 4. Thus, the choice of a chordal ring with a chord length of $\sqrt{N}+3$, instead of a mesh-torus, leads to a reduction in the number of network links, and hence in the total cable length, since the number of links in a N-node chordal ring is $3N$, and the number of links in a N-node mesh-torus is $4N$.

1. INTRODUCTION

Internet traffic is increasing due to the rise of new communication services and due to the explosive growth of the number of hosts [1]-[2]. To satisfy the bandwidth requirements that enable those new services and a higher number of users, WDM systems are being deployed world wide, which is seen as the first phase of optical networking. After this, the introduction of optical add/drop multiplexers in a linear architecture and the use of WDM protection switches are expected. This architecture will rapidly evolve to a WDM ring architecture, and a further possible evolution scenario may be the interconnection of WDM rings and mesh networks [3]. Whereas the evolution from the point-to-point WDM transmission system to interconnected rings is clear, from the point of view of physical topology, the optimal topology to be used for the mesh network is a subject less studied. In [4], a study is presented of the influence of node degree on the fibre length, capacity utilisation, and average and maximum path lengths of wavelength routed mesh networks. It is shown that average node degrees varying between 3 and 4.5 are of particular interest.

In this paper, we consider chordal rings, which are a particular family of regular graphs of degree 3 [5]. Using an analytical model, we present an assessment of the blocking performance in wavelength routed chordal ring networks, considering chord lengths of 3, $N/4$ and $\sqrt{N}+3$, where N is the number of nodes.

The remainder of this paper is organized as follows. The analytical model used to compute the blocking probability in wavelength routed chordal ring networks is briefly described in section 2. The assessment of the blocking performance in chordal rings is presented in section 3 and main conclusions are presented in section 4.

2. COMPUTATION OF BLOCKING PROBABILITY IN CHORDAL RING NETWORKS

2.1 Chordal Ring Topology

A chordal ring is a ring structured network in which each node has an additional link, called a chord. The number of nodes in a chordal ring is assumed to be even and nodes are indexed 0, 1, 2, ..., N-1 around the N-node ring. It is also assumed that each odd-numbered node i (i=1, 3, ..., N-1) is connected to a node $(i+w)$ *mod* N, where w is the chord length, which is assumed to be positive odd and, without loss of generality, we also assume that $w \leq N/2$, as in [5].

For a given number of nodes N, several chordal rings can be obtained for different values of the chord length. Figure 1 shows a chordal ring with N=20 and w=5.

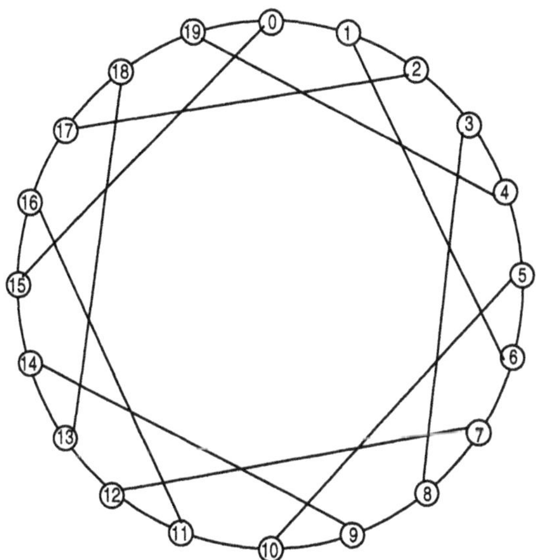

Figure 1. Chordal ring network with N=20 nodes and w=5

According to the Theorem 2 presented in [5], the diameter of a chordal ring network with N nodes, where N is a square and $N \geq 8^2$, is greater than or equal to $\sqrt{N}+3$. This equality holds when $\sqrt{N}+3 \leq w \leq \sqrt{N}+2h+1$, where h is the largest integer that satisfies $\sqrt{N} \geq 2\left(2h^2+h+1\right)$. This Theorem is particularly important since it allows the identification of chord lengths that lead to the smallest network diameter.

2.2 Evaluation of Blocking Probability

To compute the blocking probability in optical networks with wavelength interchange, we have used the model given in [6], since it applies to ring topologies, has a moderate computational complexity, and takes into account dynamic traffic and the correlation between the wavelengths used on successive links of a multi-link path.

The following assumptions are used in the model [6]: 1) Call requests arrive at each node according to a Poisson process with rate λ, with each call equally likely to be destined to any of the remaining nodes; 2) Call holding time is exponentially distributed with mean $1/\mu$, and the offered load per node is $\rho=\lambda/\mu$; 3) The path used by a call is chosen according to a pre-specified criterion (e.g. random selection of a shortest path), and does not depend on the state of the links that make up a path; the call is blocked if the chosen path can not accommodate it; alternate path routing is not allowed; 4) The number of wavelengths per link, F, is the same on all links; each node is capable of transmitting and receiving on any of the F wavelengths; each call requires a full wavelength on each link it traverses; 5) Wavelengths are assigned to a session randomly from the set of free wavelengths on the associated path.

The analysis presented in [6] also assumes that the hop-length distribution is known, as well as the arrival rates of calls at a link that continue, and those that do not, to the next link of a path. The call arrival rates at links have been estimated from the arrival rates of calls to nodes, as in [6]. The hop-length distribution is a function of the network topology and the routing algorithm, and is easily determined for most regular topologies with the shortest-path algorithm. However, for a bi-directional N-node chordal ring with chord length w, it was not possible to obtain a general expression for the hop-length distribution. We have found analytical expressions for the hop-length distribution when the chord length is 3 (w=3), when the chord length is maximum ($w=N/2$ or $w=N/2-1$), when the chord length is as close as possible to the mean chord length ($w=N/4$), and for a chord length that leads to the smallest network diameter ($w=\sqrt{N}+3$).

2.3 Performance Metrics

The performance assessment presented in this paper is based on the evaluation of the path blocking probability. However, a direct comparison of blocking probabilities is sometimes preferred in order to put in evidence some features and to quantify benefits. Besides the blocking probability, two metrics are used in this paper to quantify and to put in evidence the benefits of using chordal ring networks. We consider the blocking gain due to a better choice of chord length (G_{cl}), and the blocking gain due to an increase in the node degree (G_{nd}). Since we consider only blocking gains along the paper, G_{cl}, and G_{nd} are also called here as chord length gain and node degree gain, respectively. These two metrics are defined as:

$$G_{cl} = \frac{P_{w3}}{P_{wl}} \tag{1}$$

where P_{w3} is the blocking probability in a chordal ring network with a chord length of 3, and P_{wl} is the blocking probability in a chordal ring network with a chord length of wl (both obtained for the same number of nodes, wavelengths per link, and load per node);

$$G_{nd} = \frac{P_{(D-1)}}{P_D} \tag{2}$$

where $P_{(D\text{-}1)}$ is the blocking probability in a network with a node degree D-1, and P_D is the blocking probability in a network with a node degree D (both obtained for the same number of nodes, wavelengths per link, and load per node).

3. ASSESSMENT OF BLOCKING PERFORMANCE IN CHORDAL RING NETWORKS

In this section, we present an assessment of the blocking performance in chordal ring networks. The performance assessment is focused on chordal rings with w=3, w=N/4, and $w=\sqrt{N}+3$. The last one is very important since it leads to the smallest network diameter. If the number of nodes is larger or equal than 484, there are other chord lengths that also lead to the smallest network diameter, but here we concentrate only on the above referred chord lengths. We consider chordal ring networks with N=100 nodes. The blocking performance obtained with the maximum chord length (w=N/2-1) is similar to the performance obtained with w=3 and, therefore, it is not presented here.

Figure 2 shows the chord length gain for a 100-node chordal ring without wavelength interchange. As can be seen, the use of w=N/4, instead of w=3, clearly improves the blocking probability. The corresponding blocking probability gain, for 16 wavelengths per link, is of the order of 10^2 for a load per node of 1 Erlang, and of the order of 10^6 for a load of 0.1 Erlang, while, for 8 wavelengths per link, the blocking gain is about 2.7 for a load per node of 1 Erlang, and about 10^3 for a load of 0.1 Erlang (see Figure 2). A blocking probability improvement is also observed when the chord length of $\sqrt{N}+3$ is used, instead of w=N/4 or w=3. The blocking probability gain due to the choice of $w=\sqrt{N}+3$, instead of w=3, for 16 wavelengths per link, is of the order of 10^4 for a load per node of 1 Erlang, and of the order of 10^{10} for a load of 0.1 Erlang, while, for 8 wavelengths per link, the blocking gain is about 13.5 for a load per node of 1 Erlang, and of the order of 10^4 for a load of 0.1 Erlang (see Figure 2). This clearly shows that the chord length of $\sqrt{N}+3$ gives rise to very high chord length gains. From Figure 2, we may also see that similar chord length gains are obtained for w=N/4 with 12 wavelengths per link and for $w=\sqrt{N}+3$ with 8 wavelengths per link. Thus, for a similar blocking performance, the use of $w=\sqrt{N}+3$ reduces the number of wavelengths per link, which is important from the point of view of network dimensioning.

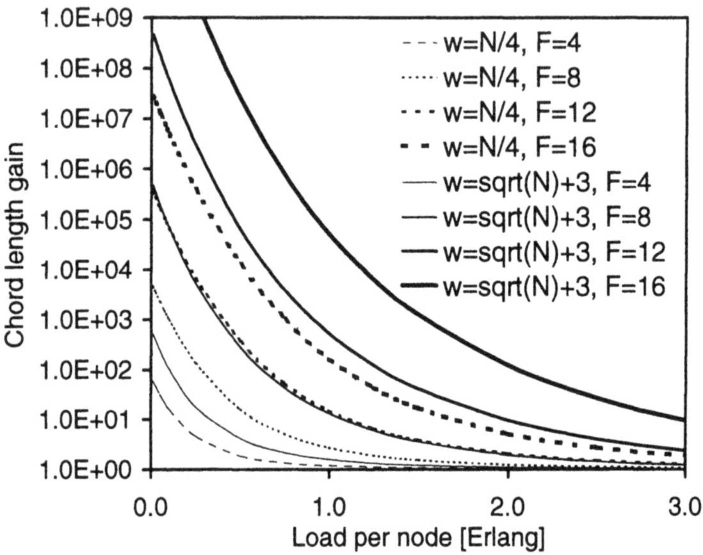

Figure 2. Blocking gain due to a better choice of chord length, as a function of load per node, for a chordal ring network with 100 nodes and without wavelength interchange.

Figure 3 shows the blocking probability, versus converter density, for 100-node chordal ring networks with chord lengths of 3, $N/4$, or $\sqrt{N}+3$, and with a load per node of 0.5 Erlang. The corresponding chord length gain is shown in Figure 4. As can be seen, there is a slight chord length gain when the number of wavelengths per link is small (4 or 8). If the number of wavelengths per link is large (12, 16), the chord length gain decreases up to a certain converter density, and then the chord length gain begins to increase slightly. Another observation that can be made from this figure is that the converter density, at which the gain starts increasing, increases with the number of wavelengths per link.

Figure 5 shows the blocking gain due to an increase of the node degree from 3 (chordal ring networks with 100 nodes) to 4 (mesh-torus networks with 100 nodes), as a function of the load per node. If $w=3$ is used with loads lower than 0.25 Erlang, the blocking gain is larger than 10^4 for 8 wavelengths per link, and larger than 10^{10} for 16 wavelengths per link.

For $w=N/4$, gains of the order of 10^2 and 10^5 were obtained for 8 and 16 wavelengths per link, respectively, with a load per node of 0.1 Erlang. For the case of $w=\sqrt{N}+3$, very small node degree gains were observed. As the load per node decreases from 5 Erlang to 0.01 Erlang, the variation of the node degree gain remains within one order of magnitude for the numbers of wavelengths per link considered: 4, 8, 12 and 16. These small gains were also observed in networks with a larger size (1600 nodes).

These very small node degree gains, obtained for mesh-torus and chordal rings with $w=\sqrt{N}+3$, may be explained by the same order of magnitude of the average hop-length.

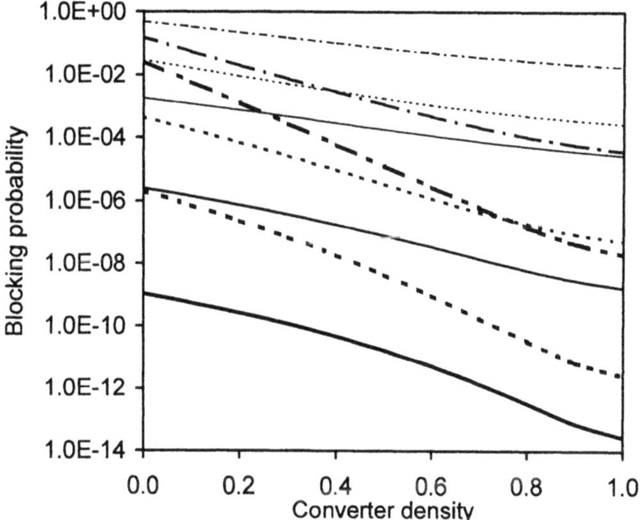

Figure 3. Blocking probability versus converter density for 100-node chordal ring networks with a load per node of 0.5 Erlang.

w=3, F=8
w=3, F=12
w=3, F=16
w=N/4, F=8
w=N/4, F=12
w=N/4, F=16
w=sqrt(N)+3, F=8
w=sqrt(N)+3, F=12
w=sqrt(N)+3, F=16

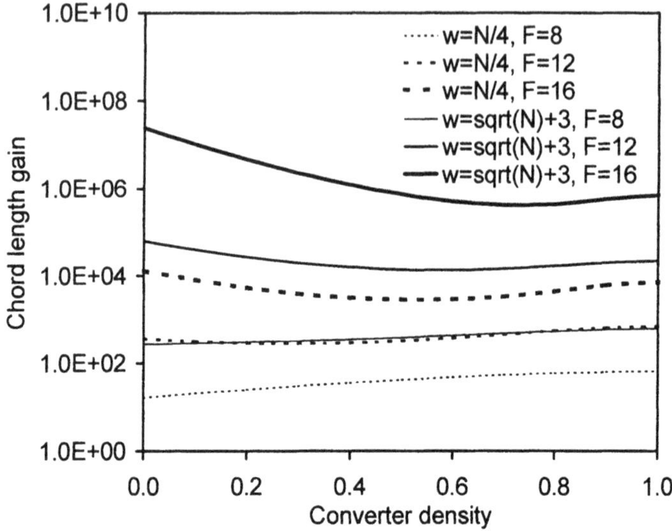

Figure 4. Blocking gain due to a better choice of chord length, as a function of the converter density, for 100-node chordal ring networks with a load per node of 0.5 Erlang.

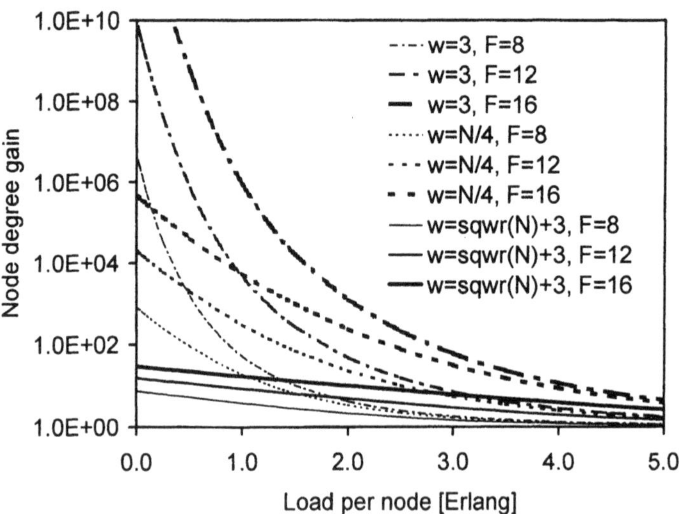

Figure 5. Blocking gain due to an increase of the node degree from 3 to 4, as a function of the load per node, for chordal ring networks and mesh-torus networks, both with 100 nodes (without wavelength interchange).

If a small blocking performance degradation is allowed, the choice of chordal ring networks with $w=\sqrt{N}+3$, instead of mesh-torus networks, leads to a reduction in the number of network links, and hence in the total cable length, since the number of links in a N-node chordal ring is $3N$, and the number of links in a N-node mesh-torus is $4N$. However, there are some restrictions that limit the practical implementation of chordal rings with $w=\sqrt{N}+3$ (as well as mesh-torus), when compared with other chord lengths. In fact, as referred in section 2, the smallest network diameter is obtained with $w=\sqrt{N}+3$ for a N-node chordal ring, where N is a square ($N=m^2$) and $N \geq 64$. The restriction associated with the square is not imposed to other chord lengths such as $w=N/4$ or $w=3$.

4. CONCLUSIONS

Using an analytical model, we have presented an assessment of the blocking performance in wavelength routed chordal ring networks with chord lengths of 3, $N/4$, or $\sqrt{N}+3$. It was shown that, in a 100-node chordal ring, the use of $w=N/4$, instead of $w=3$, clearly improves the blocking probability. An improvement in blocking performance was also observed if the chord length of $\sqrt{N}+3$ is used, instead of $w=N/4$ or $w=3$. In the latter case, the blocking probability gain due to the choice of $w=\sqrt{N}+3$, instead of $w=3$, for 16 wavelengths per link, is of the order of 10^4 for a load per node of 1 Erlang, and of the order of 10^{10} for a load of 0.1 Erlang, while, for 8 wavelengths per link, the blocking gain is about 13.5 for a load per node of 1 Erlang, and of the order of 10^4 for a load of 0.1 Erlang.

Chordal rings have also been compared with mesh-torus networks. An important feature has been revealed by this comparison: very small blocking gains were observed due to the increase of the node degree from 3 (chordal ring with 100 nodes and $w=\sqrt{N}+3$) to 4 (mesh torus with 100 nodes). In this case, as the load per node decreases from 5 Erlang to 0.01 Erlang, the variation of the node degree gain remains within one order of magnitude for the numbers of wavelengths per link considered: 4, 8, 12 and 16. Thus, the choice of a chordal ring with $w=\sqrt{N}+3$, instead of a mesh-torus, leads to a reduction in the number of network links, and hence in the total cable length.

Acknowledgements

Part of this work has been supported by Fundação para a Ciência e Tecnologia, Portugal, in the framework of project TRANSPARENT.

REFERENCES

[1] Cisco Systems, Cisco optical internetworking, available on-line at the URL: http://www.cisco.com/warp/public/779/servpro/solutions/opt/oi_brochure.html

[2] S. Harbour, "Optical networking: it's the end of the beginning", *FibreSystems*, Vol. 3, No. 10, pp. 30-32, 1999.

[3] P. Arijs, B. Caenegem, P. Demester, P. Lagasse, W. Parys, and P. Achten, "Design of ring and mesh based WDM transport networks", *Optical Networks Mag.*, Vol. 1, No. 3 pp. 25-40, 2000.

[4] D. R. Hjelme, "Importance of meshing degree on hardware requirements and capacity utilization in wavelength routed optical networks", in *Proc. ONDM'99*, Paris, France, February 8-9, 1999, pp. 417-424.

[5] B. W. Arden and H. Lee, "Analysis of chordal ring network", *IEEE Transac. on Computers*,Vol. C-30, No. 4, pp. 291-295, 1981.

[6] S. Subramaniam, M. Azizoglu, and A. K. Somani, "All-optical networks with sparce wavelength conversion", *IEEE/ ACM Transac. Networking*,Vol. 4, No. 4, pp. 544-557, 1996.

Near Optimal Design of Lightpath Routing and Wavelength Assignment in Purely Optical WDM Networks

Hong-Hsu Yen[1] and Frank Yeong-Sung Lin
Dept. of Information Management
National Taiwan University
Taipei, Taiwan, R. O. C.
Tel: +886-2-23638423
Fax: +886-2-27584773
Email: d4725001@im.ntu.edu.tw yslin@im.ntu.edu.tw

Key words: Lightpath Routing, Wavelength Assignment, Lagrangean Relaxation, WDM Network, Optimization.

Abstract: This paper proposes the optimal design of lightpath Routing and Wavelength Assignment (RWA) problem in Wavelength Division Multiplexing (WDM) networks without wavelength conversion. We formulate RWA as a mixed Integer Linear Programming (ILP) problem where the objective is to minimize the cost of wavelength assignment to the fiber links in the network. The Lagrangean relaxation technique and the optimization-based heuristics are used to solve this problem. Two sets of computational experiments are performed to test the algorithms for the maximum carried traffic and minimum wavelength requirements in three different network topologies (GTE, ARPA, OCT network). Based on solution quality to the computational experiments, the error gaps between the upper bound and the lower bound are close enough that the near optimal solutions could be obtained. On the other hand, we also show that the solution quality degrade gracefully under more and more heavy traffic network environment. By assessing solution quality and the computational time, we propose the efficient and effective optimization-based algorithms based on the Lagrangean relaxation method for the RWA problem in the WDM networks without wavelength conversion.

1. INTRODUCTION

WDM is a promising technique to utilize the enormous bandwidth of the optical fiber where the multiple wavelength-division multiplexed channels can be operated on a single fiber simultaneously [1]. A lightpath is an all-optical transmission path between two network nodes, implemented by the allocation of the same wavelength throughout the path [2]. However, how to route the wavelengths to a set of lightpaths is a challenging issue and proven to be a NP-Complete problem [6, 7, 9].

[1] Correspondence Author

A number of researchers have addressed this issue. Chlamtac [2] and Liang [3] have introduced the semilightpath technique to find the routing path. Both of the two works try to find the routing path by shortest path algorithm based on the arc weight of the auxiliary graph. And the arc weight of the auxiliary graph is fixed in order to calculate the optimal routing path for the particular Origin-Destination (O-D) pair. However, the arc weight should change dynamically in different lightpath routing assignments for each O-D pair. Furthermore, they did not address the issue of how to avoid the wavelength assignment collision between all O-D pairs.

Kim [4] model the RWA in terms of the ILP formulation where the objective is to minimize the number of OXCs. And the heuristic algorithm based on branch & bound technique is proposed. However, no solution quality data between the solutions from the heuristic algorithm and the optimal solutions are reported from the computational experiments in order to clarify the effectiveness of this heuristic algorithm.

Chen [5] propose layer-graph model for solving the routing and wavelength assignment problem. Several heuristics based on shortest path algorithm and layer-graph approaches are proposed. However, no lower bound are reported in this work to verify the solution quality of the proposed heuristic algorithms.

Banerjee [6] formulated a Linear Programming (LP) formulation with the objective to minimize the number of wavelengths and solve by approximation algorithms in order to deal with large networks. However, in this work, only at most one lightpath is considered from a source to a destination.

In this paper, we try to optimize the cost of wavelength assignment on the fiber-optic links in the WDM networks such that to route the wavelengths to their destination without violating the wavelength continuity constraint. In addition, multiple lightpaths from a source to a destination is considered. The wavelength continuity constraint means that the same wavelength must be used on all the links along the selected path for the O-D pair [1, 5]. And the algorithms that we proposed, unlike previous researches, are based on the lower bound and upper bound approaches at the same time.

This paper is organized as follows. In Section 2, mathematical formulation of the RWA problem is proposed. In Section 3, the dual approach for the RWA problem based on the Lagrangean relaxation is presented. In Section 4, the getting primal feasible heuristic is developed to get the primal feasible solution from the solutions to the dual problem. In Section 5, the computational results are reported. In Section 6, the concluding remarks are presented.

2. RWA PROBLEM FORMULATION

The WDM network is modeled as the graph $G(W, L)$ where W is the set of O-D pairs and L is the set of fiber links. Here we assume that each node has the switching capability to route the wavelength to proper links. We show the definition of the following notation.

L	the set of candidate fiber links in the WDM network
W	the set of O-D pairs in the WDM network
J	the set of admissible wavelengths in the WDM networks
λ_w	the traffic requirement (in number of wavelengths) for each O-D pair $w \in W$
P_w	a given set of simple directed paths from the origin to the destination of O-D pair $w \in W$
δ_{pl}	the indicator function which is one if link $l \in L$ is on path $p \in P_w$ and zero otherwise
$\varphi_{lj}(C_{lj})$	the cost for installing wavelength $j \in J$ on link $l \in L$

And the decision variables are depicted as follows.

x_{pj}	1 when path $p \in P_w$ with wavelength $j \in J$ is used to transmit the traffic for O-D pair $w \in W$ and 0 otherwise
C_{lj}	1 when wavelength $j \in J$ is installed on link $l \in L$ and 0 otherwise

The RWA problem is formulated as a mixed ILP optimization problem, as shown below.

$$Z_{IP} = \min \sum_{l \in L} \sum_{j \in J} \varphi_{lj}(C_{lj}) \qquad \text{(IP)}$$

subject to:

$$\sum_{p \in P_w} \sum_{j \in J} x_{pj} = \lambda_w \qquad \forall w \in W \quad (1.1)$$

$$x_{pj} = 0 \text{ or } 1 \qquad \forall p \in P_w, w \in W, j \in J \quad (1.2)$$

$$\sum_{w \in W} \sum_{p \in P_w} x_{pj} \delta_{pl} \le C_{lj} \qquad \forall l \in L, j \in J \quad (1.3)$$

$$C_{lj} = 0 \text{ or } 1 \qquad \forall l \in L, j \in J. \quad (1.4)$$

The objective function of (IP) is to minimize the total cost of wavelength assignments in the WDM networks. Constraints (1.1) and (1.2) require that the wavelength requirements for each O-D pair should be routed to its destination. Here distinct routing paths for the O-D pair with different wavelength requirements are allowed only if each wavelength is routed on one path. Hence, the wavelength continuity constraint is explicitly enforced in these two constraints.

Constraint (1.3) enforces that any wavelength should be installed on the link before assigned by the O-D pairs for routing on this link. Constraint (1.4) requires that the wavelength assignment on each link is a zero/one integer constraint, which means each wavelength could only be installed on each link for one time or none. From Constraints (1.3) and (1.4), each wavelength could only be assigned by no more than one O-D pair on every link is strictly enforced.

3. LAGRANGEAN RELAXATION FOR (IP)

In order to solve the above formulation successfully, we relax (1.3) to obtain the following (LR) for (IP).

$$Z_D(a) = \min \sum_{l\in L}\sum_{j\in J}\varphi_{lj}(C_{lj}) + \sum_{l\in L}\sum_{j\in J} a_{lj}\left(\sum_{w\in W}\sum_{p\in P_w} x_{pj}\delta_{pl} - C_{lj}\right) \qquad \text{(LR)}$$

subject to:

$$\sum_{p\in P_w}\sum_{j\in J} x_{pj} = \lambda_w \qquad \forall w\in W \qquad (2.1)$$

$$x_{pj} = 0 \text{ or } 1 \qquad \forall p\in P_w, w\in W, j\in J \qquad (2.2)$$

$$C_{lj} = 0 \text{ or } 1 \qquad \forall l\in L, j\in J. \qquad (2.3)$$

We can decompose (LR) into two independent subproblems.

Subproblem 1: for x_{pj}

$$\min \sum_{w\in W}\sum_{l\in L}\sum_{j\in J}\sum_{p\in P_w} a_{lj} x_{pj}\delta_{pl}$$

subject to (2.1) and (2.2).

Subproblem 2: for C_{lj}

$$\min \sum_{l\in L}\sum_{j\in J}(\varphi_{lj}(C_{lj}) - a_{lj}C_{lj})$$

subject to (2.3).

Subproblem 1 could be further decomposed into $|W|$ independent subproblems. For each independent subproblem, it looks like a shortest path problem but the wavelength assignment makes this subproblem slightly more complicated. We propose the Wavelength-Routing Algorithm (WR) to solve this subproblem.

Wavelength-Routing Algorithm(WR)

Step 1: For each O-D pair $w \in W$, first finding the shortest path with respect to each wavelength. Since the multiplier is positive, so the Dijkstra's Shortest Path Algorithm could be applied.

Step 2: There are total $|J|$ shortest paths for this O-D pair. Then, the optimal solutions for O-D pair $w \in W$ are λ_w number of shortest paths with the lowest costs.

Subproblem 2 could also be further decomposed into $|L||J|$ independent subproblems. For each independent subproblem, which could be solved by the Wavelength-Assignment Algorithm (WA).

Wavelength-Assignment Algorithm(WA)

Step 1: For each wavelength $j \in J$ on each link $l \in L$, calculate the value of $\varphi_{lj}(C_{lj}) - a_{lj}$.

Step 2: If this value is greater than zero, assign C_{lj} to zero else assign C_{lj} to one.

According to the algorithms proposed above, we could successfully solve the Lagrangean relaxation problem optimally. By using the weak Lagrangean duality theorem (for any given set of non-negative multipliers, the optimal objective function value of the corresponding Lagrangean relaxation problem is a lower bound on the optimal objective function value of the primal problem), $Z_D(a)$ is a lower bound on Z_{IP}. We construct the following dual problem to calculate the tightest lower bound and solve the dual problem by using the subgradient method.

$$Z_D = \max Z_D(a) \qquad \text{(D)}$$

subject to: $a \geq 0$.

Let the vector S be a subgradient of $Z_D(a)$ at (a). In iteration x of the subgradient optimization procedure, the multiplier vector $m^x=(a^x)$ is updated by $m^{x+1} = m^x + \alpha^x S^x$, where $S^x(a) = (\sum_{w \in W} \sum_{p \in P_w} x_{pj} \delta_{pl} - C_{li})$.

The step size α^x is determined by $\delta \frac{Z_{IP}^h - Z_D(m^x)}{\left\| S^x \right\|^2}$, where Z_{IP}^h is an primal objective function value at iteration k (an upper bound on optimal primal objective function value), and δ is a constant ($0 \leq \delta \leq 2$).

4. GETTING PRIMAL FEASIBLE SOLUTIONS

To obtain the primal feasible solutions to the RWA problem, solutions to the Lagrangean relaxation problems (LR) is considered. From the wavelength-routing assignment x_{pj} to determine the wavelength assignment for each link C_{lj} in order to satisfy the Constraint (1.3). The algorithm to get the primal feasible solution is proposed as follows.

Joint Wavelength-Routing and Wavelength-Assignment Algorithm(WR-WA)

Step 1: For each O-D pair, say w_b, the wavelength routing assignment x_{pj} from the solutions to the dual problem is used. If the wavelength j along the routing path x_{pj} have not been used, assign the associated C_{lj} along this routing path to be one and go to Step 4. If wavelength assignment violation occurs at any link, go to Step 2.

Step 2: Find all the wavelengths on the links where $C_{lj} = 1$ and add the associated arc weight a_{lj} with a very large number, say $G1$. Then identify the O-D pair, say w_c, that use this conflict wavelength link. Go to Step 3.

Step 3: There are two ways to resolve conflict. First to release the associated C_{lj} along the routing path selected by w_c, and let the w_b find the minimum shortest path and the associated a_{lj} is modified accordingly. Then, finding the minimum shortest path of w_c. Compute the total cost, say $t1$, by adding the cost of the shortest paths for w_b and w_c. Second is to locate the minimum shortest path of w_b with the shortest path of w_c remain unchanged. The total cost of w_b and w_c is also computed, say $t2$. If the lower cost of $t1$ and $t2$ is greater than $G1$, then it is an infeasible primal solution and stop the whole algorithm else assign the associated C_{lj} to be one and go to Step 4.

Step 4: Repeat the whole process until all O-D pairs are executed. And the primal feasible solutions are obtained. Stop the whole algorithm.

5. COMPUTATIONAL EXPERIMENTS

The computational experiments for the RWA algorithms developed in section 3 and 4 are coded in C and performed at PC with INTEL^{TM} PIII-500 CPU. We tested the algorithm for 3 network topologies--ARPA, GTE, OCT with 21, 12 and 26 nodes. The network topologies are shown in Fig. 1, 2 and 3.

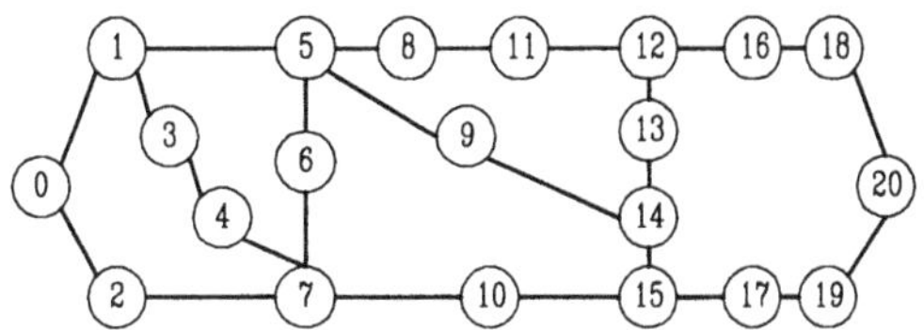

Figure 1. 21-node 52-link ARPA Netwo

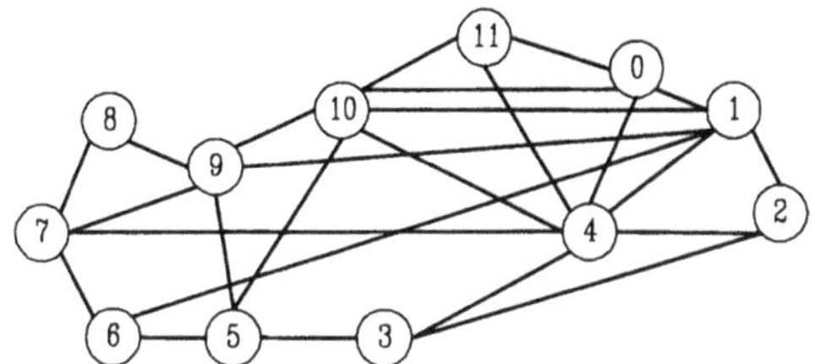

Figure 2. 12-node 50-link GTE Netw

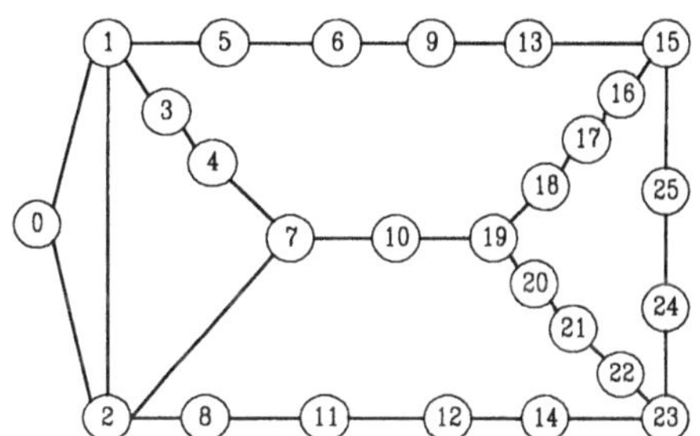

Figure 3. 26-node 60-link OCT Netw

The maximum number of iterations for the proposed dual Lagrangean algorithm are 1500, and the improvement counter is 30. The step size for the dual Lagrangean algorithm is initialized to be 2 and be halved of its value when the objective value of the dual algorithm does not improve for 30 iterations.

Two sets of computational experiments are performed. The computational time for these two sets of computational experiments are all within fifteen minutes. In these computational experiments, the cost of installing each wavelength on each link is randomly generated from one to ten. In the first set of computational experiment, we try to explore the threshold of the traffic that the networks could afford in the different network topologies under the given total number of available wavelengths. This is to maximize the carried traffic for the RWA problem, which is similar to the objectives in a number of related researches [5, 8]. In this experiment, the total number of wavelengths available for each link is 40. Table 1 summarizes the results.

Here the error gap is defined as the {(upper bound – lower bound)/lower bound} * 100%. Since the traffic requirement is randomly generated, in the second column of Table 1, it means the range of traffic requirements (in wavelengths) for each O-D pair.

As could be seen from Table 1, the error gap is tighter (all below 3%) in lightly loaded (below 0~6) GTE network environment and become loser in highly loaded environment. And when the range of traffic requirement is above 12, no feasible solution could be found. On the other hand, in the ARPA and OCT network topologies, it is not easy to find the feasible solution, since the degree of the nodes is small as compared to the GTE network. Table 1 shows that no feasible solution could be found when the range of traffic requirements is above 3 in both of these two network topologies. In other words, no primal feasible solution could be obtained even in the lightly loaded environment when the degree of the nodes is small. Hence, the degree of the nodes affects the solution quality tremendously.

In the second set of computational experiment, we try to explore the threshold of the number of available wavelengths under fixed traffic requirements. In this experiment, the GTE network topology is tested, and the traffic requirements (in wavelengths) are randomly generated from zero to four for each O-D pair. Table 2 summarizes the results.

As could be seen from Table 2, the solution quality is getting loser as the number of available wavelengths for each link is getting smaller. And for the number of wavelengths to be no greater than 13, no feasible solution could be found. It is interesting to see that as the number of wavelengths is approaching the threshold, the solution quality is becoming unstable. That is, when the number of available wavelengths is 15, no feasible solution could be obtained, but feasible solution could be found when the number of available wavelengths is 14. In addition, the error gap is only 7% at the threshold of the number of wavelengths. Some researchers try to minimize the number of available wavelengths [6], in the similar way, we have located the minimum number of wavelengths needed to support a given set of lightpaths on a GTE topology. In this set of computational experiments, the minimum number of wavelengths is 14. By assessing the error gap (7%) at this number of wavelengths, we also provide the effective algorithms to find the minimum number of wavelengths.

6. CONCLUDING REMARKS

In this paper, we successfully solve the RWA problem in which wavelength conversion is not considered. We formulate RWA problem as a mixed ILP problem and solved by Lagrangean relaxation method. We introduce two algorithms to solve each independent dual subproblem successfully, and we also propose an optimization-based heuristic to get the primal feasible solution based on the solutions to the dual problem.

Two sets of computational experiments are performed. In the first set of computational experiments, the threshold of the traffic requirements is explored in different network topologies under a given number of wavelengths. The solution quality is good (error gap are below 3 percent) in lightly loaded network and reasonably good (error gap are below 13 percent) in more highly loaded network. On the other hand, we also show that the degree of the nodes is an important factor for the algorithms to find the feasible solutions. That is, in low degree network topology, it is difficult to find feasible solution even in the lightly loaded traffic environments.

In the second set of computational experiments, we try to locate the minimum number of total wavelengths which is the objective function in the past research. The approach to locate the number of total wavelengths is by iterative decreasing the number of total wavelengths when the feasible solution could be obtained. At the time that no primal feasible solution could be found, the minimum number of wavelengths is located. Based on the computational experiments, the solution quality is still good at the minimum number of wavelengths.

As recalled from the computational time and the solution quality of the computational experiments, we propose the efficient and effective algorithms to solve the RWA problem. The further research is to solve the RWA with wavelength conversion problem.

References:

[1] B. Ramamurthy and B. Mukherjee, "Wavelength Conversion in WDM Networking," *IEEE JSAC*, Vol., 16, No. 7, pp. 1061-1073, Sep. 1998.

[2] I. Chlamtac, A. Faragó and T. Zhang, "Lightpath (Wavelength) Routing in Large WDM Networks," *IEEE JSAC*, Vol., 14, No. 5, pp. 909-913, June 1996.

[3] W. Liang, G. Havas and X. Shen, "Improved Lightpath (Wavelength) Routing in Large WDM Networks," *18th International Conf. On Distributed Computing Systems*, pp. 516-523, 1998.

[4] B. Kim, J.-H. Eom, J. Park and C.-J. Chae, "A Lightpath Routing Reconfiguration Algorithm for WDM Optical Networks," *5th Asia-Pacific Conf. On Communication and 4th Optoelectronic and Communication Conf.*, Vol. 1, pp.11-14, 1999.

[5] C. Chen and S. Banerjee, "A new model for optimal routing and wavelength assignment in wavelength division multiplexed optical networks," *Proc. of INFOCOM*, pp. 164-171, 1996.

[6] D. Banerjee and B. Mukherjee, "A Practical Approach for Routing and Wavelength Assignment in Large Wavelength-Routed Optical Networks," *IEEE JSAC*, Vol., 14, No. 5, pp. 903-908, June 1996.

[7] P. Raghavan and E. Upfal, "Efficient Routing in All-Optical Networks," *Proc. 26th ACM STOC*, pp. 384-395, 1994.

[8] N. Wauters and P. Demeester, "Design of the Optimal Path Layer in Multiwavelength Cross-Connected Networks," *IEEE JSAC*, Vol. 14, No. 5, pp. 881-892, June 1996.

[9] I. Chlamtac, A. Ganz and G. Karmi, "Lightpath Communications: An Approach to High

Bandwidth Optical WAN's," *IEEE Trans. on Communications*, Vol. 40, No. 7, pp. 1171-1182, July 1992.

TABLE 1 – COMPUTATIONAL RESULTS (THRESHOLD OF THE TRAFFIC REQUIREMENTS UNDER DIFFERENT NETWORK TOPOLOGY)

Network topology	Traffic requirements	Lower bound	Upper bound	Error gap(%)
GTE	0~2	1.454951e+002	1.490000e+002	2.408975
	0~3	3.620000e+002	3.710000e+002	2.486190
	0~4	6.159618e+002	6.270000e+002	1.792022
	0~5	9.386195e+002	9.630000e+002	2.597483
	0~6	1.278489e+003	1.313000e+003	2.699331
	0~7	1.660505e+003	1.768000e+003	6.473647
	0~8	2.198478e+003	2.338000e+003	6.346305
	0~9	2.827843e+003	2.947000e+003	4.213689
	0~10	3.379600e+003	3.736000e+003	10.545640
	0~11	4.285055e+003	4.814000e+003	12.343940
	0~12	5.387118e+003	X#	X
ARPA	0~2	2.329307e+003	2.412000e+003	3.550106
	0~3	6.698546e+003	X	X
OCT	0~2	5.146004e+003	5.471000e+003	6.315504
	0~3	1.205400e+004	X	X

means no feasible solution could be found.

TABLE 2 – COMPUTATIONAL RESULTS (THRESHOLD OF THE NUMBER OF AVAILABLE WAVELENGTHS)

Network topology	# of available wavelengths	Lower bound	Upper bound	Error gap(%)
GTE	20	1.491610e+003	1.529000e+003	2.506690
	19	1.426938e+003	1.456000e+003	2.036654
	18	1.401527e+003	1.441000e+003	2.816450
	17	1.389967e+003	1.411000e+003	1.513170
	16	1.700296e+003	1.776000e+003	4.452425
	15	2.052708e+003	X	X
	14	1.883020e+003	2.024000e+003	7.486924
	13	2.204916e+003	X	X

Part Two

Protection and restoration in WDM networks

An Intelligent and Mobile Agent-based Approach for Dynamic Protection Set-up in Future Optical Networks

Daniel Rossier-Ramuz[†°], Dr. Daniel Rodellar[†], Dr. Rudolf Scheurer[°]
[†]Swisscom Corporate Technology and [°]University of Fribourg

Key words: Intelligent wavelength services, intelligent and mobile agents, FIPA, quality of protection

Abstract:

The future Wavelength Division Multiplexing (WDM) based optical networks are expected to provide new Intelligent Wavelength Services (IWS) for the transport of all types of data (voice, multi-media, Web-based data, etc.) at bit rates exceeding 1 TBit/s. The survivability of such networks will depend on efficient protection mechanisms. Although several protection strategies will be proposed to the customers, managing the complexity of protection-related interactions between client layers and the optical network will become a considerable challenge. In this paper, we develop a number of statements related to the management of IWS and we propose an approach based on intelligent and mobile agents in order to tackle the management of protection-related client requirements. In that scope, we introduce the notion of p-groups as a possible client requirement and we define a Quality of Protection (QoP) as a new metric to assess the protection quality during the lifetime of client connections.

1. INTRODUCTION

The tremendous growth of bandwidth needs expressed by end customers has made the Wavelength Division Multiplexing optical networks very attractive. But the client demand matrix has not only increased in terms of bandwidth but also regarding the diversity of the client requirements related to the different protocols and architectures used on top of the Optical Transport Network (OTN).

The QoS-based end-to-end optical channel allocation and protection requirements for future mesh optical networks remain an important subject of research. At the University of Fribourg, the OPTIMA[1] project aims at studying new advanced solutions based on intelligent and mobile agents for the management of future OTN.

The investigations within OPTIMA will mainly focus on transparent OTN, i.e. transparent in the sense that the regenerators and wavelength converters within the OTN are fully operating in the optical domain, without electrical/optical conversion. We consider the transparent OTN as an intermediate step in the technological development towards a full-optical OTN based on optical packet switching OPTIMA is also intended to contribute to the conception of Intelligent Wavelength Services (IWS). The goal of IWS is to provide more intelligence and thus flexibility to OTN in order to integrate a multitude of client protocols as well as to cope with their specific requirements regarding protection.

In this paper, we first introduce the next generation of optical networks and the concept of Intelligent Wavelength Services (Chapter 2). In a next step we discuss the protection issues related to IWS as imposed by the complexity of dependencies between client protection requirements, and we introduce a way to formally define protection requirements and to express protection quality (Chapter 3). This will be the base for the introduction of the mixed stationary/mobile agent based architecture to manage the allocation of protection paths in IWS (Chapter 4). We finally present our conclusions so far and the plans for future work (Chapter 5).

2. NEXT GENERATION OPTICAL NETWORKS

New advances in optical technology now allow carrying signals optically in the network from end to end, and thus totally eliminating the need for electrical regeneration. Owners of the optical networks may use four network functionalities to make their transport networks more agile (protocols like IP and MPLS can also profit from these functions). The first function is the Optical Add-Drop Multiplexing (OADM) that allows inserting and removing a given wavelength from the optical fibre; the second one is wavelength routing that enables to route wavelengths from input fibres to the output ones. The third function is wavelength conversion.

[1] OPTical network management with Intelligent and Mobile Agents – project submitted to the Swiss National Science Foundation

Finally there is optical switching that enables wavelength switching capabilities from one fibre to another (the wavelength is present on one fibre or the other). The four functions are shown on *Figure 1.* Although they are all represented on a static basis for comprehension purposes, they could have a dynamic behaviour (variation as a function of time).

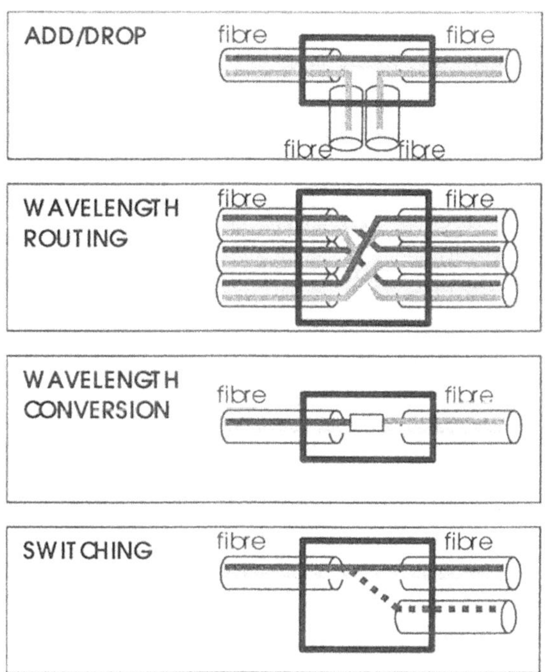

Figure 1. Four types of functionality of next generation optical network elements

The elements realising these functions can operate in the optical domain without requiring optical/electrical conversion. The WDM equipment should perform several of these functions concatenated. Full optical components capable to route whichever inputs to whichever outputs on whichever wavelengths will gradually appear on the market, going through intermediate ranges of limited equipment. Basically, the switch matrix is quite limited nowadays (in the order of 32 x 32 optical wavelengths) and the optical transponder does not provide full conversion between all the wavelengths. Wavelength continuity constraint consequently remains to be considered when new optical paths are being allocated. These constraints introduce potential blocking problems when the client demands are being allocated.

The concept of Intelligent Wavelength Services (IWS) will provide an advanced optical transport layer that will promote the Fibre To The Home

(FTTH). This paper proposes a novel agent-based architecture to tackle the dynamic protection management in an optical mesh network.

2.1 Intelligent Wavelength Services (IWS)

The future OTN will have to support all kinds of client protocols (IP, Gigabit Ethernet, PoS, clear-channel with bit rates from 2 Mbit/s to 2.5 Gbit/s, etc.) in a multi Service Providers (SPs) and multi Network Operators (NOs) environment. New business models are now emerging to support the new inter-parties interactions. We believe that this is a pre-requisite for a successful deployment of "fibre to the home" (FTTH) networks, in which the customers will dynamically allocate and release wavelength transport channels for specific data services (voice, digital TV, Internet, etc.). In this context, the elaboration of some kind of automatic Service Level Agreement (SLA) based negotiations as considered in the future UMTS networks seems to be a promising way to support the client requests.

Hereafter, we use the concept of Intelligent Wavelength Services (IWS) depicted on *Figure 2*. IWS support all the network and service management functions of the OTN, i.e. from the network element as well as from the customer viewpoint. It has to be noted that for IWS, the customer can be either a service provider or an end-user.

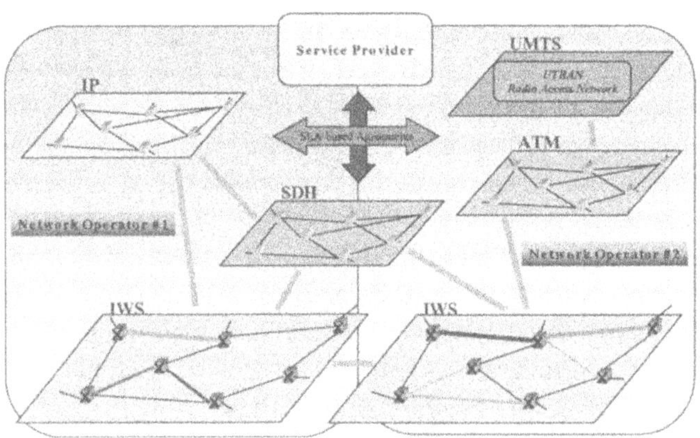

Figure 2. IWS in a multi-SP multi-NO environment and supporting multiple client protocols

The SLA in the IWS is used between the customer and the network operator to specify explicitly the traffic conditions regarding bandwidth, channel availability, protection requirements, type of interface, etc. The IWS will configure the OTN dynamically in order to meet the objectives as described in the SLA. Normally, if one of the specified objective can not be satisfied, even for a limited period of time, the NO has to pay penalties to the customer. The description of the negotiation process taking place behind the SLA is out of the scope of this paper. A similar SLA-based approach for future UMTS networks can be found in [LRDC+Oct00].

2.2 Management Issues

The functional aspects of OTN management are currently under investigation in ITU-T. The Telecommunication Management Network (TMN) [M3100May96] is usually considered as the generic framework for the definition of all information related to managed objects and to the interactions between managers and agents. The TMN platform-centred approach however entails drawbacks in scalability, reliability, efficiency, and flexibility and is consequently unsuitable for large and heterogeneous networks [ZZ1998].

In TMN, two information streams are used for management purposes: the first stream uses the Data Communication Network (DCN) of TMN and usually requires an IP-based network. The second stream consists in an Embedded Communication Channel (ECC), in which in-band signalling can be passed through between Network Elements (NE). The signalling protocol is implemented by the transport protocol itself (SDH, ATM, etc.) and requires specific bytes from the overhead.

The search for optimal protection paths attached to an optical end-to-end connection can be based upon the definition of the Routing and Wavelength Assignment (RWA) problem described in the next section. The on-line computation requires a distributed version of the algorithms used to solve the RWA. The approach we are investigating consists in a combination of TMN legacy systems with advanced distributed mechanisms to deal with the dynamic client requests as well as with their protection requirements. To do this, we use intelligent agents making complex interactions between NEs possible.

The software agent-based architecture proposed in this paper uses as much local information as possible. Local information is called the *environment* of the agents. From the implementation viewpoint, access to the information is provided by the use of a specific gateway. In the future, the transport of messages between agents or the transport of the agents themselves will be possible thanks to new structures such as Digital Wrapper or ASON [NBJan99,D697Apr00]. Such an approach is already described in [RSDec99].

2.3 Dynamic Routing and Wavelength Assignment (RWA)

We define the *lightpath* as the optical path used for a client connection, i.e. the path between an optical source node and an optical destination node. A lightpath can use one or several wavelengths along the way.

The RWA can be divided into two sub-problems: the *routing problem* that is, finding the route between two optical nodes and the *wavelength assignment problem* that is, finding the best wavelength allocation along the discovered route. Combining the two sub-problems provides a solution minimising the number of wavelengths required in an optical network for a given traffic matrix. Both of the two sub-problems are difficult to solve (NP-complete) and require the usage of heuristics. An overview of RWA algorithms and heuristics as well as the mathematical formulation of the problem can be found in [EUR709Apr99], [ZJMJan00].

The RWA problem is originally considered during the network design and planning process, when business costs are relevant. Therefore, most of the algorithms proposed to solve RWA are based on a centralised architecture in which the entire network topology is known in advance. The traffic matrix is required by the RWA and is normally issued by statistical data. At this stage, there is no dynamic consideration. The working paths and the protection paths are established in a similar way.

This is possible as long as the traffic matrix does not differ from reality too much and the client demands are static. As soon as the client demands are dynamic and the requirements regarding the protection are different from one customer to another, as will be the case in IWS, the protection paths and the working paths have to be computed dynamically. The major problem of the dynamic approach resides in the blocking problem, i.e. there is no available wavelength to satisfy a client's demand.

The agent-based algorithms firstly require the study of distributed algorithms. For RWA, and more specifically for the wavelength assignment part, we have found the Dynamic Relative Capacity Loss (DRCL) algorithm. The implementation of DRCL with FIPA agents is explained in section 4.3.

2.3.1 Dynamic Relative Capacity Loss (DRCL)

The DRCL algorithm [ZJMJan00] is the distributed version of RCL. Briefly explained, DRCL consists in calculating the *relative capacity loss* for each path on each available wavelength every time the network state changes. For each client request, the route is first found using the Bellman-Ford algorithm; the wavelength minimising the RCL value is then selected.

The RCL values are stored locally in the optical node and are updated in a distributed way after a new path has been allocated.

In the IWS, the RCL values are handled by stationary agents (see section 4). RCL tables and agents are considered as the *environment* in which M-agents (mobile agents) evolve.

2.4 Protection Management

Since several Tbit/s will be transported through the OTN and since the protection will be considered as a value-added service, protection management in IWS is a strategic key issue. The protection schemes in OTN are derived from the mechanisms as specified by ITU-T [G841Jul95] for SDH networks; the dedicated protection (1+1) and shared protection (1:n or n:m) are two essential protection modes.

As considered in our research, however, the IWS, will give the customer the opportunity to select different protection strategies according to his needs. The protection might already be present at the client layer, or might be required at both layers (client + OTN layers).

2.4.1 Protection / Restoration

Two important strategies for protecting OTN are based on protection and restoration mechanisms. The former consists in allocating protection paths so that the protection mechanism can be raised using hardware mechanism and can thus react very fast. The latter is initiated right after the failure and consists in finding alternative paths in order to preserve the running services.

The restoration can be proposed as an ultimate way to keep a service running in case of failure. When a customer claims low quality protection, it might happen that no protection paths are available when a failure occurs. In this case, the only way to preserve the connection is to launch a restoration mechanism. We have not considered this approach in our research yet.

2.4.2 Dedicated / Shared Protection

The dynamic protection set up might result in a combination of mixed dedicated and shared optical channels, depending on the availability of protected paths at a given time and on the priority of client requirements. We assume that the dynamic reconfiguration capabilities of OTN allow the configuration of protected paths to be changed over time.

The client requirements related to QoS, including the protection, can be described in the SLA before the allocation of optical resources. At the moment, we are investigating a possible description of protection-oriented requirements.

3. PROTECTION ISSUES IN IWS

In this chapter, we propose two concepts that allow us to figure out optimal protection paths: ***p-group*** and ***QoP***.

3.1 Inter-operability with Client Layers / Protected Group (p-group)

The future IWS will support all kinds of client protocols as well as advanced client requirements, particularly concerning the protection strategy. In this context, the customer could own a certain group of NEs like SDH, ATM or IP switches, for example. Virtual topology - or logical topology - is the set of connections that will be set-up by the higher-level network using the OTN. This kind of requirement also refers to the concept of Virtual Private Network (VPN).

Four different protection strategies have been identified according to the customer's wish (*Figure 3*): protection at the OTN layer, protection at the client layer, protection at both layers or no protection at all. As an additional requirement, the customer could also tell the IWS how many failures points do the client accept to have (single failure/multiple failures). Another kind of requirement consists in claiming special protection by giving a sub-topology, e.g. a critical zone, requiring better protection than the rest of the network. The customer could have experienced that a high traffic load will occur in a certain country at a certain point of time and could therefore require 100% protection for this part of the network.

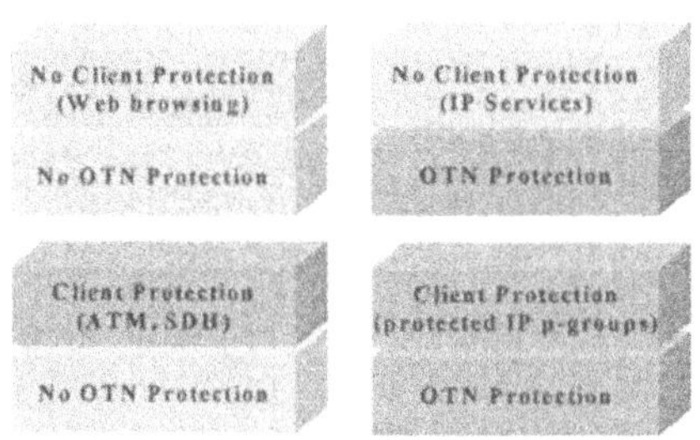

Figure 3. Possible Protection Schemes between client layers and the OTN

A possible way to express the client requirement regarding the protection strategy formally is the protected group (p-group, *Figure 4*), as described in [Cro1998]- "*the p-group represents the demands belonging to a common protected group of the higher level network. Each p-group is characterised by a number and is a list of demands*". The number refers to the level of protection (1 means that the client layer is able to support 1 failure).

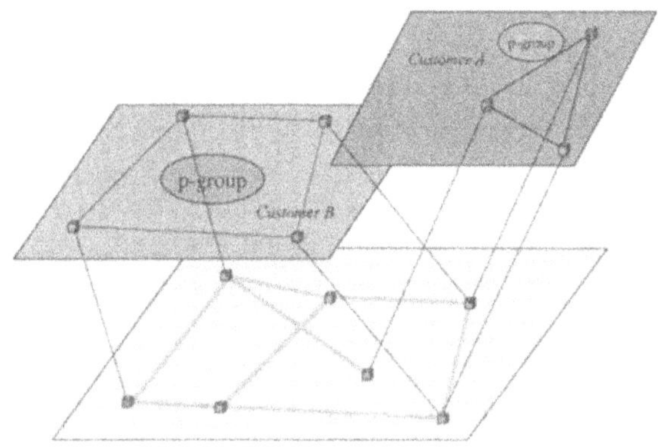

Figure 4. Two customers with their own protected VPN sharing the same physical OTN

The concept of p-group allows the customer to specify a virtual topology for the protected network components. In the future, the customer will decide which part of the VPN should be protected, and he will indicate the quality degree of the required protection. The interoperability of protection mechanisms between the client and the OTN must be considered and can lead to three potential problems: the *bottleneck*, *connectivity* and *multiple groups* problem. Details concerning interoperability issues can be found in [Cro1998].

3.2 Quality of Protection (QoP)

The second concept is based on a metric called Quality of Protection (QoP), which is being developed at our institute. Thanks to QoP, the agents will be able to assess the quality of protection dynamically and therefore to compare it with the requirements specified in the SLA. QoP(t) provides a temporal metric for an on-line measure of the protection quality.

To define a QoP metric, we first need to understand which components can fail and at which frequency. An analysis of this issue is proposed in [GRMar00]. In a first approach, we have decided to consider two components: *the entire node* and *the optical fibre*.

Let us define the following two functions: SN(t), the number of Shared Nodes along a protected path, SL(t) the number of Shared Links (fibres) along a protected path. It has to be noted that the shared link implies that the protection and working paths are on the same link but are using different wavelengths. In case of fibre failure, the connection obviously breaks down. Shared links are acceptable only in the case of optical channel malfunction within the optical node.

SN(t) and SL(t) are time dependent. The configuration of protected paths can change over time. If SL(t) is constantly equal to 0, it means that the protection path follows a dedicated (1+1) protection strategy. Moreover, if SN(t) = 0, the protection path uses disjoint nodes that those used for the working path.

Let us define $P_{NetworkFailure}$ as the probability that a failure occurs in the OTN layer (fibre cut, node malfunction) and $P_{ServiceFailure}$ as the probability of a service disruption. The *Dynamic Protection Set-up* (DPS) problem is stated as follows:

DPS: *given an end-to-end connection, find a protection path that maximizes QoP(t).*

$$QoP(t) = \begin{cases} QoP(t) = 1, & \text{if SL = 0 and SN = 0} \\ QoP(t) = \alpha\omega_1 \dfrac{1}{SN(t)} \beta(t), & \text{if SL = 0 (dedicated 1+1)} \\ QoP(t) = \alpha\omega_2 \dfrac{1}{SL(t)} \beta(t), & \text{if SN = 0} \\ QoP(t) = \alpha(\omega_1 \dfrac{1}{SN(t)} + \omega_2 \dfrac{1}{SL(t)}) \beta(t), & \text{otherwise} \end{cases}$$

with $\omega_1 + \omega_2 = 1$

$0 < QoP(t) < 1$, 1 means that the quality of protection is maximal, 0 means no protection quality.

$0 < \alpha < 1$ is to be considered in case of multiple operators (or domains) interactions. α is a constant value and refers to horizontal interactions (see *Figure 5*). This factor can be used to assess the overall influence on the protection when an end-to-end connection needs to traverse several domains.

$0 < \beta < 1$ is a time function that assesses the respect of constraints related to the interoperability between the protected groups of the client layer and OTN. The value of β depends on client demands and can change dynamically depending on the current configuration of protection paths. This factor is concerned with multi-layer interactions (see *Figure 5*). Details of this function are still under investigation.

$0 < \omega_1, \omega_2 < 1$ are two factors associated to the time function SN(t) and SL(t) accordingly. The customer can give them as part of his protection requirements and can also give information concerning the importance given to SN and SL functions (i.e. rather shared nodes, rather shared links).

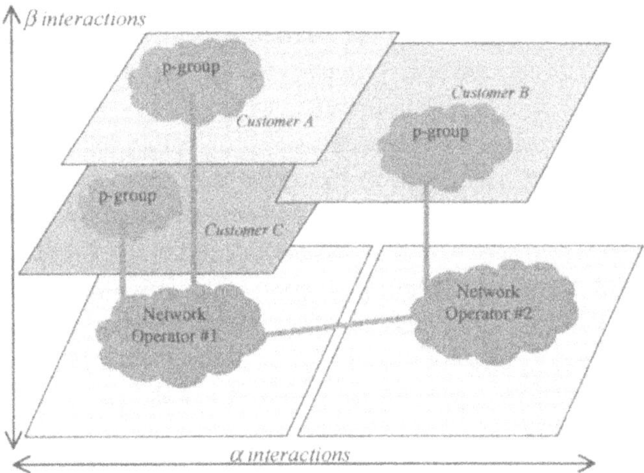

Figure 5. Two-dimensional interactions for the quality of protection (multi-layer, multi-NO)

From the customer's point of view, the protection requirement can be expressed as the probability that the service will be disrupted. This can be simply computed as follows:

$$\mathbf{P_{ServiceFailure} = P_{NetworkFailure} * (1 - QoP)}$$

The value $P_{ServiceFailure}$ can be monitored in a continuous way. If QoP is equal to 1, the protection is maximal and the probability that the service will be interrupted is reduced to 0. Without any protection, QoP is equal to 0, and the probability of being in trouble is equal to the probability to be in trouble at the OTN layer.

Having defined the p-group and QoP metric, the customer can now specify the protection level, the weight attached to shared nodes/shared links and the sets of p-group as possible protection-oriented requirements.

4. AN AGENT-BASED APPROACH

In this chapter, we describe an agent-based approach we are investigating in order to deal with the complexity of dynamic protection management based on complex requirements, as described in section 3. We introduce two families of agents, FIPA-agents – which handle DRCL tables - and M-agents – which use the QoP metric to find optimal paths.

Intelligent software agent technology can be considered as a natural extension of object-oriented technology. Although there is no widely adopted definition the agent itself, an intelligent agent should exhibit a minimal set of properties such as: *autonomy* – the agent has its own execution context (code+data), *reactivity* – the agent is able to react to external events (alarms, notification, etc.), *pro-activity* – the agent is able to make decisions according to its mental state, i.e. its internal representation of the environment, and *sociability* – the agent is able to communicate with other agents, i.e. to exchange complex knowledge in order to achieve certain objectives. The intelligence of such systems does not only refer to the agent behaviour itself but also to the intelligent behaviour that emerges from a society of agents that is, the results of co-operative work performed by all the agents. Methodologies for the design of multi-agent systems (MAS) are still under investigation. Further details about software agent can be found in [BZW1998,HBMar99].

The concept of intelligent agent in telecommunication has gained growing interest since Internet and the Java programming language appeared. The deployment of large-scale agent systems is now possible and can gradually be implemented into network devices. Standard organisations such as **FIPA**[2] or **OMG**[3] are promoting agent-based systems by focusing on interoperability between agent platforms.

In order to ensure that the message content exchanged between agents is correctly interpreted and to avoid mismatch between the different domains of management, the representation of terms and rules governing a given domain is precisely specified. This representation is called *ontology*. In the context of software agents, the ontology is the basic level of a knowledge representation scheme. It gives a meaning to terms, so that terms exchanged between agents can be correctly understood. Specifying ontology is particularly important in case of multiple domains (SPs/NOs) interactions. The definition of an ontology related to IWS is part of our research.

Based on the RWA decomposition, we are developing two families of intelligent agents. The first family of agents is constituted by stationary FIPA-compliant agents that reside in every optical node. Their objective is to

[2] Foundation of Intelligent Physical Agents – http://www.fipa.org

[3] Object Management Group – http://www.objs.org/agent

maintain RCL tables used by the DRCL algorithm in order to provide an optimal wavelength assignment for a given route. The second family contains adaptive mobile agents (M-agents) that are responsible for finding appropriate routes within the network. These agents are characterised by their ability to embed client requirements related to the protection strategy dynamically.

4.1 FIPA and FIPA-OS

FIPA agents can be implemented with the Nortel FIPA-OS[4] platform, the first publicly FIPA-compliant agent platform. FIPA-OS is an *Open Source* project and is well designed for telecommunication applications. FIPA-OS supports most of the FIPA specifications.

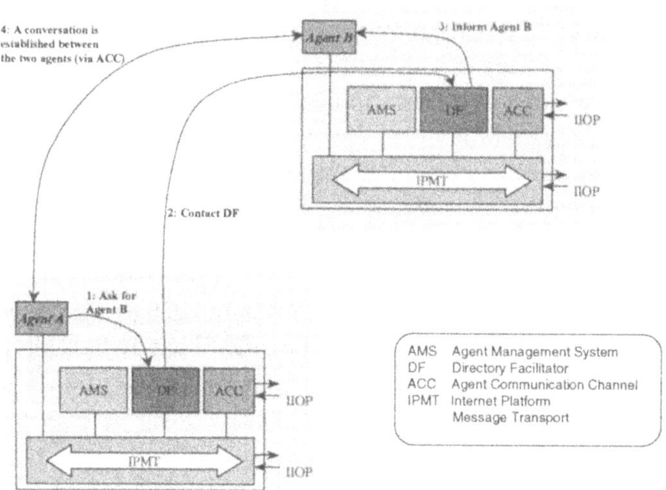

Figure 6. FIPA communication model for an agent platform

The distributed nature of agent-based applications requires advanced mechanisms for knowledge exchanges. The agents operate in an asynchronous way and can process messages belonging to several domains (SPs/NOs). FIPA has designed a communication model using the notions of *ontology* and *Agent Communication Language* (ACL). While the ontology is used to define a collection of terms and rules related to a specific domain,

[4] http://fipa-os.sourgeforge.net

the ACL gives the agents the flexibility required to establish some kind of advanced conversation. The ACL is made up of several types of messages called *performatives*. The performatives allow one agent to inform another agent of its intention. The agent then adapts its internal representation of the environment according to the message content. Examples of performatives are *request*, *query if*, *inform*, *reject*. The performative itself includes a message content expressed in a content language such as SL0 or RDF. Both languages are supported by FIPA-OS.

FIPA architecture has defined three basic agents (*Figure 6*): AMS, ACC and DF. Briefly explained, the AMS is considered as the kernel of the agent platform, the ACC is responsible for send/receive messages and the DF is a kind of "yellow page" service, which provides information about the registered agents.

4.2 Our approach: FIPA agent and M-agents

The software agent can be either *stationary* or *mobile*. Mobile agents are now easier to study since programming languages such as Java support code mobility. Several mobile agent platforms such Aglet, Odyssey, Voyager, Tacoma, and more recently Grasshopper[5] which is the first OMG MASIF[6] compliant mobile agent platform are available. At our institute, we are investigating new mobile agents (M-agent) and interaction models. We wish to use the abstraction of mobility as an efficient way of managing the protection requirements in the future IWS. Typically, M-agents will be used for the routing part of the RWA problem.

FIPA agents are stationary agents implemented into the optical nodes. The main function of a FIPA agent is to maintain the RCL table up to date and to manage the interactions with M-agents. The next section describes the implementation of DRCL into FIPA-agents using a peer-to-peer protocol for the knowledge exchanges.

4.3 Wavelength Assignment Using Stationary FIPA Agents

We now proceed to describe briefly the implementation of DRCL into FIPA agents.

[5] http://www.grasshopper.de

[6] Mobile Agent Service Interoperability Facilities http://www.omg.org

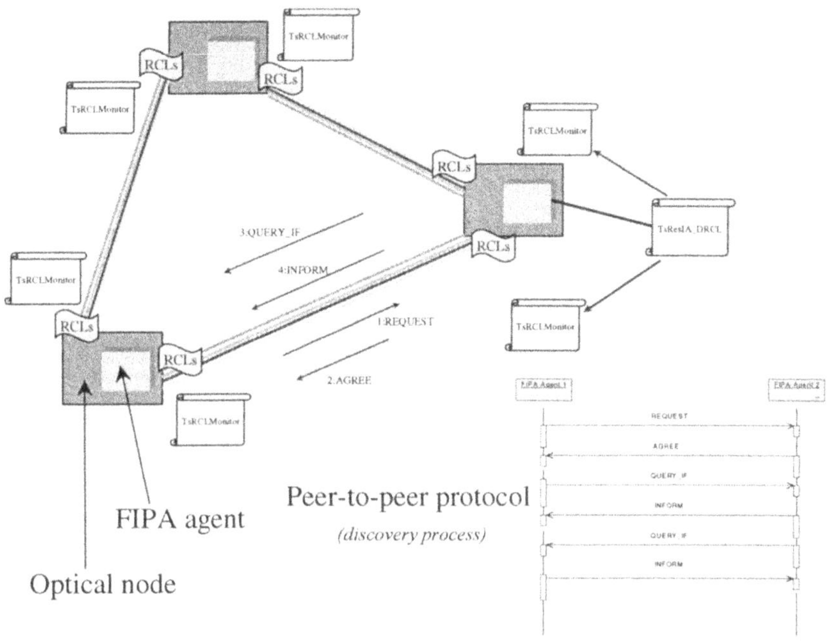

Figure 7. Implementation of DRCL with stationary FIPA-agents

Figure 7 shows the FIPA agent based architecture we are developing. The agents have to keep internal RCL tables up to date. The agent task starts with a discovery process which can be initiated everywhere in the network. The agent discovers the output ports and initiates a task *TsRCLMonitor* for each port. This task creates a RCL table (RCLs in the figure) and establishes a conversation with the remote agent according to the peer-to-peer protocol depicted on the figure. A permanent conversation is then set up and will be referred to when the agent needs to send a message to this port. The performative *inform* is used to transfer the XML-encoded RCL tables. A comparison mechanism allows avoiding cyclic updates.

The RCL table is made of triples (src, dst, rcl) on each wavelength. It is therefore possible to know which destination is reachable from a source node for a certain wavelength. Once the M-agent has discovered a satisfying protection path, the wavelength with the lowest *rcl* value is selected.

4.4 Protection Management with M-agents

In this chapter, we give a short overview of the basic agent-based mechanisms that are presently being implemented in our research group.

The conjunction of two active agents societies provides optimal configuration for protection paths. M-agents are used to capture the client protection requirements that are formulated with a rule-based language, which is not described in this paper. The M-agents then evolve in the network and try to find an itinerary that is, a route. The itinerary of M-agents is mainly based on their internal rules. One of the rules consists in finding paths to maximise the QoP metric. There is one M-agent per end-to-end connection.

Stationary FIPA agents are used to provide additional information to mobile agents regarding the wavelength assignment by computing the relative capacity loss. Once the destination node has been reached, the M-agent returns to the source node by selecting the appropriate wavelengths thanks to RCL tables.

The M-agent architecture is currently being defined. So is an interaction model relying on the abstraction of agent mobility. In the end, each protection path can be configured separately in a co-operative way.

5. SUMMARY AND FUTURE WORK

In this paper we have presented some intermediate results of our research work in the framework of the OPTIMA project. The investigations so far represent a promising basis for the further research work within OPTIMA.

Software agents provide an open, scalable and flexible way to deal with the legacy network management systems and the embodiment of specific OTN-related algorithms. On the one hand, stationary intelligent agents are able to perform reasoning and to transfer knowledge regarding local optical nodes; on the other hand, reactive and adaptive mobile agents figure out optimal routes thanks to a QoP temporal metric and interaction models. Such an approach can support future protection-oriented client requirements.

Within the context of OPTIMA, we intend to investigate in more detail the coordination models for different combination degrees of intelligent and mobile agents. From the ATM world, we are aware of ongoing research work centered on market-based approaches for resource allocation using agent technologies. We will investigate to which extent such approaches may be adopted and included in the IWS concept.

Furthermore it is planned to work on an OTN modelling tool to get a flexible framework to perform the (mainly functional) evaluations of our agent-based approaches. This tool will have to integrate the FIPA platform and the co-ordination models as developed in OPTIMA.

6. REFERENCES

[BZW1998] Walter Brenner, Rüdiger Zarnekow, Hartmut Wittig, "Intelligent Software Agents", Springer-Verlag Berlin Heidelberg, 1998.

[Cro1998] O. Crochat, "Wavelength Division Multiplexing Networks And Failure Protection", PhD thesis Nr 1851, Ecole Polytechnique Fédérale de Lausanne, 1998.

[D697Apr00] ITU-T (USA), "Work on the automatic switched optical network", Delayed contribution D.697 (WP3/15).

[EUR709Apr99] EURESCOM Project P709, "Planning of Full Optical Network", http://www.eurescom.de/

[GRMar00] Ornan Gestel and Rajiv Ramaswami, "Optical Layer Survivability: A Services Perspective", IEEE Communications Magazine, March 2000.

[G841Jul95] ITU G.841, "Types and Characteristics of SDH Network Protection Architectures", July 1995.

[HBMar99] Alex L.G. Hayzelden, John Bigham, "Software Agents for Future Communication Systems", Springer-Verlag Berlin Heidelberg 1999.

[LRDC+Oct00] Jingming Lisalina, Daniel Rossier, Manuel Dinis, Laurie Cuthbert, Laurissa Tokarchuk & John Bigham, "Agent-based resource management for 3G networks", Mobile Communications Summit, Galway, Ireland, 1-4 October 2000.

[M3100May96] ITU M.3100, "Principles for a Telecommunications management network", May 1996.

[NBJan99] G. Newsome and P. Bonenfant, "A Proposal for Providing Channel-Associated Optical Channel Overhead in the OTN", ANSI T1X1.5/99-002, Jan 1999; available at http://www.t1.org/index/0816.htm

[RSDec99] D. Rossier-Ramuz, R. Scheurer, "An Introduction to Optical Agents: Intelligent and Mobile Agents for WDM Optical Network Management", in Proceedings of IMPACT'99, Impact of Agent Technology on Telecommunications, Seattle, USA, 2-3 December 1999, pp. 131-139.

[ZJMJan00] Hui Zang, Jason P. Jue, Biswanath Mukherjee, "A Review of Routing and Wavelength Assignment Approaches for Wavelength-Routed Optical WDM Networks", Optical Networks Magazine, January 2000.

[ZZ1998] Dianlong Zhang, Werner Zorn, "Developing network management applications in an application-oriented way using mobile agent", Computer Networks and ISDN Systems 30 (1998), pp. 1551-1557.

A Framework for Service-Guaranteed Path Protection of the Optical Internet

Pin-Han Ho and H.T. Mouftah
Department of Electrical and Computer Engineering, Queen's University at Kingston, Ontario, Canada, K7L3N6

{hoph, mouftah}@ece.queensu.ca

TEL: (613) 533-2934 FAX: (613)533-6115

Key words: Protection and restoration, optical networks, DWDM

Abstract: The ordinary path-based and link-based shared protection schemes can only provide a limited spectrum of protection services with coarse protection granularity, which will not be able to satisfy the versatile requirements of multimedia applications on the Internet in the foreseeable future. In this paper we propose a framework, Short Leap Shared Protection (SLSP), for service-guaranteed end-to-end shared protection for the optical Internet. We will show that SLSP enhances the 1:N and M:N shared protection schemes in terms of scalability, flexibility and class of service.

1. INTRODUCTION

The use of DWDM (dense wavelength division multiplexing) in the Internet backbone has opened a new era for multimedia communication networks by dynamically provisioning nearly unlimited bandwidth.

Although this approach can accommodate tremendous amount of data, it may also risk a serious data loss when a fault occurs (e.g. a fiber cut or a node fault), which could downgrade the service to the customers to the worst extent. To improve the survivability, the Internet service providers (ISPs) are required to equip the networks with protection and restoration schemes that can provide end-to-end guaranteed services to their customers according to the service level agreements (SLAs). The networks with SONET or ATM as the intermediate layers (e.g., existing IP/SONET/DWDM or IP/ATM/SONET/DWDM) have provided robust and layered protection and restoration mechanisms by which the service to the customers can be recovered from a fault within tens of miliseconds[13,15]. However, the disadvantages of these multi-layered structures[14] have motivated the design for the IP/MPLS control directly over the optical transport network (OTN) that was defined by ITU-T G.872. With the trend of integrating IP/MPLS/DWDM control plane for the optical Internet, it is desirable to develop a parametric, systematic and scalable protection and restoration schemes that can work in the optical layer to cope with the more stringent requirements for the quality of service.

In this paper, we propose a framework, the Short Leap Shared Protection (SLSP), to perform end-to-end path protection of a single failure within the optical layer of IP over DWDM networks. It can also be easily extended to consider multi-fault situation. We will show that SLSP yields a wide spectrum of protection service levels with schedulable restoration granularities, and enhances the ordinary 1:N and M:N[13] shared protection in terms of flexibility, scalability, and class of service.

2. OVERVIEW OF THE ORDINARY SHARED PROTECTION SCHEMES

2.1 Basic Assumptions

This section discusses the ordinary schemes for 1:N and M:N shared path and link based protection. The basic assumptions in this paper are defined as follows. The resources along protection paths (called "protection resources", which could be the optical switching fabric, angles of prisms, etc) are assumed to be pre-computed instead of being real-time established in order to achieve service guarantees. With this, in both path- and link-based protection, the protection resources will not be configured until being informed by signaling; therefore, they can also be reserved for the protection use by the other working paths which do not violate the shared risk link group (SRLG)[1] restrictions with each other. To increase the resource utilization, the protection resources may be arranged to carry the best-effort traffic or some time-non-sensitive data flows during the normal operation. The best-effort traffic has to yield the right of way once the protected traffic subjects to any unpredictable interruption.

2.2 Path-Based Protection

For a path-based protection, the first hop node[2] of the working path w1 computes the protection path p1 that has to be diversely routed from the working path according to the SRLG information, as shown in Fig.1. If a fault occurs on the working path, LMP (link management protocol)[3] will help localizing the fault, and the first hop node will be notified to activate a traffic switchover. The fault localization and notification processes are described as follows. When a fault occurs, the downstream node(s) along the path suffered from LOL (loss of light, which is monitored in the optical layer), or loss of signal (LoS, which has to be detected in the IP layer), will send FIS (fault indicator signal)[2] to the upstream nodes. The node that does not suffer from LOL or loss of signal but receives FIS from its downstream node, will localize the fault. The fault-localizing node then sends a NIS (notification indicator signal)[2] to notify the first hop node the occurrence of the fault. Then, the first hop node will immediately send a wake-up packet to activate the configuration of the nodes along the protection path and then switch over the whole traffic on the working path.

This approach is necessary when the IP/MPLS control plane directly handles exceptions thrown by the optical layer. The fact that physical and logical paths are identical helps simplification of signaling and management. However, on the other hand, it incurs the following difficulties and problems. First, the complexity of calculation for the diverse protection route grows fast with the increasing number of nodes in the domain. Second, the protection resources cannot be shared by any other working paths that violate the SRLG restriction with the protected working path. For example, in Fig.1, p1, the protection path for w1, cannot share any of its resources to protect w2 because w2 shares the same link group with w1 only 1 out of the 17 links. Third, the restoration time strongly depends on the location of the working path at which a fault occurs. In Fig.1, it is easy to see that the recovery of a fault on link 1 will take longer period of time than the case in which a fault occurs on link 2.

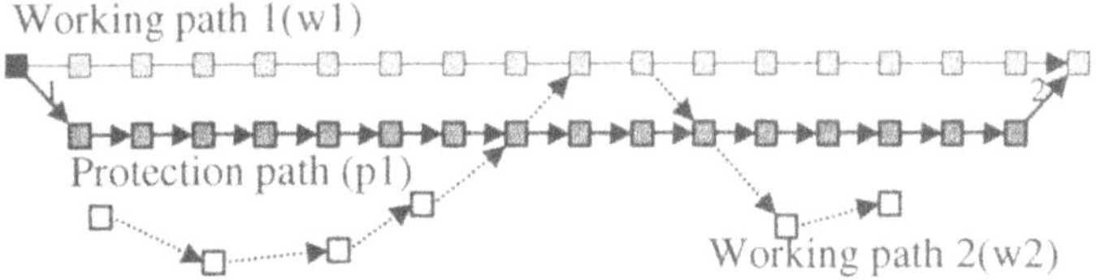

Fig.1. Ordinary path-based 1:N protection

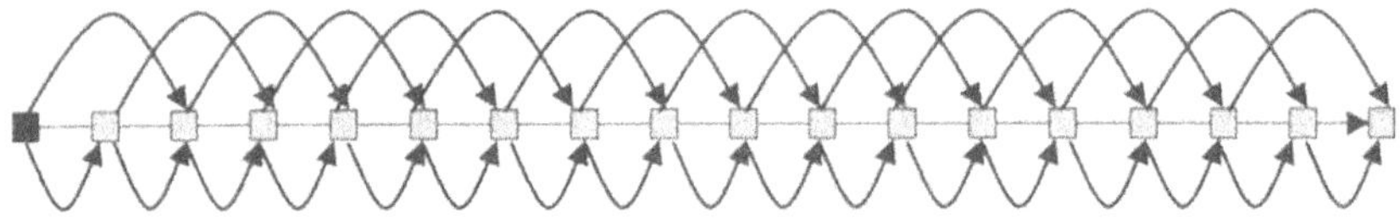

Fig.2. Ordinary link-based protection, in which every piece of resources requires a redundancy for protection use.

2.3 Link-Based Protection

For a link-based protection, restoration can be performed at the two nodes directly connected to the fault (which can be a link or node failure) so that a local restoration can be made wherever the fault is along the working path, as shown in Fig.2. The node localizing a downstream fault behaves as a PSL (path switch LSR (label switching router))[2], which only needs to notify its PML (path merge LSR)[2] before traffic can be switched over to the protection path, therefore, in this local restoration mechanism the time for transmitting NIS is totally saved. However, the protection resources consumed are much more than that by the other shared schemes because every piece of resources (node and link) along the working path must have its own corresponding protection path, in which case it may be classified as a "golden" or "premium" corresponding to the consumption of protection resources. Although link-based protection is comparable with the SONET ring mechanism[13] in terms of restoration time, it is expensive and sometimes unnecessary for time-non-sensitive services. To meet the requirement of the incoming versatile Internet applications and the demands for class of service, the future protection framework should possess a wider spectrum of service levels and finer restoration granularities so that ISPs will have more revenue-generating services to satisfy their clients from every aspect.

3. NETWORK ARCHITECTURES

3.1 Interconnection under the MPLS Control Plane

The network is IP-centric with MPLS (Multi-Protocol Label Switching)-based control plane, in which both OLXC (optical layer cross connect) and LSR are required to be under the unified service model[4] in the same administration and trust. In this case, the IP user domain and optical domain run the same signaling protocol and routing instance. The control plane for both OLXCs and LSRs is the same, in which every node is in a peer relationship. That is, the platform for SLSP is "flat"[5] within the same OSPF domain.

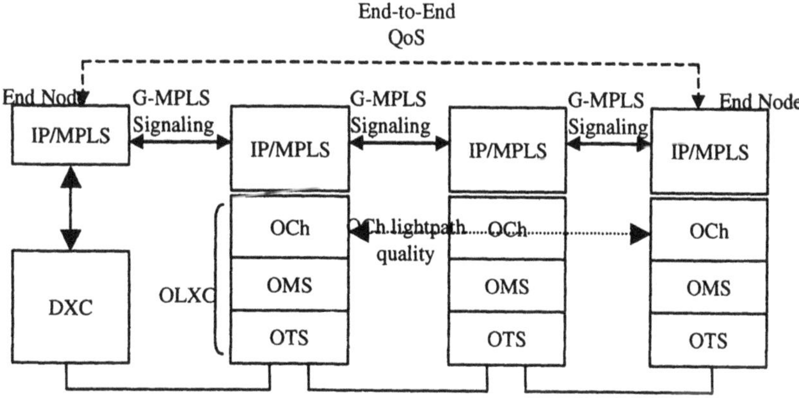

Fig.3. The IP/MPLS optical network architecture with G-MPLS signaling.

Both LSRs and OLXCs are required to contain interfaces for them to communicate with each other. The generalized MPLS (G-MPLS)[9] defined interfaces that are Packet Switching capable (PSC), Time-Division Multiplexing (TDM) capable, Lambda Switching capable (LSC), and Fiber-Switching capable (FSC) for LSRs. The transport network architecture is illustrated in Fig.3. The traffic-initiating node, with any or all capabilities mentioned above, assigns a "label" (which could be a number in the PSC case or a wavelength for LSC, etc.) to each traffic request based on its service requirements. The requirements can be the BER limitation, survivability and security, etc. Accordingly, a new forwarding equivalence class (FEC) will be created if there is not a suitable and available existing FEC to use. The issues about signaling mechanisms defined in G-MPLS that activates the distribution of the labels will be detailed in the following sub-section.

Because of the adoption of the MPLS-based control plane, the QoS-aware layer mentioned in [19] is no longer needed. The FEC mapping and label assignment have well performed the function of class of service for each traffic flow.

3.2 Functional Architecture of Network Nodes

The functional diagram of OXCs and LSRs are shown in Fig.4. The IP/MPLS control box is configured with the MPLS control plane functions for both OXC and LSR. The transport layers for OXC and LSR are optical layer cross connect (OLXC) and digital cross connect (DXC) respectively, which are controlled by their IP/MPLS control boxes. There are three main functions provided by the IP/MPLS control box.

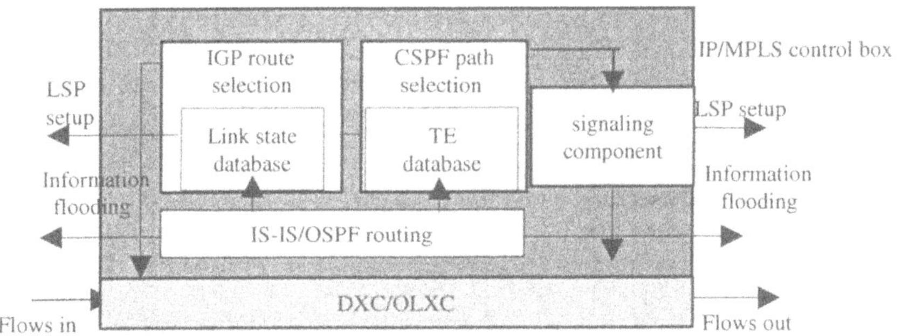

Fig.4: The functional architectures for OXC and LSR

First, the constrained shortest path first (CSPF) path selection module can calculate both work and protection paths upon different constraints such as reservable bandwidth of a link, wavelength availability, and diversity requirement, according to the TE (traffic engineering) database and SRLG information. Second, the extended RSVP or CR-LDP[6, 7] signaling component will be activated to send a Path Message to setup up a path based on the calculation result from the CSPF path selection. Besides, it can also respond to the requests from the other nodes in the network for establishing connections. Third, there is an ISIS/OSPF routing protocol module with ISIS/OSPF optical and TE extensions[8] for disseminating link states and TE metrics, so that the other nodes in the domain can verify/update the information in their link state and Traffic engineering (TE) database. All the signaling mechanisms specifically for the optical networking environment have been included into the G-MPLS, which summarizes the optical and TE extensions from the MPLS protocol, and can handle inter-domain signaling (i.e., between optical and user IP domains) and TE link states dissemination for the IP user and the optical domain.

4. SLSP DESCRIPTION

The new protection scheme, SLSP, is a scalable end-to-end service-guaranteed shared protection scheme, which accommodates the characteristics of both path-based protection and link-based protection. It provides finer and more service granularities than the traditional shared protection schemes. The main idea of SLSP is to divide the protected working path into several overlapped segments, each of which is assigned by the first hop node a protection domain ID (PID) at the routing stage, as shown in Fig.5. The calculation of the protection path for each protection domain can be done either by the first hop node alone, or decentralized to PSLs in each sub-domain, depending on how the SRLG information is configured in each node and how heavy workload the first hop node can afford at that time. With the CSPF path selection module, every node (LSR or OXC) is able to do a diverse routing within its protection domain. Although the decentralized selection of the protection paths lays down part of the burden of the first hop node, it may also subject to a problem that some of the PSLs fail to find qualified protection paths. How to determine which way to go is still under study.

Fig.5 illustrates how a path under SLSP is configured and recovered when a fault occurs. Node A is the first hop node and node N is the last hop node[2], which could respectively be the source node and the destination node of this path. The first protection domain (PID = 1) starts at node A and ends at node F. The second protection domain (PID = 2) is from node E to node J, and the third is from node I to node N. In this case, (A, F), (E, J) and (I, N) are the corresponding PSL-PML pairs for each of the protection domains.

Since each protection domain is overlapped with its neighboring protection domains by a link and two nodes, a single failure on any link or node along the path can be recovered. For example, a fault on link 4 or node E is localized by node D. A fault on link 5 or node F is localized by node E. In the former case, node D will send a FIS to notify node A that a fault occurred in their protection domain. In the later case, node E is itself a PSL. In each of the cases, the PSL (i.e. A or E) immediately sends a wake-up packet to activate the configuration of each node along the corresponding protection path, and then the traffic can be switched over to the protection path. A Tell-And-Go (TAG) [12] strategy can be adopted at this moment so that the PSL (i.e., node A or node E) may switch the traffic to the protection path before an acknowledgement packet is received from the PML (i.e., node F or node I). At the completion of the switchover, the information associated with this reconfiguration has to be disseminated to all the other nodes. By doing this, the situation that the other traffic sources send best-effort traffic to these newly occupied resources can be avoided.

Under the single-failure assumption, the other working paths that share the same protection resources are not supposed to be interrupted during the time when their protection paths are used by the switched working path. However, for the environment where multiple failures are considered, a working path has to possess two or more sets of partially or totally disjoint protection paths to prevent from the possibility that its protection resources are busy while it needs them.

When the fault on the working path is fixed and a switchback to the original working path is required, a packet for releasing the protection resources can be sent by the PSL right after the traffic is switched. With this, the protection resources can be reported as "free" again to all the other nodes.

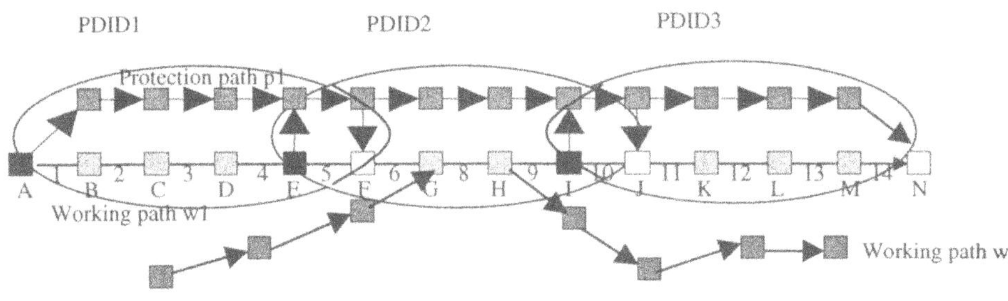

Fig.5. SLSP protection scheme divides the working path into several overlapped protection domains. Node A, E, I are the PSLs, and node F, J, N are the PMLs.

To implement the protection information dissemination, the association of the protection resources in each sub-domain with corresponding working path segments (PDID) has to be included in its forwarding adjacency (FA) [4]. The other working paths must know this association before they can reserve any piece of protection resources for protection use. An example is shown in Fig.5. w1 and w2 possess the same SRLG on link 8. However, w2 can share all the protection resources of w1 except those in the second protection domain (PDID2). In addition, a signaling protocol, e.g. RSVP or CR-LDP, needs further extensions for the Path Message and Label Request Message to carry object to assign PSLs and PMLs for implementing SLSP.

The average restoration time and longest restoration time are two service indexes of interest for a path. The average restoration time will be determined by averaging the restoration time of all the protection domains. The longest restoration time will depend on the size and node number of the largest protection domain along the path.

4.1 Inter-Domain Protection

Each of the ASON (automatically switched optical networks)[10,18] and the ODSI (optical domain service interconnect)[11,16,17], which is based on the viewpoint of a domain service model[4], has defined user-user interface (UNI) and/or network-network interface (NNI) to handle the inter-domain connection. Since the relationship between domains (In ASON, for both IP-optical and optical-optical domains; in ODSI only for IP-optical domain) is not treated as a peer, an end-to-end protection across UNI/NNI is hard to be purely implemented in the optical layer. Instead, the layered approach will be adopted at the expense of the complex signaling mechanisms. According to the drafts associated with the two protocols, the promising candidates of interfacing carrier frame are SDH, Gigabit Ethernet and SONET, etc.[10,11,16,17,18], in which protection mechanisms has been well arranged.

For a pair of trusted domains, the inter-domain protection can be done with SLSP as shown in Fig.6, which would not differ from the situation where the working path is protected within a single domain.

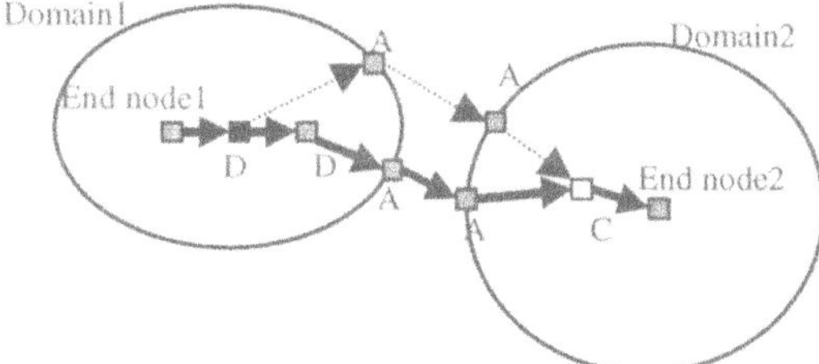

Fig.6. SLSP for inter-domain protection. The solid bold lines are for the working path, the dotted lines are for the inter-domain protection path. Nodes A are border routers, node B is a PSL, node C is a PML, and node D is an intermediate node within the protection domain.

5. ADVANTAGES OF SLSP

The advantages of SLSP framework over the ordinary path protection schemes are stated below. First, the complexity of calculating a diverse route under the constraint of whole domain's SRLG information can be segmented and largely diminished to several protection domains, in which the provisioning latency for dynamic path selection can be reduced. Second, since both notification and wake-up are performed on a very limited number of nodes, therefore, the restoration time is reduced according to the size of the protection domain. Third, the service is easier to be guaranteed since the restoration time does not vary with the length of the whole path, instead, the average size of the protection domains will be the dominant factor, which can be one of the items with which the service providers bill their customers. Fourth, SLSP can work well in any type of topologies. Examples are shown in Fig.7 and Fig.8. Fig.7. shows two implementations of SLSP on the Pan-American Network. Fig.7(a) demonstrates the protection service with a longer recovery time than that in Fig.7(b) since the protection domains in Fig.7(a) are larger. The solid bold lines are for the working path, and the thick dash lines and dotted lines are for the inter-segmented protection paths.

It can also be applied to a path traversing several domains to achieve an end-to-end path protection as shown in Fig.6. Fifth, the computation complexity of protection domain allocation is simplified. The decentralized allocation scheme has largely decreased the computation efforts for the first hop node.

Compared with the pure link protection approach, SLSP provides flexibility in compromising restoration time and protection resources required, with which the class of service can be achieved with more granularities. The ISPs can put proper constraints on the path selection according to the SLA with each of their customers. The constraining parameters for the selection of protection path can be those related to the restoration time along the working path, such as the diameter of each protection domain and the physical distance between each PSL-PML pair, etc.

6. CONCLUSION

This paper proposed a new framework, SLSP, for end-to-end shared path protection, and qualitatively compared SLSP with the ordinary path-based and link-based shared protection schemes. In addition, the network service architecture and signaling mechanisms upon which SLSP can be implemented were demonstrated. SLSP can provide a wider spectrum of protection services with finer restoration granularities, which enhances the 1:N and M:N shared protection in terms of scalability, restoration time and class of service. It will satisfy the versatile requirements of the multimedia applications and may be one of the best candidates strategies for ISPs to guarantee their services to the customers.

Fig. 7(a). An implementation of the SLSP on the Pan-American Network with large protection service domains.

Fig. 7(b). An implementation of the SLSP on the Pan-American Network with small protection service domains.

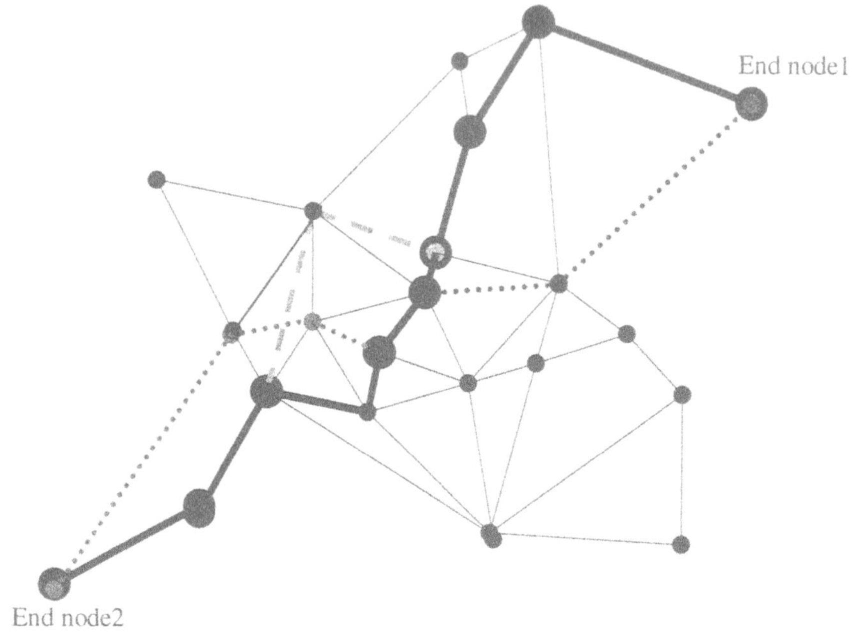

Fig.8. An implementations of SLSP on the European Network.

7. REFERENCES

[1] A. Chiu and J. Strand, "Unique Features and Requirements for the Optical Layer Control Plane," Internet Draft, <draft-chiu-strand-unique-olcp-00.txt>, July 2000.

[2] S. Makam, V. Sharma, K. Owens, C. Huang, F. Hellstrand, J. Weil, L. Andersson, B. Jamoussi, B. Cain, S. Civanlar, and A. Chiu, "Framework for MPLS-based Recovery," Internet Draft, <draft-makam-mpls-recovery-frmwrk-01.txt>, work in progress, Junn 2000.

[3] J. P. Lang, K. Mitra, J. Drake, K. Kompella, Y. Rekhter, L. Berger, D. Saha, D. Basak, and H. Sandick, "Link Management Protocol (LMP)," Internet Draft, <draft-ietf-mpls-lmp-00.txt>, work in progress, July 2000.

[4] N. Chandhok, A. Durresi, R. Jagannathan, R. Jain, S. Seetharaman, and K. Vinodkrishnan, "IP over Optical Networks: A Summary of Issues," Internet Draft, <draft-osu-ipo-mpls-issues-00.txt>, work in progress, July 2000.

[5] D. Pendarakis, B. Rajagopalan, and Debanjan, "Routing Information Exchange in Optical Networks," Internet Draft, <draft-prs-optical-routing-00.txt>, work in progress, Feb. 2000.

[6] J. P. Lang, K. Mitra, and J. Drake, "Extensions to RSVP for optical networking," Internet Draft, <draft-lang-mpls-rsvp-oxc-00.txt>, work in progress, March 2000.

[7] Z. B. Tang, D. Saha, and B. Rajagopalan, "Extensions to CR-LDP for Path Establishment in Optical Networks," Internet Draft, <draft-tang-crldp-optical-00.txt>, work in progress, March 2000.

[8] D. Fedyk, A. Ghanwani, and R. Balay, "Multiple Metrics for Traffic Engineering with IS-IS and OSPF," Internet-Draft,<draft-fedyk-isis-ospf-te-metrics-00.txt>, work in progress, March 2000.

[9] P. Ashwood-Smith, Y. Fan, A. Banerjee, J. Drake, J. P. Lang, L. Berger, G. Bernstein, K. Kompolla, E. Mannie, B. Rajagopalan, D. Saha, Z. B. Tang, Y. Rekhter, and V. Sharma,

"Generalized MPLS - Signaling Functional Description," Internet Draft, <draft-ashwood-generalized-mpls-signaling-00.txt>, work in progress, June 2000.
[10] M. Mayer, "First Draft of G.ason," Contribution to T1 Standards Project, March 2000.
[11] G. Bernstein, J.Weiss, R. Coltun, J. moy, A. Sodder, and K. Arvind, "Optical Domain Service Interconnect (ODSI) Functional Specification," ODSI Documents, Aug. 2000.
[12] C. Qiao, "A High Speed Protocol for Bursty Traffic in Optical Networks," SPIE's All-Optical Communication Systems, vol. 3230, Nov. 1997.
[13] Rajiv Ramaswami, Kumar N. Sivarajan, "Optical Networks - A Practical Perspective", Morgan Kaufmann Publishers, Inc, 1998.
[14] D. O. Awduche, Y. Rekhter, J. Drake, and R. Coltun, "Multi-Protocol Lambda Switching: Combining MPLS Traffic Engineering Control with Optical Crossconnects," Internet-Draft, < draft-awduche-mpls-te-optical-02.txt>, work in progress, July 2000.
[15] K. Owens, V. Sharma, and M. Ommen, "Network Survivability Considerations for Traffic Engineering IP Networks," Internet Draft, <draft-owens-te-network- survivability-00.txt>, work in progress, March 2000.
[16] S. Chaudhuri, G. Hjalmtysson, and J. Yates, "Control of Lightpaths in an Optical Network," Internet Draft, <draft- chaudhuri-ip-olxc-control-00.txt>, Feb. 2000.
[17] O. Duroyon, E. Hoebeke, and H. D. Neve, "Triggering and advertising lightpaths in an IP over optical network," Internet Draft, <draft-duroyon-te-ip-optical-00.txt>, work in progress, July 2000.
[18] S. J. Chen, "ASON - the UNI and its Computational Model," Contribution to T1 Standards Project--T1X1.5, July 2000.
[19] T. D. N. Nada Golmie, David H. Su, "A Differentiated Optical Services Model for WDM Networks," IEEE Communication Magazine, vol. 38, pp. 68-73, Feb. 2000.

Multiple Objective Heuristic for Ring Loading and Logical Wavelength Assignment in OCH-SPRings

Raf Meersman, Wim Van Parys, Peter Arijs
Comsof N.V.
Ketelvest 16, B-9000 Gent, Belgium
Tel: +32-9-234.13.56 - Fax: +32-9-233.30.02
E-mail: raf.meersman@comsof.com

Piet Demeester
Department of Information Technology, Ghent University - IMEC
Sint-Pietersnieuwstraat 41, B-9000 Gent, Belgium
Tel: +32-9-267.35.91 - Fax: +32-9-267.35.99

Eugenio Iannone, Matteo Pierpaoli, Filippo Bentivoglio
Cisco Systems Inc., Photonics Transport Business Unit
Viale Sarca, 222, 20126 Milano, Italy

Key words: optical ring networks, network design, OCH-SPRing, routing and wavelength assignment algorithm

Abstract: The dimensioning process of a multi-ring network based on optical channel shared protection is discussed. This dimensioning starts with a multi-ring routing algorithm to decide on the rings traversed by a connection, followed by "Ring Loading"- and "Logical Wavelength Assignment"-algorithms applied to the individual rings. In this paper we focus on this second aspect, for which we introduce an integrated approach. The performance and applicability of this heuristic is evaluated for different objectives.

1. INTRODUCTION

The use of Wavelength Division Multiplexing (WDM) rings has been heralded as the next step in the evolution process from point-to-point WDM links towards optical transport networking (OTN) [1]. Two main types of WDM rings are envisaged: dedicated and shared protection rings. In this paper we focus on the *Optical Channel Shared Protection Rings* (OCH-SPRing).

The design process of an OCH-SPRing network can be divided in multiple consecutive sub-problems, as illustrated in *Figure 1*.

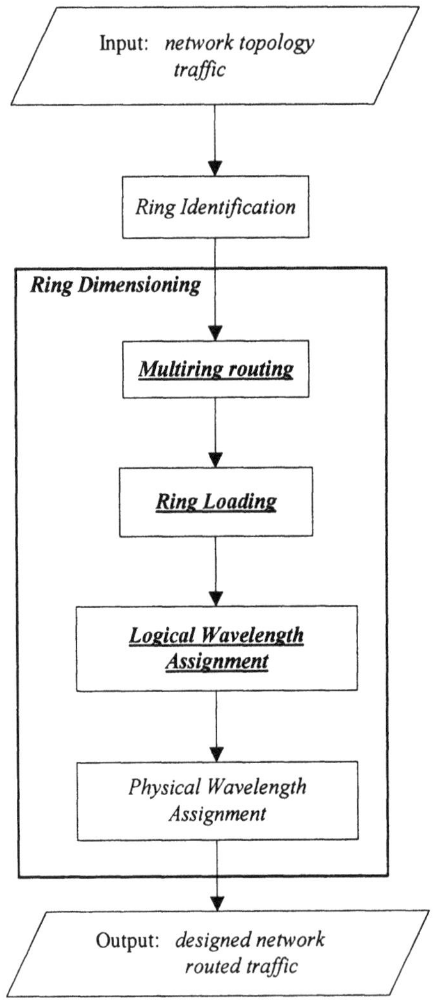

Figure 1. multiple network design steps

In case the rings are not defined up-front, the "***Ring Identification***" *(RI)*-process will try to identify a proper set of rings. Afterwards, these rings have to be dimensioned in order to accommodate the traffic forecast. We propose to split this dimensioning problem in 4 sub-problems. First the traffic is distributed over the different rings, so that the remainder of the dimensioning problem can be solved as a set of independent single-ring problems. This step is called "***Multi-ring Routing***"*(MR)*. Then the so-called "Ring Loading"- and "Logical Wavelength Assignment"-problem [2] are solved on a ring-per-ring base. These problems can be defined as follows:

- ***Ring Loading*** *(RL):* decide for each connection whether to route it clockwise or counter-clockwise on the ring, not taking into account wavelength assignment.
- ***Logical Wavelength Assignment*** *(LWA):* decide which connections to be grouped in a virtual ring (i.e. connections grouped on the same wavelength).

Finally a *"**Physical Wavelength Assignment**"(PWA)* step is executed in order to minimise the cost of the required equipment. This step is especially interesting when the ring has a modular structure, i.e. built-up of stacked fibres (or wavelength bands), and different types of OADM's are available. Suppose for example that there are three types of OADM's, dropping 0, 4 or all wavelengths of a fibre (or a wavelength band). In each node, the packing of logical wavelengths will then influence the choice of OADM for each stacked fibre (or band). For more details about this problem and the solution techniques proposed, we refer to [3].

In the remainder of this paper we focus on the first three sub-problems of the ring dimensioning. Paragraph 2 describes how we deal with the multi-ring routing problem. In paragraph 3 we introduce a new integrated solution method for the RL and LWA problem and finally paragraph 4 discusses the results obtained with this new heuristic.

2. MULTI-RING ROUTING

In the "multi-ring routing"-phase, we try to determine for each connection which rings are best suited to use and through which nodes the connections enter and leave each of the visited rings. Although these early decisions limit the freedom in the next steps of the design, they are required to split up the global problem into several single-ring problems. Note that this step only decides on the rings and interconnections visited. The routing on the ring – clockwise or counter-clockwise – is left to next steps. As such the decisions made in this phase have only limited impact on the following steps.

Our solution method is based on balanced shortest paths.

We use the shortest path algorithm to determine the shortest path in the network, build up of all rings and the interconnections between them. We associate a cost to each of the links. Note that there can be multiple links between 2 nodes, because each ring has its own links. This cost can be fixed or link-dependent. We have chosen for the first option, but extensions can be made easily. Using the same cost for all the lines in the network implies "least hops"-routing.

The ties that occur when paths having the same hop length are found, is a drawback of using hop count for the shortest path routing. In order to obtain a better spreading of the traffic over the rings in the network, we extended the cost-model to obtain a 'balanced routing'. Initially, all links of the rings have a cost 1000. Each time a link is used by a connection with a capacity C, C is added to the cost of this link. This way, we select the path with the least used links in case there are multiple "least-hop" paths.

The influence of the balanced cost-model on the dispersion of the traffic, depends on the way the connections are routed. We order the connections according to their demand. The ones with the highest demand are routed first. This way, connections with lower capacity can be used to eliminate the relative big difference in cost between the used and the unused links of the first connections.

Although each connection has been routed at the end of this phase, these routes are not yet final, because this would be too restrictive for the logical wavelength assignment. Part of the information will be eliminated to allow more possibilities for the sharing of wavelength resources. *Figure 2* shows a possible representation of a route in a multi-ring network. Each route can be completely described by means of an ordered list of visited rings, with for each ring the nodes in which the connection enters and leaves the ring and the direction of the connection over the ring (clockwise (CW) or counter-clockwise (CCW)).

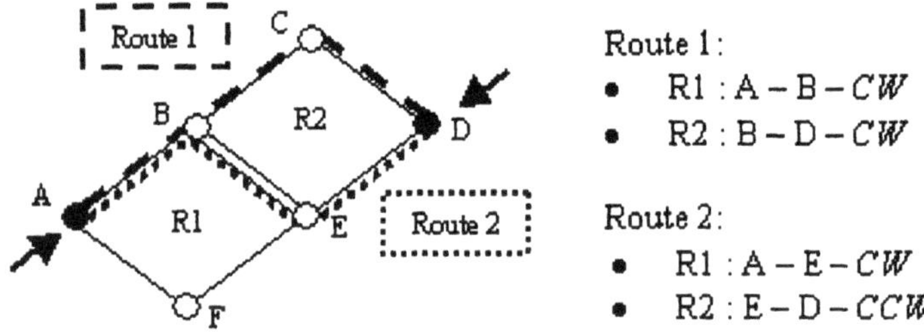

Figure 2. Example of route representation

From the routes, only the rings and end-points are retained. Removing all the directions in this representation, means that this decision is postponed. This offers additional freedom to the logical wavelength assignment algorithm. Eliminating the directions, leaves just enough information to split up the multi-ring dimensioning problem into multiple single-ring problems. Indeed, with the remaining information, the global traffic matrix can be split up into traffic matrices per ring. From this point on the dimensioning process is continued ring per ring by the algorithm discussed in the next paragraph.

3. RL+LWA: MULTI-CRITERIA HEURISTIC

The "Ring Loading"- (RL) and "Logical Wavelength Assignment"-problem, have been subject to previous studies in literature. Both problems can be solved independently and consecutively, as in [2][4]. An integrated approach based on linear programming has been presented in [5]. To deal with both problems in an integrated way in a computationally efficient way, we present a heuristic approach. This heuristic iteratively assigns connections to wavelengths in a greedy manner according to a set of different criteria. As such the ring can be optimised according to different objectives. An obvious objective is to minimise the amount of required wavelengths, but it could also be interesting to minimise the number of access points, in order to share a protection card in an OADM for two connections. A possible secondary objective could be to maximise the number of free links in the final configuration, in order to offer some flexibility for future traffic. Even a combination of different objectives could be interesting for the network designer. Our heuristic approach called "Multi-criteria heuristic", is able to take these multiple objectives into account.

In this paragraph we explain this method in detail. In the next paragraph, the results for different objectives are discussed.

This algorithm is based on the construction of "Shared Connection Sets" (SCS's). A SCS is defined as a set of non-crossing routed connections, such that there is no other connection available that can be added to form a new SCS. A simple example (*Figure 3*) will clarify this. The set of connections 1-2 and 2-3 is not a valid SCS since there is still room on the ring to route the connection between node 3 and 4. By adding connection 3-4 we end up with a valid SCS. For this small example there are 3 valid SCS's.

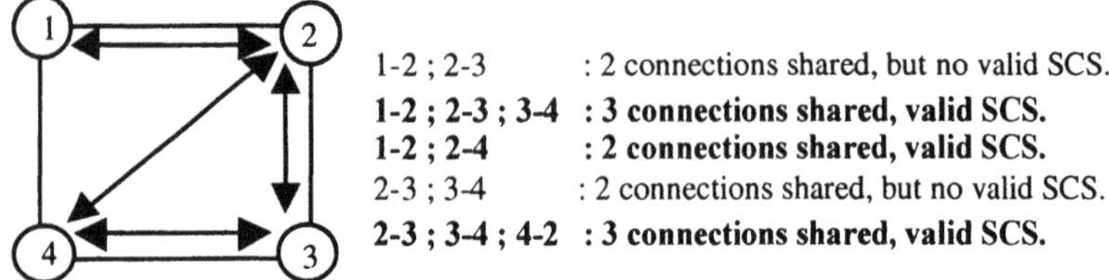

Figure 3. Example of ring with all valid SCS's

Figure 4 gives an overview of the different steps of the algorithm, which works in an iterative way.

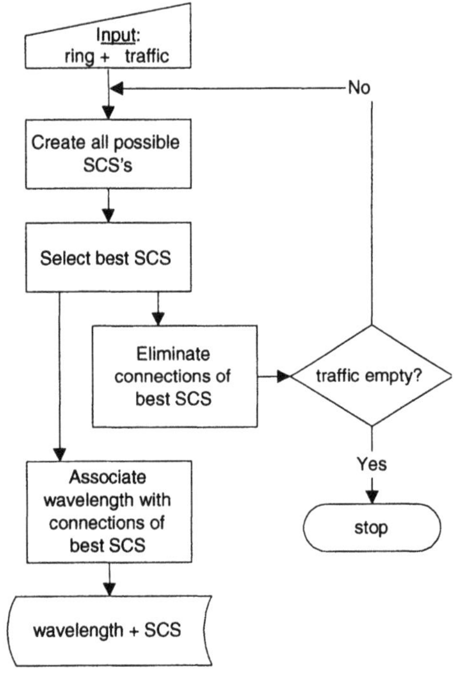

Figure 4. Multi-criteria heuristic

Each iteration starts with the creation of all possible SCS's, given the ring and its (remaining) traffic demand. With each of these SCS's, we can associate some values, reflecting the quality of the SCS, e.g. the number of connections in the set or the number of links occupied on the ring. Based on these values, we can identify one of these sets as the best, according to a certain criterion. The connections of the best SCS are then associated with one virtual ring and we start over with the remaining connections, until they are all assigned to a virtual ring.

The advantage of this solution method is that the selection criterion for the best SCS can be chosen according to the designer's preferences. When trying to minimise the number of wavelengths, good results are generally found when we select in each iteration the set with the maximum number of connections. Often there will be more than one such set. If we favour for instance the set that uses the least number of links, we will end up with a solution that maximises also the number of free links. On the other hand, we could try to minimise the number of shared protection cards by selecting the set with the most adjacent connections. To take into account different possible choices we developed five criteria, having different optimisation objectives. They select the best SCS based on following ideas:

- *Criterion 1*: best SCS contains highest number of connections, and least number used links in case of a tie.
- *Criterion 2*: best SCS occupies most links without using longest path.
- *Criterion 3*: best SCS has minimum weighted sum of number connections and number access points
- *Criterion 4*: best SCS has the least number of access points per connection.
- *Criterion 5*: best SCS has the highest value, obtained by a weighted function of the number of connections, number of used links and the number of access points per connection.

An additional advantage of the multiple criteria, is that the overall performance of the algorithm gets better when we combine more than one criterion, and select for each case the best result according to the preferred objective(s). In the results described in paragraph 4, it is shown that different criteria can complement each other.

4. RL+LWA: PERFORMANCE EVALUATION

4.1. Multi-criteria

To illustrate the use of multiple criteria, we performed the RL+LWA on a set of single ring networks with the amount of nodes varying from 3 to 16. To simulate the traffic of a multi-ring network, we consider a mix of random (intra-ring) and hubbed (inter-ring) traffic in different proportions. The validated objective is the number of required wavelengths. In the first three cases, the result is obtained with a single criterion, respectively 1,3 and 5. In the last case the best result of all three criteria is always selected. *Figure 5* gives an overview of the percentage of the average amount of extra wavelengths required by the four different solutions, compared to the minimum, obtained by means of a de-coupled ILP algorithm [2]. The chart shows that using multiple criteria results in a considerable improvement compared to a single criterion (depending on the traffic, an improvement from 10 to 50%). If we always use all the criteria, obviously the results will be the best that can be generated with the heuristic. Calculation time on the other hand is proportional to the number of used criteria.

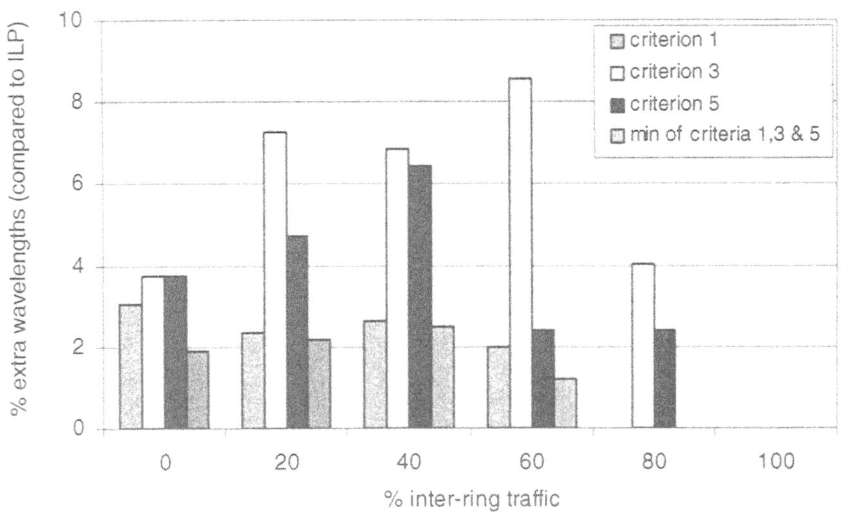

Figure 5. Comparison single criterion / combined approach

4.2. Multi-objective

With the 5 presented criteria, we can find good solutions for the three objectives we considered: minimum number of wavelengths, minimum amount of access points, maximum number of free links. Even combinations of objectives can be considered. In *Figure 6* we give an overview of the average results that were obtained with the different criteria for the different objectives in a series of experiments (each bar is average for rings with the number of nodes from 3 to 16, with five different traffic matrices per ring). Based on this chart, we can choose the best criterion corresponding with our (multi-) objective(s). E.g. if the amount of wavelengths is the most important objective, criterion 1 performs well, but if also "% free links" is important, criterion 3 is better. Considering only free links and access points favours criterion 4, etc…

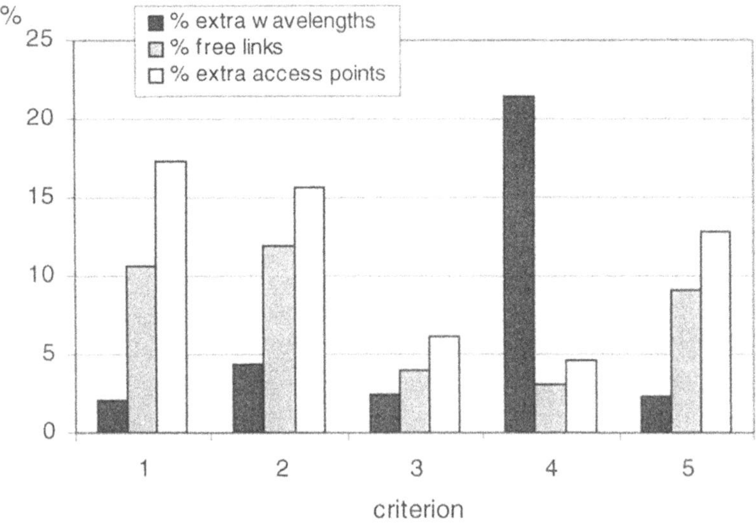

Figure 6. Performance multiple criteria for multiple objectives

4. CONCLUSIONS

In this paper we present a modular design approach for multi-ring networks based on optical channel shared protection. We discuss in detail the approach to split up the global problem into independent single-ring problems. A multi-objective heuristic, based on multiple selection criteria is introduced. This approach offers a fast way to solve the 'Ring Loading'- and 'Logical Wavelength Assignment' problem in a proper way. In addition, upon the designer's preferences, multiple objectives can be selected, which is less obvious with optimised algorithms based on e.g. integer linear programming techniques. Furthermore, this method can be easily extended for new objectives, simply by designing other criteria.

5. REFERENCES

[1] J. Manchester, P. Bonenfant, C. Newton, "The Evolution of Transport Network Survivability", IEEE Comm. Mag. **37** (8), 44-51 (1999).

[2] P. Arijs, M. Gryseels, P. Demeester, "Planning of WDM Ring Networks", Photonic Network Communications **2** (1), 33-51 (2000).

[3] P. Arijs et al., "Architecture and Design of Optical Channel Protected Ring Networks", Journal of LightWave Technology **19** (1), to appear January 2001.

[4] D. Marcenac, "Benefits of Wavelength Conversion in Optical Ring-Based Networks", Optical Networks Magazine **1** (2), 29-35 (2000).

[5] T. Carpenter, S. Cosares, I. Saniee, "Demand Routing and Slotting on Ring Networks", DIMACS Technical Report 97-02, January 1997.

Part Three

Optical packet and burst switching

Optical Packet Switching over Arbitrary Physical Topologies using the Manhattan Street Network: An Evolutionary Approach

Olufemi Komolafe*, David Harle*, David Cotter^
** Dept. of Electronic & Electrical Engineering, Strathclyde University, Glasgow G1 1XW, U.K.*
^ Corning Research Centre, Adastral Park, Martlesham Heath, Ipswich IP5 3RE, U.K.
Tel.: (+44) 141 248 2090. Fax.: (+44) 141 552 4968. Email: femi@comms.eee.strath.ac.uk.

Key words: Optical Packet Switching, Virtual Topology Design, Manhattan Street Network, Genetic Algorithms, Optical Processing

Abstract: Optical packet switching over arbitrary physical topologies typically mandates complex routing schemes and the use of buffers to resolve the likely contentions. However, the relatively immature nature of optical logic devices and the limitations with optical buffering provide significant incentive to reduce the routing complexity and avoid optical domain contentions. This paper examines how the Manhattan Street Network (MSN) and a particular routing scheme may be used to facilitate optical packet switching over arbitrary physical topologies. A novel approach, genetic algorithms (GA), is applied to the problem of deploying the MSN (near) optimally in arbitrary physical topologies. A problem encoding is proposed and different implementations of GA described. The optimum GA parameters are empirically selected and GA is successfully used to deploy the MSN in physical topologies of up to 100 nodes. Favourable results are obtained. GA are also seen to out-perform other heuristics at deploying the MSN in arbitrary physical topologies for optical packet switching.

1. INTRODUCTION

The proliferation of the Internet has provided significant incentive and impetus to develop and implement improved packet switching architectures. It is widely accepted that optical packet switching will be one of the ways the demand for bandwidth is met. One approach to optical packet switching is to assume that the switching may be controlled electronically, although the data packets will remain in the optical domain throughout the switching. However, situations may arise in which it is desirable to control the switching of packets in the optical domain. Optical logic devices will read the packet's header, process the information and switch the packet accordingly [1, 2]. An advantage of optically controlling the switching of packets is that optical processing devices are faster than their electronic counterparts. A disadvantage, however, is the fact that the algorithmic processing undertaken must be extremely simple given the relatively undeveloped nature of optical logic devices. Regardless of whether the header processing is implemented optically or electronically, another important constraint is the need to minimise or avoid optical domain contention due to the non-availability of static random access optical memory devices and the limited scalability of travelling-wave optical buffers. Nevertheless, it is often the case that real-life (hence arbitrary) physical topologies necessitate complex routing schemes and the need for buffers to resolve the likely contentions.

This paper investigates how the Manhattan Street Network (MSN) and a particular routing scheme may be used to realise optical packet switching over arbitrary physical topologies, avoiding complex routing operations and contentions in the optical domain. Section 2 introduces the MSN and the routing scheme. The MSN is deployed as a virtual topology over the arbitrary physical topology. Section 3 presents the fundamentals of the technique used to attempt to deploy the MSN optimally in the physical topologies; genetic algorithms (GA). Section 4 discusses the alternative problem-specific implementations of GA considered in this paper. Section 5 presents the results obtained when selecting the optimum parameters for the different GA, compares the results obtained for each GA implementation, and compares the performance of GA against other heuristics. Section 6 contains the conclusions.

2. THE MANHATTAN STREET NETWORK WITH THE *CLOCKWORK ROUTING* SCHEME

2.1 *Clockwork Routing* Scheme Operation

The Manhattan Street Network [3] (MSN) with the *Clockwork Routing* scheme [4, 5] has been suggested for use in optical packet switched networks for a number of reasons. The routing processing is extremely simple and suitable for optical implementation, no optical domain buffering is required, no resequencing is needed at the destination nodes and throughput is comparable with conventional store-and-forward packet switching [4, 5].

Clockwork Routing requires that all the nodes in an *NxN* MSN be synchronised to a global clock. The timeslots are arranged in a modulo-*N* sequence of frames. Each node consists of a simple 2x2 cross-bar switch. All nodes are in the cross state for the first *N*-1 timeslots in each timeframe, and in the bar state for the last timeslot in the timeframe. Figure 1 shows a 4x4 MSN. By correctly inserting a packet into a timeslot on a particular output link, it is automatically routed to the destination. No additional processing is required at intermediate nodes – these nodes need only determine whether the packet has reached its destination or not. Such a simple "for me or not for me" evaluation may be readily implemented in the optical domain [1, 2]. Optical domain contentions are avoided through the use of peripheral electronic buffers.

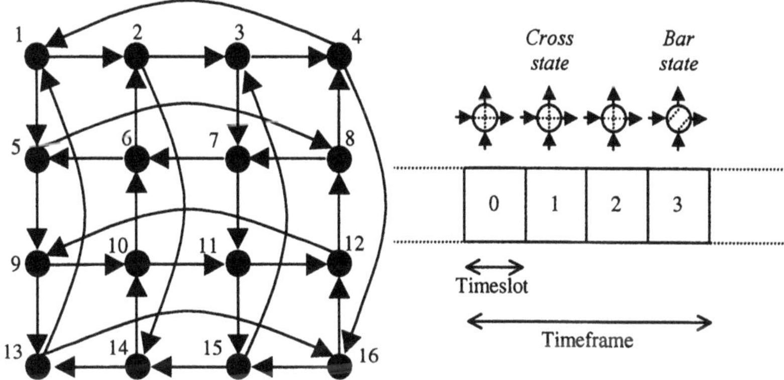

Figure 1. 4x4 MSN

2.2 Deployment of the MSN with Clockwork Routing in Arbitrary Physical Topologies

Outside the local area network, the MSN may be deployed (or embedded) as a virtual topology in WDM networks. Nodes of the MSN are mapped onto those of the physical topology. Links in the MSN are implemented as lightpaths between the corresponding nodes [6, 7]. The length of these lightpaths, in terms of concatenated physical topology hops, will vary according to which MSN nodes are mapped onto which physical topology nodes. A shortest path routing algorithm is used to route the lightpaths over the physical topology. Each node contains wavelength selective devices, in particular an optical cross-connect (OXC), that route and manipulate the lightpaths as desired. Packets are optically switched at the terminal of lightpaths with each node implementing the *Clockwork Routing* scheme.

There are several issues associated with deploying the MSN in arbitrary physical topologies; this paper focuses on one of the main issues - the (near) optimal embedding of the MSN in the physical topology. The virtual topology design problem and several of its sub-problems are known to be *NP*-complete [8] and so heuristics must be adopted to find (near) optimal solutions in reasonable time.

There are up to $N^2!$ ways to deploy an NxN MSN in a given physical topology - for a 8x8 MSN this corresponds to up to 1.27×10^{89} different embeddings! The quality of these embeddings will differ significantly. The cost considered in this paper is the mean lightpath length. (Throughout this paper, distance is in terms of concatenated hops over a topology, and not in terms of physical length.) The lightpath length is important for several reasons: it affects the number of wavelengths needed, impacts the dimensions of the wavelength selective devices, indicates the number of OXCs a packet has to traverse between any two adjacent nodes in the MSN and would impact the deployment of optical amplifiers and the consumption of other network resources. Furthermore, lightpath length corresponds to a figure of merit often used in graph embedding theory [9].

Packets will typically need to traverse more than one lightpath between source and destination. The aggregate number of physical hops traversed by a packet corresponds to the number of OXCs crossed. It is desirable to minimise the length of lightpaths and, hence, the aggregate number of OXCs that packets encounter between source and destination. The number of physical hops a packet can traverse is limited by the characteristics of the fibre coupled with distortions such as crosstalk and insertion loss introduced by the OXCs. In [10] it was observed that the aggregate source to destination distance packets encounter is simply the product of the mean lightpath length and the mean inter-nodal distance over the MSN - the latter quantity is known [4, 5] and this paper proposes the novel use of genetic algorithms to minimise the former.

3. GENETIC ALGORITHM THEORY

Genetic algorithms (GA) are a family of adaptive computational models inspired by evolution in nature. The *survival of the fittest* criterion provides evolutionary pressure for the population to develop increasingly fit individuals with time. GA were originally proposed in [11] and although several different nuances have been developed, they all adhere to the same basic principles. The first step in an implementation of GA is to create a randomly generated initial *population*. The *individuals* in the population each represent a possible solution to the problem being tackled. The individuals are normally coded as a binary string. A *fitness* (or cost) function needs to be defined and evaluated for each individual in the population. Individuals are copied from the population into what is known as the *mating pool*. The mating pool will be the same size as the population. However, the number of times an individual is copied into the mating pool is correlated to its fitness. Pairs of individuals from the mating pool are randomly selected to *reproduce*. Reproduction or mating typically yields two new *offspring*. The offspring ought to have inherited attributes from both *parents*. It can be seen that by ensuring that fit individuals appear more frequently in the mating pool, good properties ought to propagate down the *generations*. The offspring produced become the new generation and the whole process in repeated. The process is repeated until ideally all the individuals in the population converge into an extremely fit individual [12, 13]. GA work because of what is termed *implicit parallelism*. Fitness evaluation, selection and reproduction ensure that numerous areas of the search space are sampled simultaneously.

4. APPLICATION OF GENETIC ALGORITHMS TO DEPLOYING THE MSN IN ARBITRARY PHYSICAL TOPOLOGIES

GA have been applied to a plethora of complex mathematical problems [13] and to numerous telecommunications problems [14]. Nevertheless, other heuristics are typically applied to the virtual topology design problem [8]. This paper applies GA to virtual topology embedding, and compares the results obtained with some other approaches which may be adopted.

4.1 Problem Encoding

The problem encoding adopted in this paper is based on approaches taken for the application of genetic algorithms to the (in)famous Travelling Salesman Problem (TSP). The TSP problem is to find the visitation order of a list of cities that the salesman should choose in order to minimise the aggregate distance covered. Several of the issues encountered when attempting to apply GA to the TSP are pertinent when using GA to deploy the MSN in arbitrary physical topologies.

GA may be readily applied to problems that are accurately encodable as a population of individuals represented by binary strings in which the frequency and position of values in individuals is inconsequential. This is evidently not the case with the TSP or the problem of deploying the MSN in arbitrary physical topologies. The search space for the TSP may be considered to be the permutation of the set of cities. Intuitively, the cost or fitness of each individual will be the distance that results from touring the cities in the order that the individual specifies. In the problem encoding adopted in this paper, the position of an element in an individual in the population is indicative of which MSN node is mapped onto which physical topology node. Consider the example of an individual when a 4x4 MSN is to be deployed in a 16-node physical topology:

4 7 6 12 8 1 9 3 13 16 10 14 15 2 11 5

This particular individual implies that node 4 in the MSN is mapped onto node 1 in the physical topology, MSN node 7 is mapped onto node 2 in the physical topology and so on. The fitness or cost of this particular individual is the resulting mean lightpath length when all the MSN nodes are mapped onto the prescribed physical topology nodes. Note that this encoding places a slightly different emphasis compared to the TSP. In the TSP, it is the order of the elements that matter since the tour is cyclic. However, when deploying the MSN in arbitrary physical topologies, it is the absolute position of the elements that is most significant.

Several techniques exist for selecting individuals from the population to be placed in the mating pool so that fitter individuals have a greater chance of being selected than less fit individuals. The technique used in this paper is *Binary Tournament Selection* [13]. In this method, pairs of individuals are randomly selected from the population and the fitter individual is copied into the mating pool. Both individuals are replaced into the original population. This process continues until the mating pool is full.

Mutation is an operation that is often applied in GA. The premise is that it is advantageous to introduce some controlled randomness in the formation of offspring in order to allow a better exploration of the search space. Thus some elements in an offspring are randomly perturbed with a certain probability. Mutation is not explicitly applied in this paper, however it will be seen that mutation occurs implicitly in several of the GA implementations considered.

4.2 Crossover Techniques

The method used to produce offspring from two parents in the mating pool is known as *crossover*. Simple crossover consists of selecting a random point in the individuals and then swapping their "tails". Thus the resulting offspring will inherit features from both parents. Such a method of crossover is well suited for the traditional applications of genetic algorithms where the problem is readily encodable as a binary string in which the frequency and position of values is unimportant. This is evidently not the case when GA are applied to MSN deployment. Consider the crossover of Parents A and B to give Offspring P and Q illustrated below. The crossover point is randomly chosen.

A	4	7	6	12	8	1	9	3	13	16	10	14	15	2	11	5
B	3	4	11	15	14	1	13	12	7	2	6	10	9	5	8	16

P	4	7	***6***	12	***8***	1	***9***	3	13	***16***	***6***	10	***9***	5	***8***	***16***
Q	3	4	***11***	***15***	***14***	1	13	12	7	***2***	10	***14***	***15***	***2***	***11***	5

It may be readily observed that the both Offspring P and Offspring Q are invalid. The duplications and omissions that exist in both offspring have been highlighted. Recalling the fact that the elements and position indicate the mapping of MSN nodes onto physical topology nodes, it can be seen that duplications and omissions are a significant problem. Four alternative crossover techniques are considered in this paper to prevent the formation of invalid offspring.

4.2.1 Crossover and Correct (XC)

Crossover and Correct is the simplest crossover technique considered in this paper. In XC, a randomly chosen crossover point is selected and the tails of the parents are swapped. Any duplication or omissions in the two offspring produced are detected and corrected. The correction of the offspring is done in such a manner that the elements and positions from one parent are left unaltered - this will hopefully ensure better propagation of characteristics down the generations compared to the alternative of correcting arbitrarily from the portion copied from both parents. Using the example previously given, it can now be seen that both Offspring P and Offspring Q contain neither duplications nor omissions. The elements that have been changed in order to correct the offspring have been highlighted. It can see that all the correction is done only on elements from one parent. The correction operation may be considered as implicitly introducing mutation into the offspring produced.

A	4	7	6	12	8	1	9	3	13	16	10	14	15	2	11	5
B	3	4	11	15	14	1	13	12	7	2	6	10	9	5	8	16

P	4	7	***14***	12	***11***	1	***15***	3	13	***2***	6	10	9	5	8	16
Q	3	4	***8***	***9***	***6***	1	13	12	7	***16***	10	14	15	2	11	5

4.2.2 Partially Mapped Crossover (PMX)

Partially Mapped Crossover [15] is implemented by selecting two random crossover points. The elements of both parents between both parents are exchanged. In each offspring, elements that have been displaced in the crossover region are exchanged with elements that have been introduced into the crossover region. Alternatively, PMX may be considered as specifying a number of swap operations to be executed in each parent to yield the offspring. In the example below, PMX implies elements 6 & 11, 12 & 15, 8 & 14, and 9 & 13 are swapped in both parents to produce new offspring. Again it may be seen that mutation is being implicitly applied by the swapping operations.

A	4	7	6	12	8	1	9	3	13	16	10	14	15	2	11	5
B	3	4	11	15	14	1	13	12	7	2	6	10	9	5	8	16

P	4	7	11	15	14	1	13	3	9	16	10	8	12	2	6	5
Q	3	4	6	12	8	1	9	15	7	2	11	10	13	5	14	16

4.2.3 Order Crossover (OX)

Order Crossover [16] is so called as it attempts to preserve the order that elements appear in parents without producing illegal offspring. Two crossover points are randomly selected. The elements between the crossover points are copied from Parent A into Offspring Q, and from Parent B into Offspring P. The remaining elements in Offspring P are obtained by copying the elements from Parent A. Starting just after the crossover region, the elements are copied in a cyclic manner in the order they appear in Parent A. Any elements already in the crossover region are skipped in order to produce legitimate offspring. Offspring Q is formed analogously.

A	4	7	6	12	8	1	9	3	13	16	10	14	15	2	11	5
B	3	4	11	15	14	1	13	12	7	2	6	10	9	5	8	16

P	8	9	3	16	14	1	13	12	7	2	6	10	15	11	5	4
Q	15	12	7	2	6	1	9	3	13	16	10	14	5	8	4	11

4.2.4 Cycle Crossover (CX)

Cycle Crossover [16] ensures that each position in an offspring is occupied by a corresponding element from one of the parents. Therefore, in order to produce legitimate offspring, groups of elements that must be taken from the same parent have to be found. Essentially, selecting an element from a particular parent compels other elements to be chosen from that parent. These groups of elements that must be selected from the same parent are called *cycles*. Since every element in the offspring must occupy a position it occupies in either parent, in the example below it is only possible to have either 4 or 3 in position 1 in the offspring. For Offspring P, say 4 is selected from Parent A, this means that 7 must be selected from Parent A for position 2, since selecting 4 from Parent B implies that the offspring will contain 4 in two positions. Furthermore, 13 must be selected in position 9 in order to prevent duplication of 7. Consequently, positions 1,2 and 9 must be chosen from the same parent, hence are part of the same cycle. The other elements in this cycle and any other cycles must be found. The cycles obtained from the exemplar parents are given below. *X* denotes the occurrence of the same element in the same position in both parents. Offspring are formed by choosing cycles from different parents. In this paper, Offspring P always chooses the elements in Cycle 1 from Parent A and the other cycles' elements from Parent B. Offspring Q selects elements in Cycle 1 from Parent B and the remaining elements are chosen from Parent A. It can be observed that each element in both offspring occurs in a corresponding position in one of the parents.

A	4	7	6	12	8	1	9	3	13	16	10	14	15	2	11	5
B	3	4	11	15	14	1	13	12	7	2	6	10	9	5	8	16
Cy	1	1	2	1	2	X	1	1	1	3	2	2	1	3	2	3
P	4	7	11	12	14	1	9	3	13	2	6	10	15	5	8	16
Q	3	4	6	15	8	1	13	12	7	16	10	14	9	2	11	5

5. RESULTS

Cases in which there is a mismatch in the number of nodes in the physical and virtual topologies are outside the scope of this paper. This paper considers examples of embedding a 4x4, a 6x6, an 8x8 and a 10x10 MSN in exemplar arbitrary physical topologies of 16, 36, 64 and 100 nodes respectively. The network with 16 nodes is a topology of the UK [17]. The other networks were chosen to illustrate the difference, if any, that the network size makes to the best embeddings found, and to investigate the impact problem size has on GA performance. The networks were chosen so that they all have comparable mean degrees and are all illustrated in Figure

2. It can be seen that packet switching over such topologies will require the storage of large and complex routing tables. Furthermore, the arbitrariness of the connectivity implies that contentions are likely to occur. Embedding the MSN is a possible way to alleviate both these problems and allow optical packet switching.

A conventional approach when designing virtual topologies for wavelength-routed WDM networks is to use the traffic matrix as an input in the optimisation process - for example [8]. It may be argued that this is an *ad hoc* approach since the traffic matrix is dynamic. The viewpoint adopted in this paper is that it is better to deploy the virtual topology using invariable parameters (such as the connectivity of the physical and virtual topologies) as the inputs to the optimisation process. Furthermore, it is assumed all fibres have equal physical length, i.e. unity. This is done in order to realise a direct correlation between the mean lightpath length and the number of OXCs encountered by a packet or lightpath.

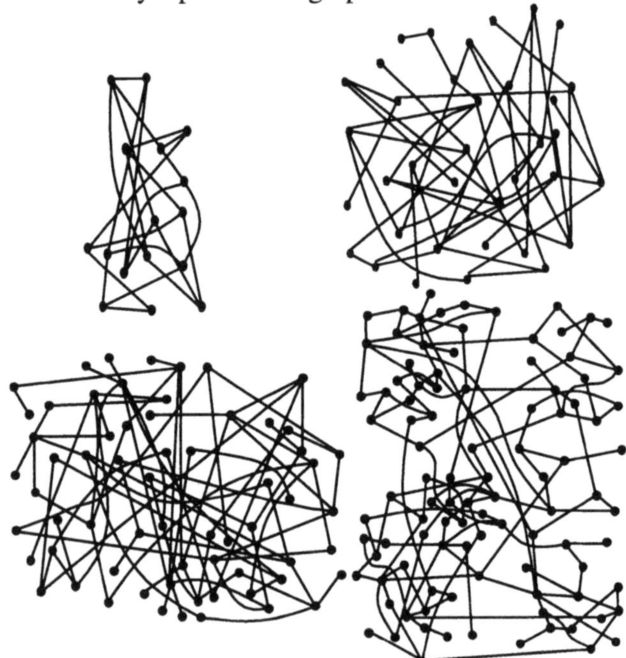

Figure 2. Exemplar arbitrary physical topologies of 16, 36, 64 and 100 nodes

5.1 Selection of Optimal GA Parameters

There are several critical discretionary parameters in the implementation of GA. The different crossover techniques and selection method have been discussed. In this paper, the population size and the probability of crossover are two parameters that were selected empirically. The population size is the number of individuals that are considered in each generation. Intuitively it seems desirable to have as large a population as possible. However, this must be balanced by the need to obtain results relatively quickly and memory limitations. Allowing the probability of crossover to be less than unity means that it is possible for two parents to survive intact into the next generation - once selected from the mating pool, the two parents are copied directly into the next generation. Population sizes of 250, 500, 1000 and 2000 were considered. The probability of crossover was either 0.5 or 1. These eight alternatives were all investigated for each crossover technique. Figures 3 to 6 show results for the example of deploying the 6x6 MSN in the 36-node physical topology. The mean of the mean lightpath lengths for the population for different generations is shown. Note that in this section, emphasis is placed not on the performance of the different crossover techniques relative to each other, but rather on the results obtained when certain parameters are varied for each crossover technique.

For Crossover and Correct (XC) and Cycle Crossover (CX) (Figures 3 and 6 respectively) it is evident that a population size of 2000 and a probability of crossover of 1 gives the best results. For XC convergence occurs just after 100 generations. It takes around 70 generations for convergence to occur with CX. For the Partially Mapped Crossover (PMX), it can be observed in Figure 4 that having a probability of crossover of 0.5 results in significantly faster convergence than always carrying out crossover. However, having the probability of crossover as 1 gives the better results. The impact of different population sizes is less conclusive and appears slightly surprising. Further investigation was carried out and it was found that a population size of 1000 gives the best results. Note that PMX takes significantly longer to converge than the other crossover techniques. For the Order Crossover (OX) from Figure 5 it is evident that no improvement to the mean lightpath length is observed when the probability of crossover is 1. When the probability of crossover is 0.5, a population size of 2000 gives the best results.

Consequently, the population sizes and probability of crossover have been determined and are used for the results presented in the rest of the paper. The population size for XC, CX and OX is 2000. PMX has a population of 1000. The probability of crossover is always 1, with the exception of OX where the probability of crossover is 0.5.

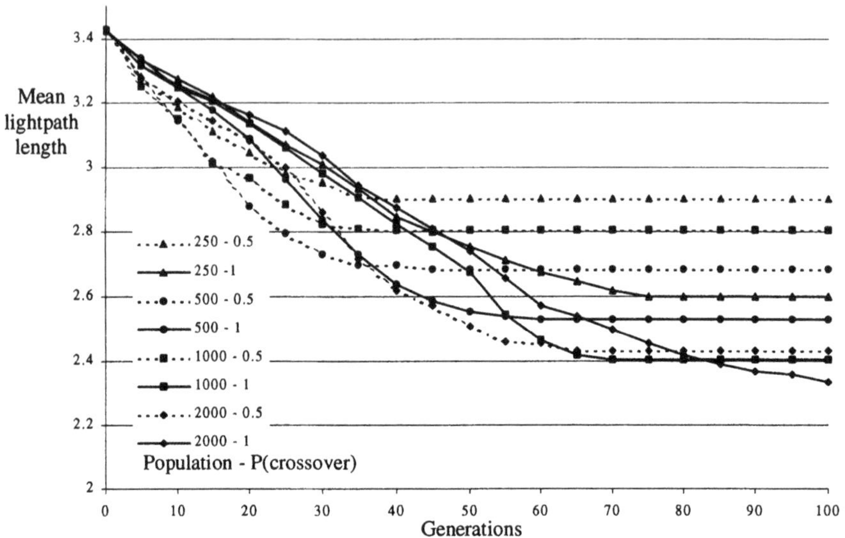

Figure 3. Effect of varying population size and crossover probability on performance of Crossover and Correct (XC)

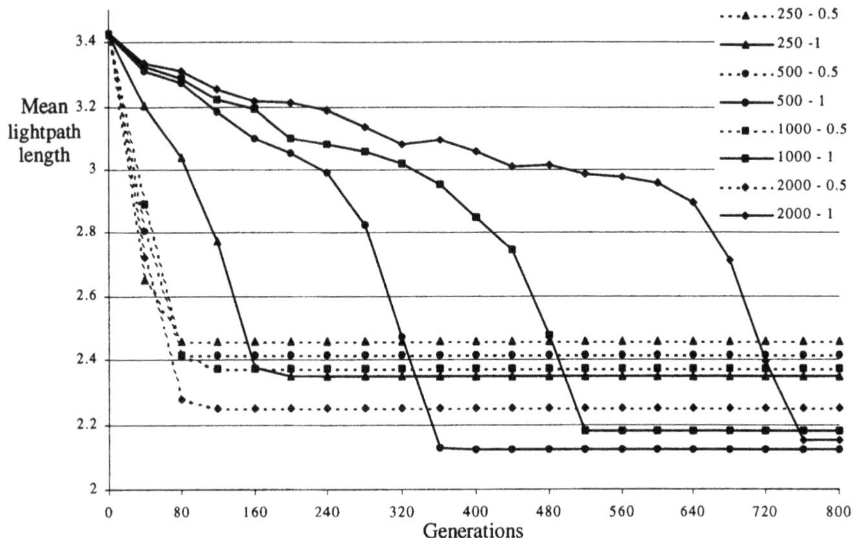

Figure 4. Effect of varying population size and crossover probability on performance of Partially Mapped Crossover (PMX)

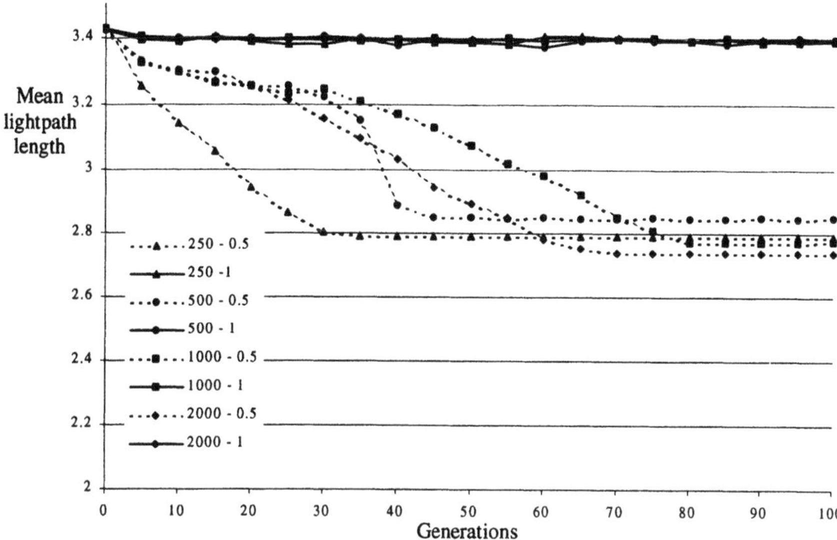

Figure 5. Effect of varying population size and crossover probability on performance of Order Crossover (OX)

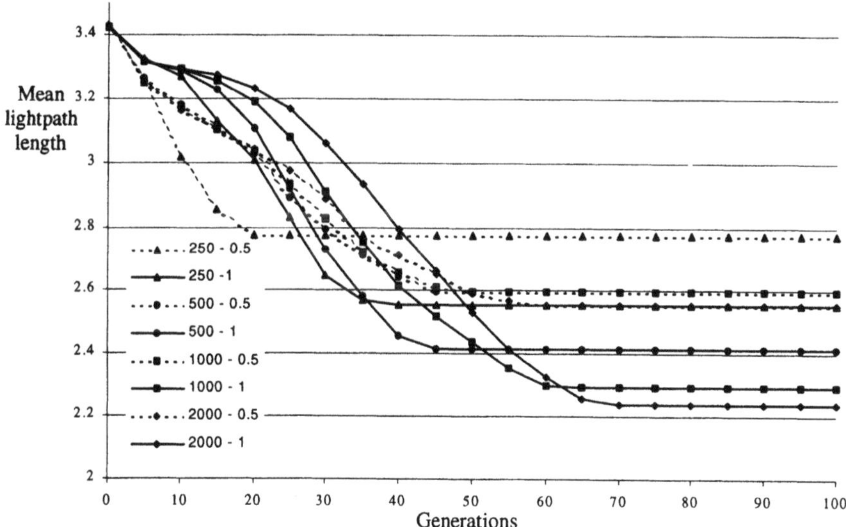

Figure 6. Effect of varying population size and crossover probability on performance of Cycle Crossover (CX)

5.2 Performance of Different Crossover Techniques

5.2.1 Results for Different Network Sizes

The different crossover techniques with the appropriate parameters were used to deploy the MSN in the physical topologies of different sizes. XC, OX and CX all have a population size of 2000 and for fairness the same randomly generated initial population was used. PMX's population is 1000, nevertheless, the implementation ensured that its initial population is a subset of the initial population for the other crossover techniques. In all cases, the overall mean lightpath length of the initial population were very similar.

Figure 7 shows the results for embedding the 4x4 MSN (Figure 1) in the 16-node physical topology shown in Figure 2. It may be observed that PMX gives the best overall result. OX gives the worst result - a final mean lightpath length is 1.78. PMX, CX and XC obtain a mean lightpath length of 1.56, 1.59 and 1.63 respectively. It is encouraging to observe that the different GA implementations all reduce the mean lightpath length significantly, making the prospect of using the MSN to achieve optical packet switching over this arbitrary topology feasible.

The results for the deployment of the 6x6 MSN in the 36-node physical topology are shown in Figure 8. It may be seen that Partially Mapped Crossover (PMX) gives the best overall results - a mean lightpath length of 2.15 after 500 generations. Cycle Crossover (CX) obtains a mean lightpath length of 2.19 after just 69 generations - this corresponds to a reduction in the mean lightpath length of almost 40% from the initial result by considering less than 140,000 different embeddings. Crossover and Correct (XC) obtains a mean lightpath length of 2.28 after 100 generations. Order Crossover (OX) yields a mean lightpath length of 2.74 after roughly 60 generations.

Once again, OX gives the worst result when deploying the MSN in the 64-node physical topology. From Figure 9, it is apparent that OX yields a mean lightpath length of 3.83 after 70 or so generations. XC comes next, obtaining a mean lightpath length of 3.15 after 140 generations. The mean lightpath length that CX obtains is 2.66, obtained after less than 100 generations. The lowest mean lightpath length obtained was achieved by PMX - 2.45 after approximately 600 generations. PMX almost halves the mean lightpath length by considering only 600, 000 different embeddings out of a search space of up to 1.27×10^{89}.

A similar pattern is observed when the 10x10 MSN is embedded in the 100 node physical topology. The results are shown in Figure 10. The mean lightpath length in the initial population is 4.96. OX manages to reduce this to 4.38 after approximately 75 generations. XC achieves a mean lightpath length of 3.63 after 160 generations. CX reduces the mean lightpath length to 2.88 - a reduction of over 40% in only 120 generations. Once again, PMX obtains the best result but takes the longest to converge. PMX's final mean lightpath length is 2.56. The mean lightpath length has been reduced by almost 50% in approximately 1700 generations.

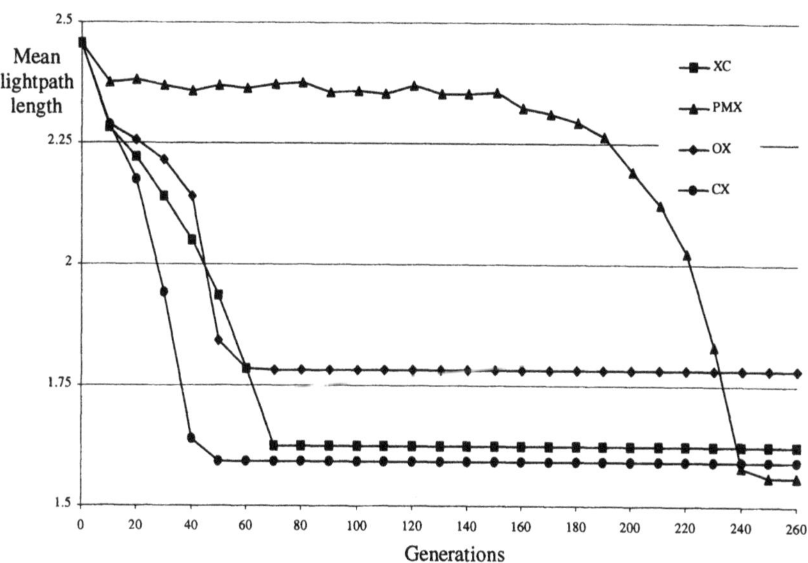

Figure 7. Deployment of 4x4 MSN in physical topology using different crossover techniques

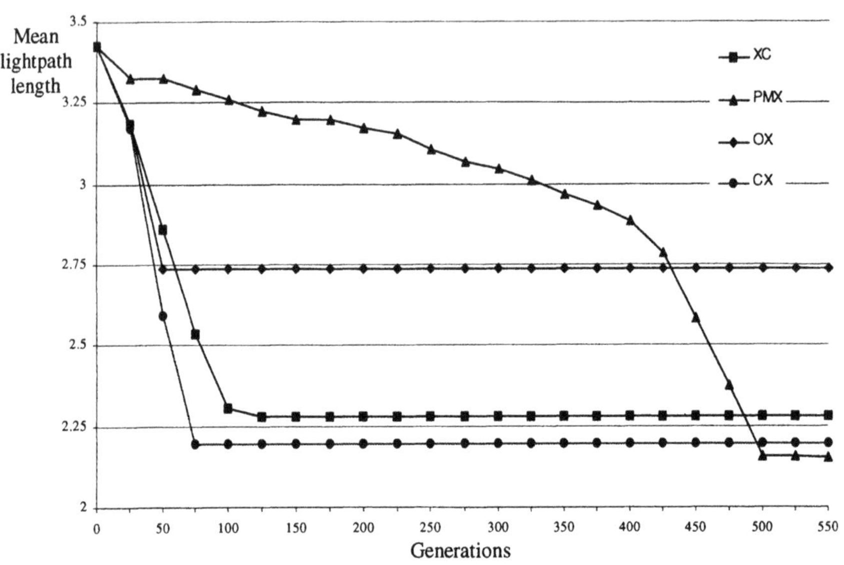

Figure 8. Deployment of 6x6 MSN in physical topology using different crossover techniques

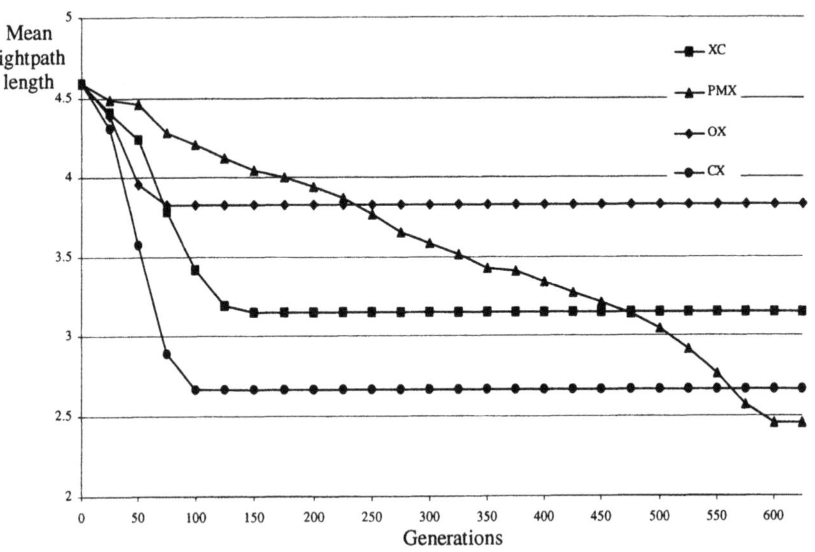

Figure 9. Deployment of 8x8 MSN in physical topology using different crossover techniques

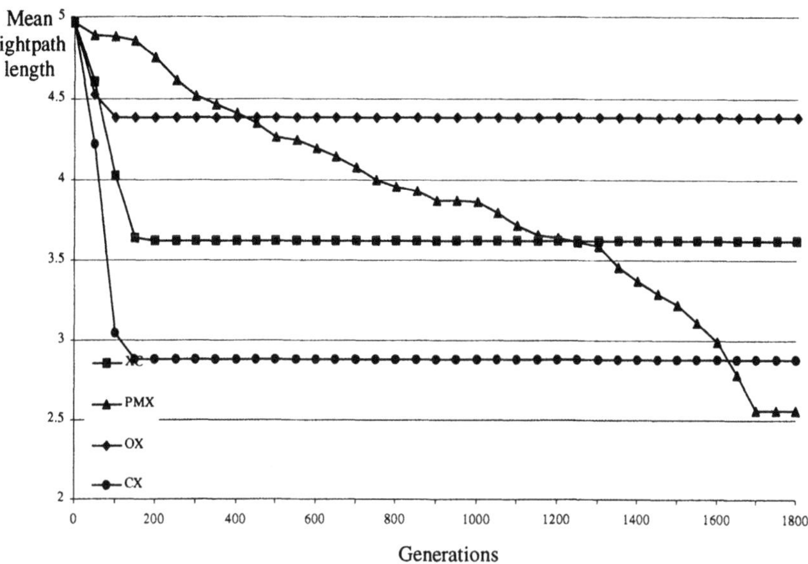

Figure 10. Deployment of 10x10 MSN in physical topology using different crossover techniques

5.2.2 Comments on Results

Several patterns emerge from the results obtained. The meritocratic ordering of the different crossover techniques is PMX, CX, XC and OX. PMX always took significantly longer to converge than the other crossover techniques but always gave the best results. The variation in the performance of the different crossover techniques appears to increase with increasing network size. For example for a 4x4 MSN, PMX is only 12% better than OX but the difference is greater than 40% for a 10x10 MSN.

The reason for the discrepancy in the performance of the different crossover techniques may be understood in terms of the two fundamental operations of optimisation or search techniques - *exploration* and *exploitation* [13]. A good search algorithm should explore the search space whilst exploiting the information gained.

Order Crossover (OX) consistently gives the worst result of all the crossover techniques considered in this paper. From the description of OX given in Section 4.2.3 it is evident that OX does a lot of exploration but very little exploitation - or to be more accurate exploits the wrong information. OX seeks to preserve the ordering of the nodes in the individual - this is valid information for the Travelling Salesman Problem (TSP) for which OX was originally proposed. However, in the problem encoding described in this paper, it is the absolute position of elements in the individual that is most important, and OX does not exploit this information adequately. This hypothesis is buttressed by the observation from Figure 5 that a crossover probability of 0.5 and not 1 gives the better results. This indicates that if crossover is always carried out, relevant characteristics of parents are not passed onto the offspring. However, adopting a crossover probability less than one allows the preserving of important characteristics of parents since the parents have a chance of appearing unaltered in the subsequent generation.

Cycle Crossover (CX) is arguably the best crossover technique at preserving and exploiting the important characteristics of the parents - each position in the offspring is filled with elements from the same position in one of the parents. CX gave results that were only slightly worse than PMX but required significantly fewer generations. For example, for the 8x8 MSN, CX gave a result that was 8.6% worse than PMX but did so in over 500 generations less. This suggests that CX is extremely good at exploitation. It would be interesting to try to improve the amount of exploration done, by allowing the offspring to contain a few elements not occurring the same position in the parent. The impact of introducing mutation to CX will be extremely interesting to investigate.

Partially Mapped Crossover (PMX) obtained the best results. From this it may be concluded that PMX best combined exploration and exploitation synergistically. In [15] it was proven that PMX is extremely good at exploring different areas of the search space for the TSP with implicit parallelism. Crossover and Correct (CX) was an implementation of the simple crossover with the creation of any invalid individuals being prevented. Consequently, it should not be expected to perform as well as PMX and CX which have been specially formulated and shown to tackle the TSP excellently. Nevertheless, CX still achieves reductions on the starting mean lightpath length - an impressive feat considering its simplicity.

5.3 Comparison of GA with Other Heuristics

The different implementations of genetic algorithms have been compared to each other for varying network sizes. There are other techniques that may be used to attempt to deploy the MSN in arbitrary physical topologies, and this section compares the best GA results with the results for alternative optimisation techniques. The three methods to which GA are compared are simulated annealing [18], hill climbing and random search.

SA has been successfully used for deploying hypercubes in NSFnet [19], deploying the MSN in different physical topologies [10] and successfully implemented for similar combinatorial optimisation problems. The SA parameters that were found to obtain good results in [10] are used in this paper. The hill climb was started at a random point in the search space and the number of hill-climbing steps executed is comparable to the aggregate number of individuals considered in the GA implementation that achieved the best result. Random searches of 5 million points were conducted.

The results obtained for the different heuristics is shown in Table 1. The lowest mean lightpath length obtained for each network size has been highlighted. From Table 1 it can be observed that GA out-performs simulated annealing, hill climbing and random search for all network sizes with the exception of a 6x6 where SA and hill-climbing give the best result. (The factors that affect the performance of the heuristics are currently under investigation.) The difference between the GA result and the other techniques' results increases with the network size, suggesting that GA are well-suited to finding high quality embeddings as the search space increases.

Evaluating the cost function is relatively time consuming, consequently the number of solutions examined by each heuristic is indicative of the time taken. For say $N = 4$, GA and HC consider a similar number of solutions, whereas SA and the random search consider approximately 10 and 20 times more solutions respectively. Consequently, not only does GA produce better results than the other heuristics, it does so significantly quicker.

Table 1. Comparisons of GA with other heuristics

N	Genetic algorithms	Simulated annealing	Hill climbing	Random search
4	***1.56***	1.63	1.63	1.72
6	2.15	***2.11***	***2.11***	2.81
8	***2.45***	2.57	2.70	3.89
10	***2.56***	3.11	3.00	4.45

6. CONCLUSIONS

The Manhattan Street Network and a particular routing scheme have been seen to avoid complex routing algorithms and prevent optical domain contentions making it attractive for optical packet switching for some applications. This paper has examined how the MSN may be deployed as a virtual topology in several exemplar physical topologies. A new technique, genetic algorithms, has been successfully applied to the virtual topology design problem. Different implementations of GA have been introduced and their optimum parameters empirically determined. These GA implementations have been compared to each other in terms of the lowest mean lightpath they yield when used to deploy the MSN in the physical topologies. GA are also seen to out-perform other heuristics at deploying the MSN in arbitrary physical topologies for optical packet switching.

REFERENCES

[1] "Nonlinear Optics for High-Speed Digital Information Processing", D. Cotter *et al*, Science, Volume 286, November 1999.

[2] "Ultra-High-Bit-Rate Networking: From the Transcontinental Backbone to the Desktop," D. Cotter, J. K. Lucek, D. D. Mercenac, IEEE Communications Magazine, April 1997.

[3] "The Manhattan Street Network," N. F. Maxemchuk, Proceedings of IEEE Globecom '85, December 1985.

[4] "A New Packet Routing Strategy for Ultra-Fast Photonic Networks," F. Chevalier, D. Cotter, D. Harle, Proceedings of IEEE Globecom '98, November 1998.

[5] "Routing and Fairness Issues Associated with a Self Routing Control Strategy in Regular Mesh Photonic Networks," F. Chevalier, D. Cotter, D. Harle, Proceedings of the International Teletraffic Conference (ITC 16), June 1999.

[6] "Lightnets: Topologies for High Speed Optical Networks," I. Chlamtac, A. Ganz, G. Karmi, IEEE/OSA Journal of Lightwave Technology, Volume 11, Number 5/6, May/June 1993.

[7] "Deployment of Architectures for Ultrafast Optical Packet Switching," O. Komolafe, D. Harle, D. Cotter, Proceedings of IEE UKTS 17, May 2000.

[8] "A Survey of Virtual Topology Design Algorithms for Wavelength Routed Optical Networks," R. Duttra, G. N. Rouskas, TR-99-06, Dept. of Computer Science, N. Carolina State University.

[9] "Data Encodings and Their Costs," A. L. Rosenberg, Acta Informatica, Vol. 9, 1978.

[10] "Ultrafast Optical Packet Switching over Arbitrary Physical Topologies using the Manhattan Street Network," O. Komolafe, D. Harle, D. Cotter, Submitted to IEEE ICC 2001.

[11] "Adaptation in Natural and Artificial Systems," J. H. Holland, University of Michigan Press, 1975.

[12] "A Genetic Algorithm Tutorial," D. Whitley, CS-93-103, Dept. of Computer Science, Colorado State University.

[13] "An Overview of Genetic Algorithms: Part 1, Fundamentals," D. Beasley, D. R. Bull, R. R. Martin, University Computing, Volume 15, Number 2, 1993.

[14] "Evolutionary Telecommunications: A Summary," M.C. Sinclair, Proceedings GECCO '99, July 1999.

[15] "Alleles, Loci, and the Travelling Salesman Problem," D. E.Goldberg, R. Lingle, in Genetic Algorithms & their Applications: Proc. of an International Conf. on Genetic Algorithms, Grefenstette (Ed), L. Erlbaum Associates, 1983.

[16] "A Study of Permutation Crossover Operators on the Travelling Salesman Problem," I. M. Oliver, D. J. Smith, J. R. C. Holland, in Genetic Algorithms & their Applications: Proc. of the 2nd International Conf. on Genetic Algorithms, Grefenstette (Ed), L. Erlbaum Associates, 1987.

[17] "Dynamic Topology Approach to Network Resilience," M. Jorgenson, B Turton, A. Goode, and A. Hume, Proceedings of IEE UKTS 15, March 1998.

[18] "Optimization by Simulated Annealing," S. Kirkpatrick, C. D. Gerlat, M. P. Vecchi, Science, Volume 220, Number 4598, 1983.

[19] "Some Principles for Designing a Wide-Area Optical Network," B. Mukherjee, S. Ramamurthy, D. Banerjee, A. Mukherjee, Proceedings of IEEE Infocom 1994.

Packet-selective photonic add/drop multiplexer and its application to ultrahigh-speed optical data networkings in LAN and MAN

Ken-ichi Kitayama[1)] and Masayuki Murata[2)]
Osaka University 85
1) Department of Electronics and Information Systems, 2-1 Yamadaoka, Suita, Osaka 565-0871, Japan, (E-mail: kitayama@comm.eng.osaka-u.ac.jp)
2) Cybermedia Center, 1-30 Machikaneyama, Toyonaka, Osaka 565-0043, Japan, (E-mail: murata@ics.es.osaka-u.ac.jp)

Key words: packet transmission, asynchronous access, optical code correlation, performance modelling, optical implementation

Abstract: We propose packet-selective photonic add/drop multiplexing (P-ADM), which can determine the source and destination addresses in the packet header based upon optical code correlation and its novel application to ultrahigh-speed optical data networking in LAN and MAN. It is distinct from conventional wavelength ADM (λ-ADM) having the wavelength granularity, in that P-ADM provides a packet granularity by individually handling optical packets on a single wavelength. As an application of P-ADM to optical data networking, a *rate-controlled asynchronous access* is studied, and the performance analysis and numerical examples are presented. The optical implementation of P-ADM is also presented, in which the photonic label selector, structured with a fiber Bragg grating (FBG), is a key to the ultrahigh-speed processing capability up to a few tens of Gpps (packet per second).

1. INTRODUCTION

There is a growing demand for high-bandwidth and differentiated data services. Given the inherently bursty nature of data traffic, the fixed-bandwidth TDM links is not an efficient solution. Optical data networking, which uses

WDM links as the optical channels [1, 2], is an alternative approach to the TDM-based data networking. Optical data networking eliminates unnecessary protocol stack in each network layer, thus drastically reducing the cost and complexity of networking [3, 4].

In optical data networking, the optical add/drop multiplexer (ADM), located at access nodes, will be key device. A wavelength add/drop multiplexer (λ-ADM) enables each wavelength to either add/drop and electronically processes at the node or to optically bypasses the node electronics (Fig.1(a)). This capability of optically bypassing a node's electronic layer can be exploited to reduce overall electronic processing, resulting in the reductions of both the transfer delay and the number of electronic digital crossconnect switches and routers. λ-ADM can be adopted for different types of traffics such as static or dynamic circuit switching, packet switching, and flow switching [5]. However, λ-ADM is of granularity of wavelength, and this imposes a limit that it can only handle traffic on an optical channel, that is, a wavelength path, and it is unable to handle traffic on a packet-by-packet basis. This may waste the wavelength resource and limit the usage of optical channels. On the other hand, conventional SONET ADM has a finer granularity for handling traffic. It picks out low-speed streams from a high-speed stream and likewise, adds low-speed streams to a high-speed stream [6]. However, the speed of electronic processing will eventually become a bottleneck to the overall processing capability as the bit rate of the WDM link goes higher. Hence, there would be a growing demand for high-speed multi-functional ADM in the future optical data networking, which can handle traffic in finer granularity.

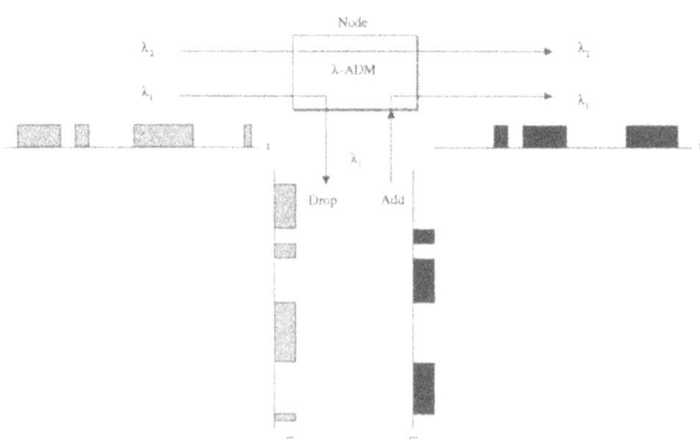

Fig. 1(a) Wavelength add/drop multiplexer (λ-ADM)

There have been a variety of optical packet switching schemes, in which the packet label are mapped in optical domain and processed optoelectronically [7-12]. Very recently, photonic router for Internet Protocol, IP packets has been proposed and demonstrated [13-15]. In the photonic IP router, the photonic labels are mapped onto a family of optical codes, which have been utilized optical code division multiplexing (OCDM) technique [16-18]. OCDM is a class of transmission and multiplexing techniques other than TDM, WDM, and space division multiplexing (SDM). OCDM inherits unique features from the wireless code division multiple access (CDMA) [19]. It was confirmed that as the encoding and decoding are performed in optical domain, the packet processing capability exceeds an order of giga-packet per second(Gpps).

In this paper, packet-selective photonic add/drop multiplexing (P-ADM), which can read the source and destination addresses in the packet header in optical domain, and its novel application to ultrahigh-speed optical data networking in LAN and MAN are proposed. It is distinct from conventional wavelength ADM (λ-ADM), in that P-ADM provides a packet granularity by individually handling optical packets on a single wavelength. In the application of P-ADM to optical data networking, a *rate-controlled asynchronous access* is studied, and the performance analysis and numerical examples are presented. The optical implementation of P-ADM is also presented, in which the photonic label selector, structured with a fiber Bragg grating (FBG), is a key to the ultrahigh-speed processing capability up to a few tens of Gpps (packet per second).

2. CONCEPT OF PACKET-SELECTIVE PHOTONIC ADD/DROP MULTIPLEXER (P-ADM)

2.1 Principle of operation

P-ADM is located at a node, and it drops to the node or bypasses the node the arriving packets or adds the generated packets from the node (Fig.1(b)). It handles traffic on a single WDM link on a packet-by-packet basis. The basic architecture is illustrated in Fig.2. The fundamental functions of P-ADM include photonic label selection and optical switching. The input and output ports are connected to a single WDM link. The photonic label in the header of arriving packet is processed, and according to the label the optical switch is driven by the control signal to direct the packet to the outgoing port either for dropping or bypassing.

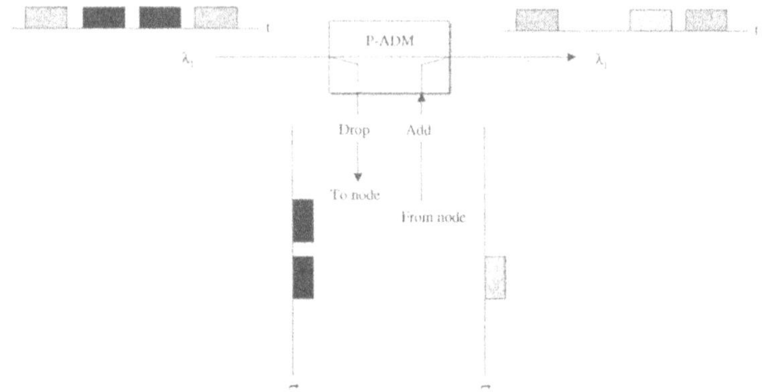

Fig. 1 (b) Proposed packet-selective photonic ADM (P-ADM)

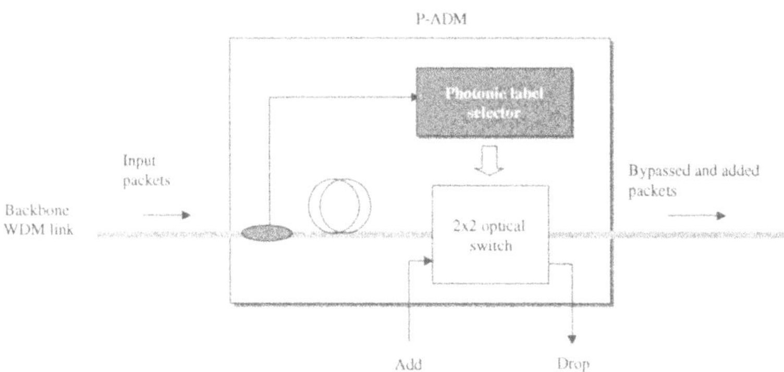

Fig.2 Basic architecture of packet-selective photonic ADM (P-ADM) module

2.2 Photonic label selector

The optical packet is structured as shown in Fig.3. The photonic label in the header is simply a relatively short, fixed-length identifier that bears the information of the destination address and source address. At the end of the packet a flag that indicates the end must be attached. The other overhead information could be included in the label upon the request. The remaining network control information will be encapsulated in the frame of higher layers.

As shown in Fig.4 each node is uniquely assigned with a label, and each label is mapped onto an optical code [13]. The optical code is a sequence of optical pulses, so-called chip pulses. The chip itself is a short pulse, and the time duration of the sequence has to be accommodated within a bit time duration. The photonic label selection is based upon the optical correlation between the optical code of arriving packet and the assigned code of the node. The optical correlation is performed by matched filtering in time domain. For example, a family of optical codes of 4-chip long codes and their auto- and cross-correlation waveforms are illustrated in Fig.4. The code is bipolar, in which the phase of optical carrier of individual chip pulse takes two states of either 0 or π, representing binary value of 1 or -1, respectively. As increasing the number of nodes, the longer code will be required. Unique to the label selector is that no optical logic operation is involved, and this is a key to the ultrahigh-speed operation of P-ADM.

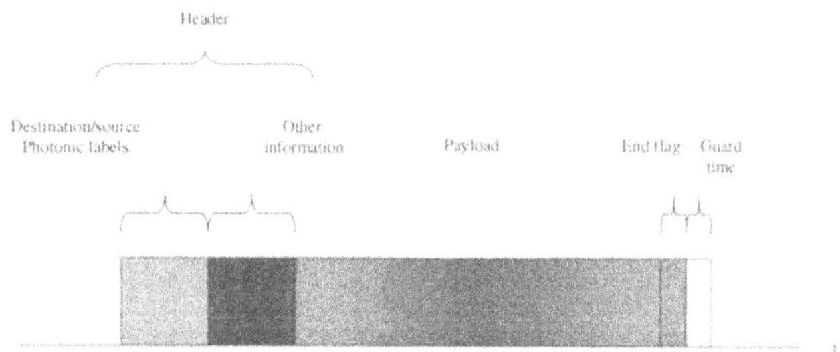

Fig.3 Optical packet

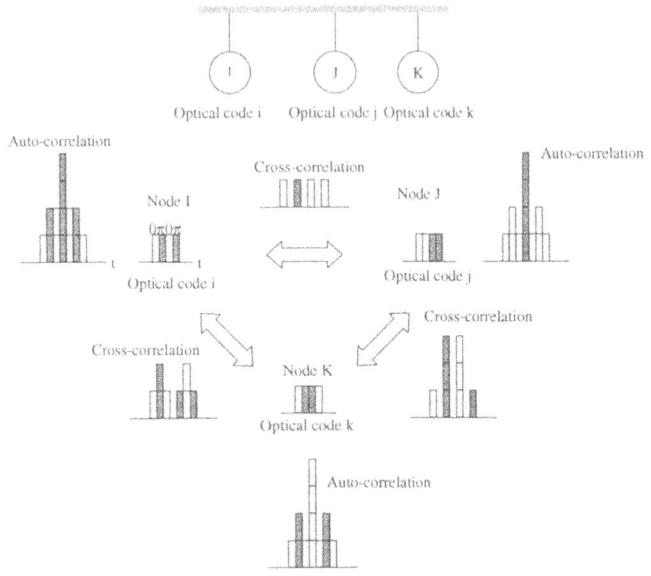

Fig.4 Address assignment to each node. A family of 4-chip long, bipolar optical codes and their auto- and cross-correlation functions

2.3 Key building block: P-ADM module

We will discuss the building blocks of P-ADM shown in Fig.2. First, the photonic label selector is structured with an optical correlator, a time-gate, an O/E conversion, and thresholding (Fig.5(a)). Promising device for the optical correlator is a fiber Bragg grating (FBG) (Fig.5(b)). An advantage is that a single FBG can select the photonic label as well as the wavelength, if the grating period is tuned to a desired wavelength. A super-structured FBG can generate bipolar optical codes [20]. The identical optical device must be used for the optical encoder.

Several high-speed off-the-shelf optical space switches have now become available. The switching speed of semiconductor optical amplifier (SOA) gate switch is in the sub-nanosecond regime, which is well beyond the requirement for the switching speed. Unfortunately, there is no random access optical memory, equivalent to electronic RAM. The optical fiber delay line is only an optical device that temporarily buffers an optical packet.

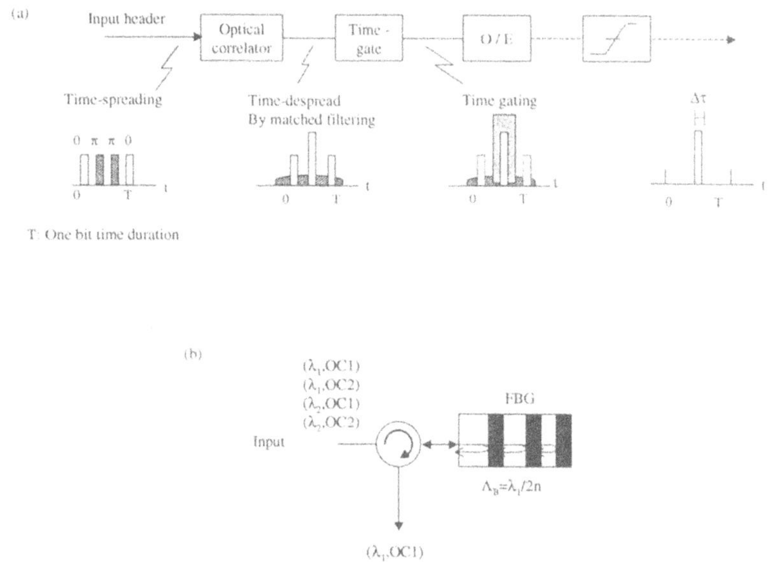

Fig.5 (a) Configuration of photonic code selector. (b) FBG-based optical correlator

3. APPLICATION TO OPTICAL DATA NETWORKINGS IN LAN AND MAN: LINE ACCESS BY RATE-CONTROLLED ASYNCHRONOUS PACKET TRANSMISSIONS

3.1 Operation mechanisms

For data transmissions we want to allow the bursty transmission of packets in LAN and MAN, by which a flexible use of bandwidth is expected. For this purpose, we will propose a rate-controlled line access method, and we will implement this by using P-ADM. We use a leaky-bucket mechanism [21] to control the packet emission onto the line at each node.

The leaky-bucket is a virtual device to control the burstiness of the flow. As shown in Fig.6, tokens arrive at the token pool with rate ρ and are buffered in the token pool if the token pool is not full. If an arriving packet finds the token pool nonempty, it departs the node instantaneously, and at the same time one token is removed from the token pool. The packet is discarded if an arriving packet finds the token pool empty. When a token is generated, and the packet buffer is nonempty, the packet on the top of queue departs the buffer immediately by removing one token. The depth of the token pool determines the burstiness of the flow; the larger size of the token pool allows more bursty packet transmissions.

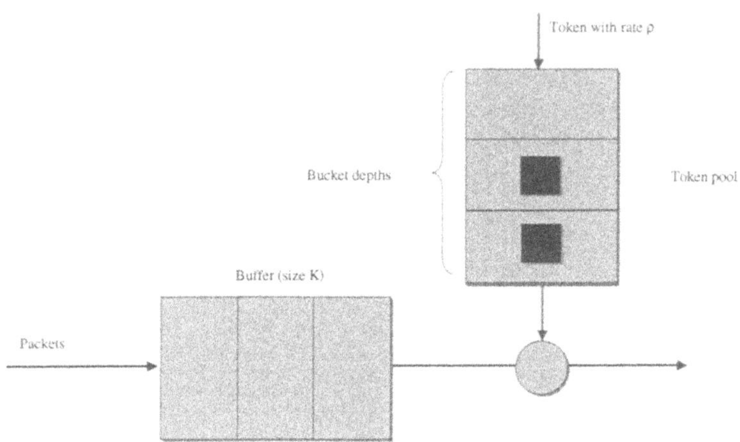

Fig.6 Logical structure of leaky bucket

3.1.1 Rate control using leaky-bucket

For a realistic scenario of the rate-controlled line access, we need to take account of the dynamics of the leaky-bucket. The leaky-bucket parameters should be determined for a given distribution of packet arrival rate. Since the packet buffer is finite, the packet loss probability should be estimated when determining the leaky-bucket parameters. For this purpose, we introduce N and C that represent the line capacity and the number of terminals connected to the line, respectively. Assuming that the packet length is fixed, the following procedure then determines the leaky-bucket parameters. Here, our simulation will forcus on the bus topology, however, P-ADM will also be applied to a ring topology.

1. Assume that the traffic load distribution is known. Let the mean rate be $\lambda_{i,j}$ for the packet flow from node i to node j, and let $\lambda_i = \sum_{j=i+1}^{N-1} \lambda_{i,j}$.
2. The admissible traffic load at each node is then determined by taking into account the relative use of the line capacity. In the case of the balanced traffic, it is simply determined as

$$C \times \frac{\lambda_i}{N^2/4} \quad (1)$$

where $N^2/4$ is the number of connections on the most congested link. Note that other bandwidth allocation can also be considered by, e.g., considering the fair allocation of packet loss probabilities among nodes. For this purpose, however, we need to formulate some optimization problem.

3. The admissible load given by Eq.(1) determines the leaky-bucket parameter ρ_i for node *i*.
4. Analyze the leaky-bucket to determine the packet loss probability and waiting time at the packet buffer for given packet arrival distributions.

Note that we need more strict arguments in performing above procedures. Those will be described in the next subsection by considering the case where the packet arrivals follow a Poisson distribution.

In the above, we still have a freedom in determining the burst size specification σ_i. For larger σ_i, the packet loss probability will decrease, which leads to the performance improvement. However, the upstream node *j* is allowed the bursty packet then, bursty arrivals of packets from the upstream nodes inhibit the packet transmission at downstream node. Once we are given a permissible packet loss probability at each node, however, we can determine the bucket size and then obtain the maximum packet delay at the node in order.

3.1.2 Analysis of leaky-bucket

In this subsection, we present an analysis for the leaky-bucket at each node, which is a slight extension of the approach shown in [21]. Packets arrive at the packet buffer according to a Poisson process with rate λ. Note that we will omit the subscript representing the node number in the below.

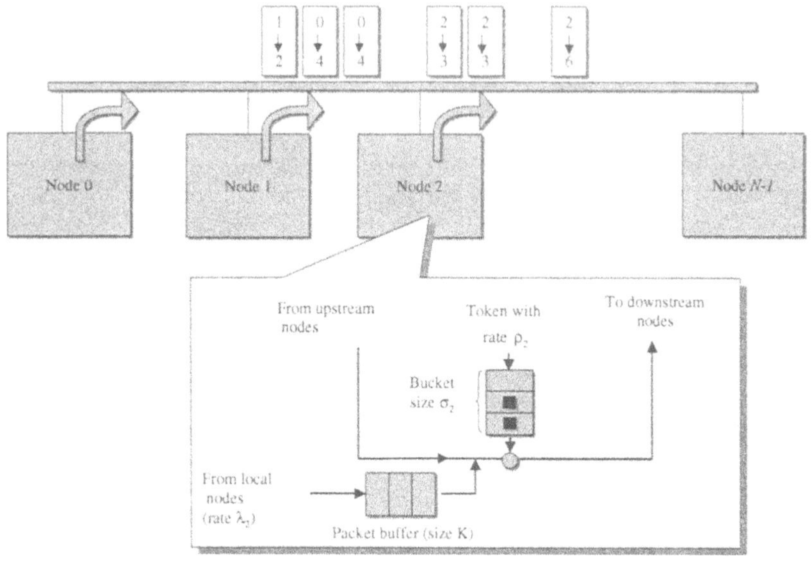

Fig.7 Flow enforcement by leaky-bucket

See Fig.7. Consider a slotted time axis where each slot is of length $1/\rho$, and a new token is generated at each slot boundary. A newly arriving token joins the token pool if the pool contains fewer than σ tokens; otherwise it is discarded. We will derive the steady–state joint probability distribution of the queue size and the token pool occupancy at slot boundaries, just prior to the generation of a new token. Note that we neglect an asynchronous operation of our system, but it is only for modelling purpose.

We introduce *P(m, i)* the steady–state probability of having m tokens in the token pool and *i* packets in the packet buffer. The number of tokens cannot exceed σ, and the packet should wait in the buffer if and only if there exist no tokens in the token pool. Thus, we have

$$P(m,i)=0, \quad m>\sigma; 1\leq m\leq\sigma \text{ and } i\geq 1. \quad (2)$$

For convenience, we introduce p_i as

$$p_i=\begin{cases} P(\sigma-i,0), & 0\leq i\leq\sigma \\ P(0,i-\sigma), & \sigma\leq i \end{cases} \quad (3)$$

As described in the above, packet arrivals at the local node are assumed to follow a Poisson process. We further assume that arrivals of transit packets from the upstream nodes destined for the downstream nodes also follow a Poisson process with rate λ_u, which is given by

$$\lambda_u = \sum_{j=0}^{i-1} \sum_{k=i+1}^{N-1} \lambda_{j,k} \quad (4)$$

for node *i*. We note here that in actual, we need to take account of the departure processes from the leaky-bucket at the upstream nodes, which is treated as an input process at the current node. However, since the variance of the departure process tends to become smaller than the input Poisson process by the leaky-bucket, we can estimate the worst case by the above assumption. We introduce a_i and b_i to represent the probabilities of *i* arrivals during a slot from the local node and from the upstream nodes, respectively, i.e.,

$$a_i = \frac{e^{-\lambda/\rho}(\lambda/\rho)^i}{i!}; \; b_i = \frac{e^{-\lambda_u/\rho}(\lambda_u/\rho)^i}{i!} \; . (5)$$

Further, we introduce $\bar{a}_i$ and $\bar{b}_i$ as

$$\bar{a}_i = 1 - \sum_{j=0}^{i} a_j; \; \bar{b}_i = 1 - \sum_{j=0}^{i} b_j \quad (6)$$

Since we consider the finite capacity of the packet buffer (denoted by *K*), a packet that arrives and finds the buffer full is discarded. The transit packets are given higher priority since we assume to have no means to store the packets in an optical domain. Accordingly, we consider that the transit packet gets the token from the token pool at the local node and leaves the node. The steady–state equations are then given by

$$p_i = p_0 \sum_{k=0}^{i} b_k a_{i-k} + \sum_{j=0}^{i} p_{j+1} \sum_{k=0}^{i-j} b_k a_{i-j-k}, \quad (7)$$
$$0 \le i \le \sigma + K - 1.$$

The above equations can be solved by first assuming a value for p_0, and recursively computing p_i's ($1 < i < \sigma + K$) using

$$p_i = \frac{1}{a_0 b_0}\left(p_{i-1} - p_0 \sum_{k=0}^{i-1} b_k a_{i-1-k} - \sum_{j=0}^{i-2} p_{j+1} \sum_{k=0}^{i-1-j} b_k a_{i-1-j-k}\right). \quad (8)$$

All quantities are finally normalized with

$$\sum_{i=0}^{\sigma+K} p_i = 1. \quad (9)$$

The throughput at the designated node is then obtained as

$$T = p_0 \sum_{k=0}^{\sigma+K-1} b_k \left\{ \sum_{i=1}^{\sigma+K-k} b_k + (\sigma + K - k)\bar{a}_{\sigma+K-k} \right\}$$
$$+ \sum_{j=1}^{\sigma+K} p_j \sum_{k=0}^{\sigma+K-1} b_k \left\{ \sum_{i=1}^{\sigma+K-j-k+1} i a_i + (\sigma + K - j - k + 1)\bar{a}_{\sigma+K-j-k+1} \right\} \quad (10)$$

Finally, we obtain the packet loss probability at the node as

$$P_{loss} = 1 - \frac{T}{\lambda / \rho}. \quad (11)$$

We can also derive waiting times for the packet stored at the buffer by applying the approach taken in [21], although we omit the result. In numerical examples, we will only show the maximum waiting time of the packets that are stored in the buffer without lost. It is simply given as

$$W_{max} = \sigma / \rho. \quad (12)$$

We last note that we can also consider the bursty traffic such as an ON–OFF process, but for that purpose, we need a more extension. See, e.g., [22]. The numerical examples for the packet loss probability will be shown in Subsection 4.1.

3.2 Optical implementation

The protocol for accessing the line is shown in Fig.8. If the arriving packet at node j has the destination label of node j, the packet is dropped from switch 1. If the arriving packet has a different destination from node j, and if the token pool is nonempty, it is bypassed to downstream nodes. However, the arriving packet having a different destination from node j is discarded when the token pool is empty. We will put a higher priority on the bypassing traffic than the traffic to be added from the node *j*, therefore, the packet transmission from node *j* is awaited during the burst from node *i* (<*j*) passes by node *j*.

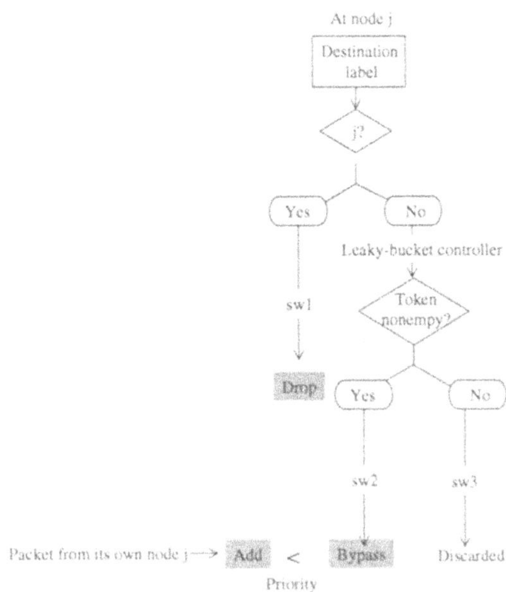

Fig.8 Access protocol of asynchronous rate-controlled packet transmission

The implementation of the P-ADM is shown in Fig.9. It consists of a photonic label selector, three 1x2 optical switches, two optical delay lines, and an electronic leaky-bucket controller. The photonic label selector directly controls optical switch 1, and the control signal from it is also fed to the leaky-bucket controller. The leaky-bucket controller controls optical switches 2 and 3. The delay time τ of the optical delay line has to be set equal to the processing time required for photonic label selector, while, the delay time τ_1 is set equal to the processing time required for the leaky-bucket controller. Optical switch 1 drops the arriving packet when it turns to the cross-state, otherwise it stays in the bar-state to bypass the packet. Optical switch 3 stays in the bar-state to bypass the arriving packet, and it turns to the cross-state to discard the arriving packet. Optical switch 2, controlled by the leaky-bucket controller, is allowed to turns to the cross-state to add the packet from the buffer only when there is no bypassing packet, otherwise it stays in the bar-state to bypass the packet.

It is noted that the leaky-bucket control is very simple; only register fetch and add operations are necessary to implement the leaky-bucket mechanism. When we consider 100 Gbps transmission line and 1,000 bit-long packet as described in the numerical examples in 4.1, a packet transmission time becomes 10ns. Thus, it is sufficient to execute several instructions for performing the leaky-bucket algorithm even by the current processor technology.

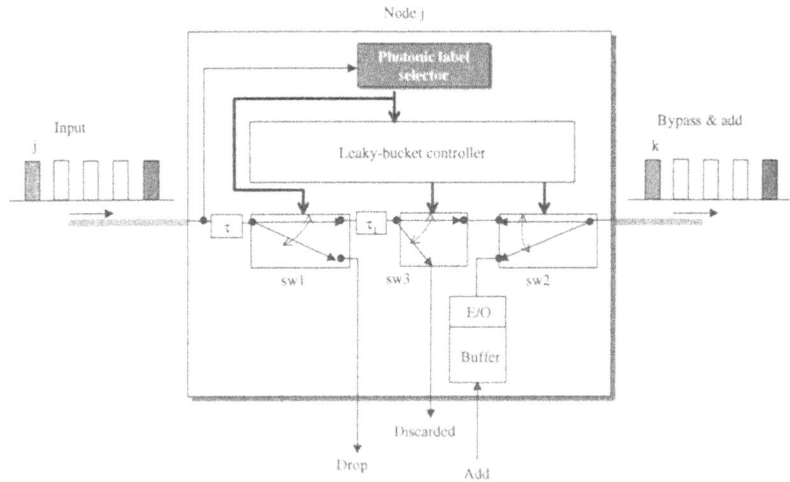

Fig.9 Optical implementation P-ADM for asynchronous rate-controlled packet transmission

4. DISCUSSIONS

4.1 Numerical examples of rate-controlled asynchronous packet transmissions

By using the analysis method in Subsection 3.1, we now provide several numerical examples. We consider the balanced traffic, i.e., $\lambda=\lambda_{ij}$ for all (i, j) throughout this subsection. The line capacity C and the packet length are fixed at 100 Gbps and 1,000 bits, respectively. The token generation rate at node i is determined by taking into account two kinds of traffic flows; $(N-I-1)$ connections are originated at node i and destined for the downstream nodes. Other i (N-i-1) connections are transit flows which are originated at the upstream nodes and destined for the downstream nodes.

In Fig.10, we first show the packet loss probabilities dependent on the node position of the line for $N = 12$. The target traffic load for each connection is set to be $C/36$ where 36 is the total number of transit and originating connections at node 5. Three cases for the buffer size are shown; $K = 10$, 15 and 20. The token pool size (or bucket depth) σ is set to be 10. The traffic load is 2.08 Gbps for each connection, which corresponds to 75% of the bandwidth allocated to each connection in the pseudo TDMA system. The packet loss probabilities are slightly increases at downstream nodes. It is because in the downstream node, the traffic from the upstream node becomes increased, and it suppresses the packet transmission from the local node. For example, in the case of the buffer size $K = 20$, the maximum waiting time at the buffer ranges from 0.100 μs (at node 5) to 0.327 μs (at node 0 and 10). By considering the time duration of 1,000 bit-long packet is 0.01 μs for the bit rate of 100 Gbit/s, the waiting time would be reasonable.

We next show the packet loss probabilities dependent on the traffic load in Fig.11. The horizontal axis shows the traffic load at each node. We plot the worst case, i.e., the packet loss probabilities at (N-2)nd node. The buffer size K is changed as 10, 15 and 20 in the figure. Last we show the packet loss probabilities dependent on the number of nodes on the line in Fig.12. For that purpose, we again fix the traffic load at 1.0 Gbps; identically set for all connections. Thus, as the number of total nodes is increased, the traffic load at each link gets large and the performance is degraded in both two cases. The packet buffer size is set to be 10, 15 and 20. The token pool size of all nodes is identically set to be 10. All results are for (N-2)nd node.

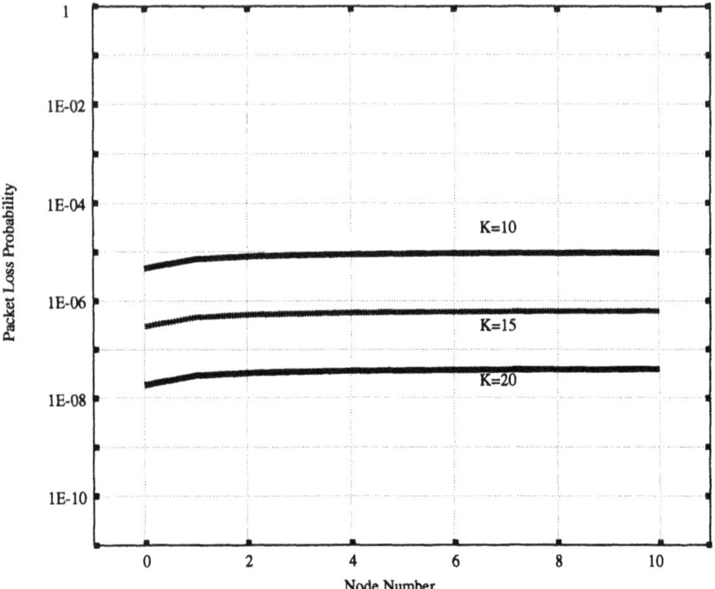

Fig.10 Packet loss probabilities at nodes

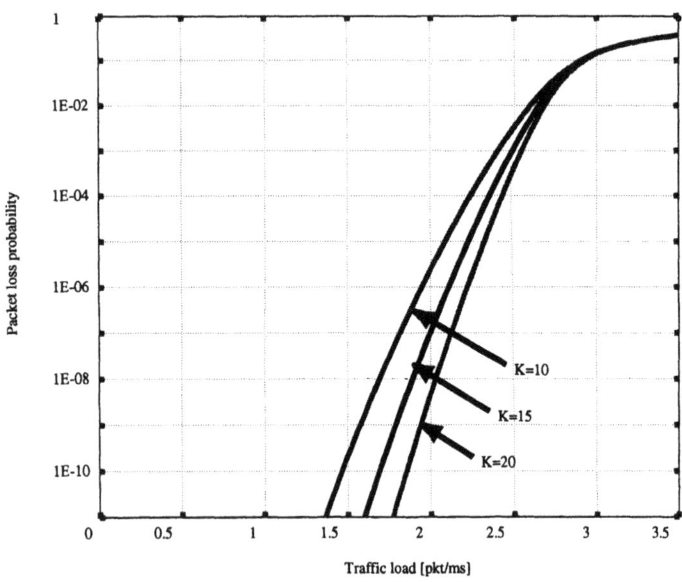

Fig.11 Packet loss probabilities vs. traffic load

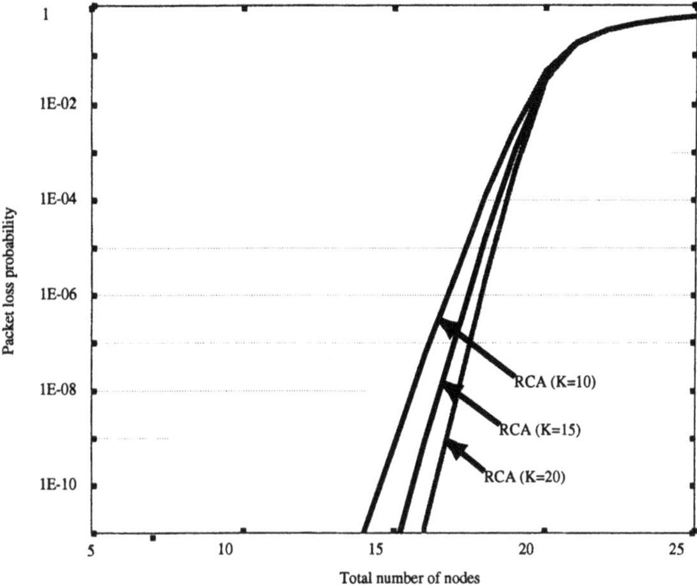

Fig.12 Packet loss probabilities vs. the number of nodes

4.2 Perspective for ultrahigh-speed optical data networking

The rate-controlled access has a distinct advantage over conventional multiple access protocols because it adopts a sort of so-called open-loop control [23]. Another is a closed loop control where the control line (or control slot) is used to reserve the next transmission on the line. Such a control method can be found in, e.g., Distributed Queue Bus (DQDB). The minimum packet size in Ethernet has to be at least twice the propagation delay in bits so that the node will hear a collision before it stops sending and mistakenly assumes the transmission is successful. As increasing the bit rate, more and more bits can be packed into the fiber. For example, the minimum packet size for 100Gbps is 10^5bits for the propagation delay of only 100m-long optical fiber. Such a large packet size is impractical and therefore, the access techniques such as GbE (E: Ethernet), 10GbE, and DQDB will have a difficulty in achieving high performance for a long distance because their performance can be easily influenced by the transmission length. On the contrary, the performance of the rate-controlled access is by no means liable to the transmission length, and this characteristic can be exploited for ultra high-speed access.

There is another an advantage in the rate-controlled access compared with optical TDMA (OTDMA), which is a multiple access protocol in optical layer [24]. An ultrahigh-speed OTDM has been realized up to a hundred Gbps [25]. All-optical regenerative OTDM add/drop multiplexing at 40Gbps has been materialized [26]. OTDMA is not as efficient for bursty data transmissions because it uses a time-slotted access protocol.

Ultimate performance of the rate-controlled asynchronous access will be predicted from a viewpoint of its optical implementation. The overall processing speed of P-ADM is limited by the capability of the photonic label selector, which is governed by the propagation delay required for the optical correlation. It has been experimentally demonstrated that optical packet consisting of photonic label using 8-chip-long optical code is processed, and optical packet of 64bit-long payload data at 10Gbit/s is routed via optical switch [14]. The processing capability reaches 29Gpps with the time duration of photonic label of 35ps. Assume that the packet length is 1,000 bit, P-ADM can be interfaced with the link of 29Tbps. The accessible number of nodes depends upon the length of optical code. For the chip pulsewidth of 1ps the maximum code length will become 5ps for the bit rate of 100Gbps. For 5-chip long optical code, the number of available optical codes are $2^{5-1}=16$. Note that the factor of -1 is due to the elimination of symmetry code sequences.

5. CONCLUSION

Packet-selective photonic add/drop multiplexing (P-ADM) using photonic label selector based upon optical code correlation and its novel application to ultrahigh-speed optical data networking have been proposed. P-ADM provides a finer granularity of traffic than conventional λ-ADM, and it can handle the traffic on a packet-by-packet basis. The optical implementation P-ADM has been presented by using off-the-shelf optical devices such as FBG and optical gate switch, and the ultrahigh-speed processing capability as high as 29Gpps has been predicted. The application to optical data networking in LAN and MAN, the rate-controlled asynchronous access has been studied, and the numerical examples have confirmed performances of low packet loss probability.

Finally, another promising application of the packet-selective P-ADM is to the network protection and restoration. P-ADM will be able to provide the protection to individual traffic on a wavelength path. Hence, this will provide with differentiated class of protections; each traffic has different priority of protection and restoration.

References:

[1] C. A. Brackett, "Dense wavelength division multiplexing networks: Principles and applications," *IEEEJ. Select. Areas in Commun.*, vol. 8, pp. 948-964, 1990.

[2] K. Sato, S. Okamoto, and H. Hamada, "Network performance and integrity enhancement with optical path layer technologies," *IEEE J. Select. Areas in Commun.*, vol.12, pp.159-170, 1994.

[3] P. Bonenfant and A. R.-Moral, "Optical data networking," *IEEE Commun. Magazine*, pp.63-70, March 2000.

[4] M. Murata, "Challenges for the next-generation Internet and the role of IP over photonic networks," to appear in *IEICE Trans. Commun.* vol.E83-B, No.8, 2000.

[5] E. Modiano and A. Narula-Tam, "Mechanism for providing optical bypass in WDM-based networks," *Optical Networks*, vol.1, pp.11-18, 2000.

[6] R. Ramaswami and K. N. Sivarajan, *Optical Networks -A Practical Perspective-*, S. F.: Morgan Kaufmann Publishers, Inc, 1998, ch.8.

[7] Y. Takushima and K. Kikuchi, "Photonic switching using spread spectrum technique," *Electron. Lett.*, vol.30, pp.436-438, 1994.

[8] M.D. Vaughn and D.J. Blumenthal, "All-optical updating of subcarrier encoded packet headers with simultaneous wavelength conversion of baseband payload in semiconductor amplifiers," *IEEE Photonic Technol. Lett.*, vol.9, pp.827-829, 1997.

[9] A. Carena, M.D. Vaughn, R. Gaudino, M. Shell, and D.J. Blumenthal, "OPERA: An optical packet experimental routing architecture with label switching capacity," *J. Lightwave Technol.*, vol.16, pp.12135-2145, 1998.

[10] X. Jiang, X. P. Chen, and A.E. Willner, "All optical wavelength independent packet header replacement using a long CW region generated directly from the packet flag," *IEEE Photonic Technol. Lett.*, vol.10, pp.1638-1640, 1999.

[11] M. C. Cardakli, D. Gurkan, S. A. Havstad, A. E. Willner, "Variable-bit-rate header recognition for reconfigurable networks using tunable fiber-Bragg-gratings as optical correlators," *2000 Optical Fiber Conference (OFC2000)*, TuN2 (Baltimore, 2000).

[12] W. I. Way, Y.-M. Lin, and G. -K. Chang, "A novel optical label swapping technique using erasable optical single-sideband subcarrier label, " *2000 Optical Fiber Conference (OFC'99),* WD6 (Baltimore, 2000).

[13] K. Kitayama and N. Wada, "Photonic IP routing," *IEEE Photonic Technol. Lett.*, vol.11, pp.1689-1691, 1999.

[14] N. Wada and K. Kitayama, "Photonic IP routing using optical codes: 10Gbit/s optical packet transfer experiment," *2000 Optical Fiber Conference (OFC'99),* WM51 (Baltimore, 2000).

[15] H. Sotobayashi and K. Kitayama, "Optical code based label swapping for photonic routing," to appear in *IEICE Trans. Commun.* vol.E83-B, No.8, 2000.

[16] J. Salehi, "Code division multiple access techniques in optical fiber networks-Part I: Fundamental principles," *IEEE Trans. Commun.*, vol.37, pp.824-833, 1989.

[17] P. Prucnal, M. Santro, and T. Fan, "Spread spectrum fiber optic local area network using optical processing," *IEEE/OSA J. Lightwave Technol.*, vol.4, pp.307-314, 1986.

[18] K. Kitayama, "Code division multiplexing lightwave networks based upon optical code conversion," *IEEE Selected Areas in Commun.*, vol.16, pp.1309-1319, 1998.

[19] K. Kitayama, H. Sotobayashi, and N. Wada, "Optical code division multiplexing (OCDM) and its applications to photonic networks (INVITED)," *IEICE A.*, vol.E82-A, pp.2616-2626, 1999.

[20] P.C. The, P. Petropoulos, M. Ibsen, and D. J. Richardson, "A 10Gb/s, 160Gchips/s OCDMA coding: decoding system based on superstructured fiber gratings," *2000 Optical Fiber Communication Conference (OFC2000)*, PD9 (Baltimore, March 2000).

[21] M. Sidi, W.-Z. Liu, I. Cidon, and I. Gopal, "Congestion control through input rate regulation," *IEEE Transactions on Communications*, vol. 41, March 1993.

[22] M. Murata, Y. Ohba, and H. Miyahara, "Analysis of flow enforcement algorithm for bursty traffic in ATM networks," *Proceedings of IEEE INFOCOM '92* (Florence, Italy), pp. 2453–2462, May 1992.

[23] H. Ohsaki, M. Murata, H. Suzuki, C. Ikeda, and H. Miyahara, "Rate-based congestion control for ATM networks," *ACM SIGCOMM Computer Communication Review*, vol. 25, pp. 60–72, April 1995.

[24] S-. W. Seo, K. Bergman, and P. R. Prucnal, "Transparent optical networks with time-division multiplexing," *IEEE J. Selected Areas in Commun./J. Lightwave Technol.*, vol.14, pp.1039-1051, 1996.

[25] S. Kawanishi, "120Gbit/s OTDM system prototype," *24th European Conference on Optical Communications (ECOC'98))*, PD, p.41 (Madrid 1998)

[26] A. Buxens, A. T. Clausen, H. N. Poulsen, P. Jeppesen, St. Fischer, M. Dulk, E. Gamper, W. Vogt, W. Hunziker, E. Gini, and H. Melchior, "All-optical regenerative OTDM add/drop multiplexing at 40Gbit/s using monolithic InP Mach-Zehnder interferometer," *CLEO2000*, CWD3, 2000.

Bandwidth Utilisation and Wavelength Re-Use in WDM Optical Burst-Switched Packet Networks

Michael Düser and Polina Bayvel
Optical Networks Group
Department of Electronic and Electrical Engineering
University College London
Torrington Place
London WC1E 7JE
United Kingdom
Tel: +44 207 679 3843, Fax: +44 207 388 9325
Email: pbayvel, mdueser@ee.ucl.ac.uk

Key words: Optical packet networks, burst switching

Abstract: Results describing the design trade-offs in bandwidth utilisation and wavelength re-use in optical burst-switched networks are reported. The effects of traffic statistics are analysed, and a set of bounds for the network design and lightpath set-up time for dynamic network control is derived. The round-trip time required for signalling is identified to be a key parameter in optimising the network performance.

1. INTRODUCTION

In future telecommunication networks traffic with different performance requirements will be merged in the same physical layer. The challenge in the design of future packet networks is, therefore, not only the increase in traffic volumes, but also the diversity of performance measures, such as packet delay, delay jitter and packet loss rate (PLR) for the different traffic types and dynamically varying network loads.

Although static wavelength-routed optical networks (WRONs) [1, 2] are relatively simple to design and operate, they are not easily adaptable to rapidly dynamically varying traffic, in sub-millisecond timescales. In contrast, optical burst-switched (OBS) networks [3-9] can accommodate dynamic traffic variations and differential quality of service (QoS) requirements by aggregating packets at the network edge to a burst, which is assigned to a free wavelength and sent into the network once pre-defined performance parameters have been exceeded. In this paper, an OBS architecture is described that allows to achieve deterministic latencies, together with a description of design parameters. New analysis and results are reported on the trade-offs in the design of edge delay, core-to-input bit-rate ratio and the round trip time, which allow to predict bandwidth utilisation and dynamic wavelength allocation gain. These results can be used to define the bounds on the round-trip time required for the setting up of lightpaths to achieve a dynamic network control scheme.

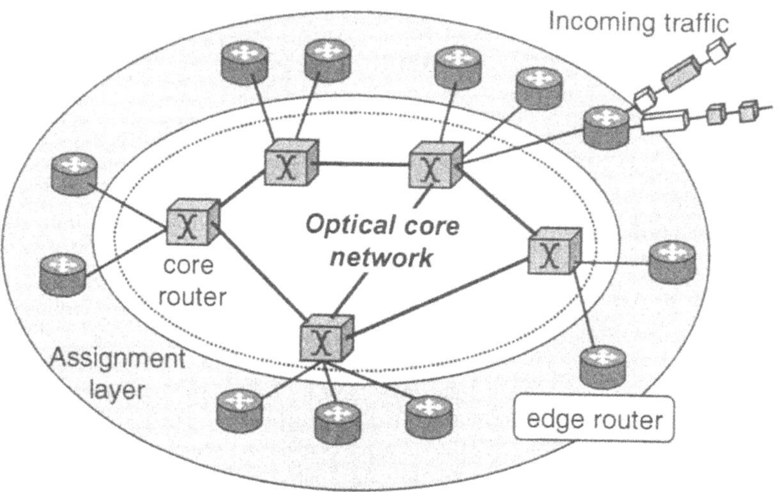

Fig. 1. Optical burst switching network architecture with electronic edge routers and optically transparent optical core

2. OPTICAL NETWORK ARCHITECTURE AND TRAFFIC MODELLING

The analysis in this work describes the network performance results presented in [7] in more detail and presents a detailed description of the assumptions made for the traffic modelling. The model is based on the OBS architecture shown in Figure 1, where electronic edge routers are connected to an optical core router. The optical core is assumed to be transparent by using dense wavelength-division multiplexing (DWDM) transport and passive routers to avoid the processing of header information and temporary buffering in the core routers. It is assumed that traffic at the edge consists of streams of packets, which are pre-sorted in the electronic edge-routers, according to their class of service (CoS) and destination into separate buffers as shown in Figure 2. Packets are not forwarded on a hop-by-hop basis as in IP networks, but are aggregated to flows, and dynamically assigned to an available wavelength, either when packets are dropped due to buffer overflow, or when a timeout signal indicates that time-critical packets have to be released to meet latency requirements. The incurred delay in the edge router, t_{edge}, is, thus, both deterministic and adjustable to meet the specific latency requirements of different traffic classes.

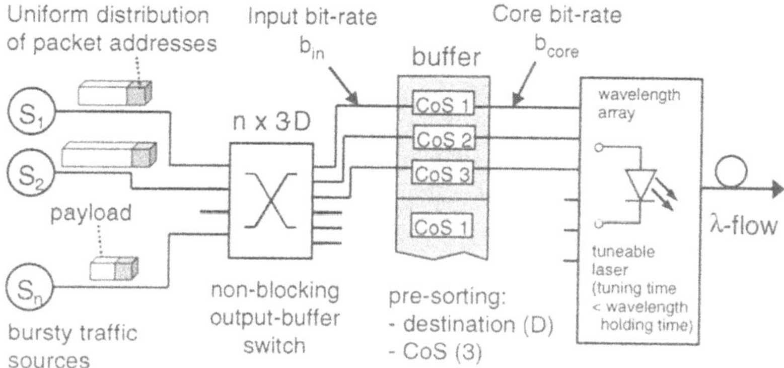

Fig. 2. Edge router model with electronic pre-sorting, non-blocking output-buffered switch and tuneable wavelength section

The implementation of the burst aggregation functionality within the network edge router requires minor changes compared to the operation of electronic IP routers. The destination address is extracted from the packet header, and the output port determined from a table look-up. Using an output-queued switch to avoid loss from blocking of the switching matrix, the IP packet would be forwarded to the respective output queue. Although purely output-queued switches are difficult to realise, combined input-output queued (CIOQ) switches with moderate internal speed-up (≤ 2) have been shown to achieve a throughput identical to output-queued switches [10].

Comparison to other OBS schemes

Optical burst switching combines the processing and buffering capabilities of electronics at the network edge with the advantage of optical DWDM transport and wavelength routing in the core. The exploitation of the advantages of both schemes allows to implement QoS differentiation in the core network. Guaranteed delays, delay jitter and PLR as QoS parameters have to be ensured even for dynamically variable traffic. A comparison with other OBS schemes, shown in Table 1, reveals that they will experience difficulties to ensure QoS guarantees.

Different schemes for the implementation of OBS networks have been suggested, but except the architecture discussed in this paper and the "Just-enough-Time" (JET) scheme [8], without allowing for different CoS. Table 1 shows a summary of the functions of different implementations schemes with respect to admission, predictability of delay over the core network, burst delivery guarantees, and QoS considerations. The following abbreviations apply:

- Horizon: 'Horizon' scheduling algorithm
- TAG: tell-and-go
- JET: just-enough-time
- IBT: in-band-terminator
- RFD: reserved-fixed-duration
- JIT: just-in-time
- MPLS: multi-protocol label switching

Except the OBS architecture with admission control discussed in this paper, all schemes suffer from potential blocking of bursts in the optical core, resulting in packet loss and non-deterministic latencies.

Table 1. Comparison of different OBS implementation schemes

Scheme	Admission	Delay	Delivery	QoS	Ref
Discussed here	Limited	Deterministic	Guaranteed	Yes	[7]
Horizon	Full	Variable	< 100%	No	[4]
TAG	Full	Variable	< 100%	No	[6]
JET	Full	Deterministic	< 100%	Potentially	[6]
IBT	Similar to TAG				[6]
RFD	Full	Variable	< 100%	No	[6]
JIT	Full	Deterministic	< 100%	No	[8]
MPLS-based	Limited	Deterministic	< 100%	Yes	[9]

3. MODELLING RESULTS AND DISCUSSION

A uniform distribution of packets addressed to other edge routers and negligible loss of packets in the output-queued switch are assumed, based on the assumption of very large buffers. The time duration before a burst is assigned to a free wavelength and released into the network is defined as the edge delay t_{edge}. As the traffic statistics defines the build up of bursts, an earlier paper focused on analysing the effects of packet inter-arrival times and size statistics [7]. Simulations for traffic loads from 0.1 to 0.8 were carried out, showing that the worst case PLR was achieved for Pareto distributed packet lengths (400 bits - 320,000 bits) and Pareto distributed inter-arrival times, the best case was achieved with fixed length packets (400 bits) and Poisson inter-arrival statistics. It was shown that for the case of Pareto statistics the queues have to be emptied with at a faster rate to avoid buffer overflow. For the given simulation parameters this rate increase in the emptying of queues was approximately a factor of 2.

Fig. 3. Source for traffic modelling with strictly alternating ON and OFF states

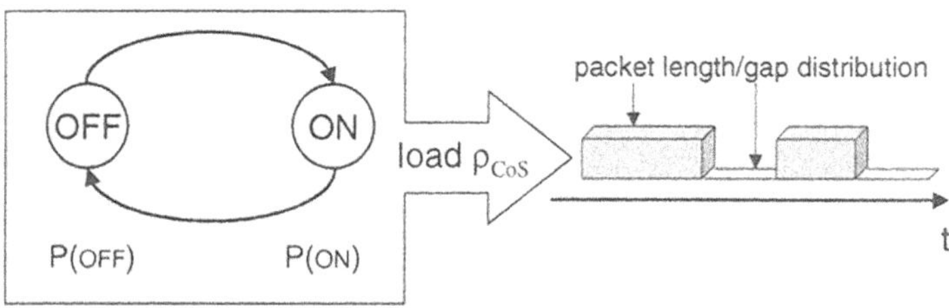

The incoming traffic is generated using ON-OFF sources as shown in Figure 3 with different probability density functions (PDF) for both the ON-state, P(ON), and the OFF-state, P(OFF) [11] to allow both variation of packet length and packet inter-arrival time. The source continuously alters between the ON and the OFF state, where different PDFs are applicable for both states to simulate a variety of traffic models.

The modelling of regular voice traffic can be described by Poisson inter-arrival time and exponential call holding times, but this model does not hold for the description of data traffic [12]. Although the correct model of data traffic is much under debate and strongly depends on the implemented protocol (e.g. IP, Ethernet), investigations report on the self-similar nature of traffic in deployed networks [13], showing that one possible implementation of this type of traffic is achieved when heavy-tailed Pareto distributions are multiplexed into the same queue. A Pareto distributed inter-arrival time t is described as

$$P(t) = \frac{\alpha \cdot A^{\alpha}}{t^{\alpha-1}} \qquad (1)$$

where $1 < \alpha < 2$, $A > 0$ and $t \geq 0$. For detailed characteristics of a Pareto distribution, see [14]. For $1 < \alpha < 2$, the resulting statistical process at the output of the buffer is *asymptotically second-order self-similar*, where self-similarity is characterised by the Hurst parameter defined as [15]:

$$H = \frac{3-\alpha}{2} \qquad (2)$$

so that $0.5 < H < 1$. The importance of the self-similar traffic characteristics in the context of networking results from its influence on the PLR and the edge delay in buffers used for burst aggregation. The edge delay, t_{edge}, is defined as the time that a burst spends in the buffer before being assigned to a free wavelength, so the queueing delay of packets entering the buffer is $\leq t_{edge}$.

To investigate the impact of traffic statistics on the queueing behaviour, simulations for both Poisson and Pareto inter-arrival statistics are shown to determine the distribution of the flow size, L_{flow}, and the resulting PLR for a finite length buffer. Here, the results for two different traffic source characteristics are applied, and their impact on the flowsize and PLR as function of tedge is analysed for:

I) fixed length packet sizes, Poisson inter-arrival distribution

II) Pareto ($\alpha = 1.5$) packet length distribution, Pareto ($\alpha = 1.5$) inter-arrival time distribution

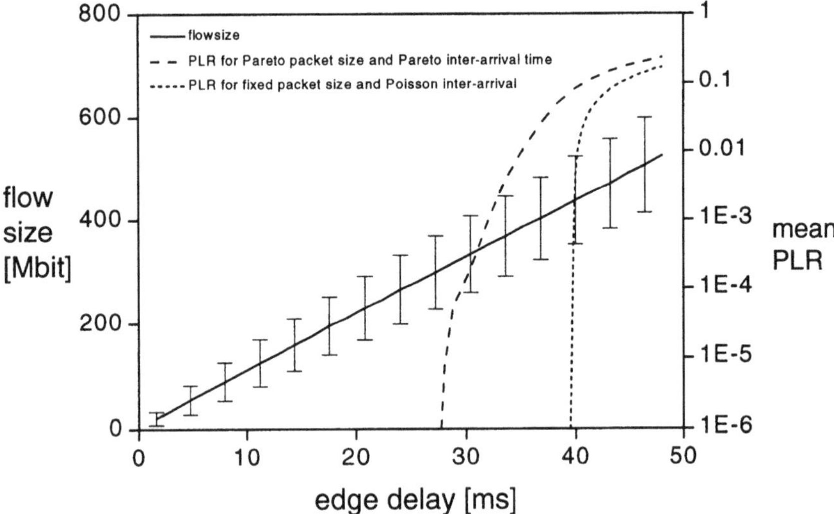

Fig. 4. Simulation results for flow size and PLR as a function of the edge delay for B = 400 Mbit and a mean input bit-rate b_{in} = 10 Gb/s. Flow size (solid) and a mean PLR for fixed packet, Poisson inter-arrival (dotted) and Pareto packet size, Pareto inter-arrival (dashed) distribution

From Figure 4 it can be seen that the mean flowsize increases linearly with t_{edge} for both traffic models. Although the standard deviation of the flowsize is negligible in case I, a significant standard deviation is observed for case II of very bursty traffic, e.g. a standard deviation of 85.73 is observed for t_{edge} = 40 ms. In the case of cbr traffic, the buffer is filled for a value of t_{edge} = 40 ms, and a PLR > 0 is expected for t_{edge} > 40 ms. The impact of the very burst traffic on the related PLR is severe: Although in case I a significant PLR is observed (> 10^{-6}) for edge delays of approximately 40 ms, i.e. nearly completely filled buffer, the same PLR is observed in case II for edge delays < 30 ms.

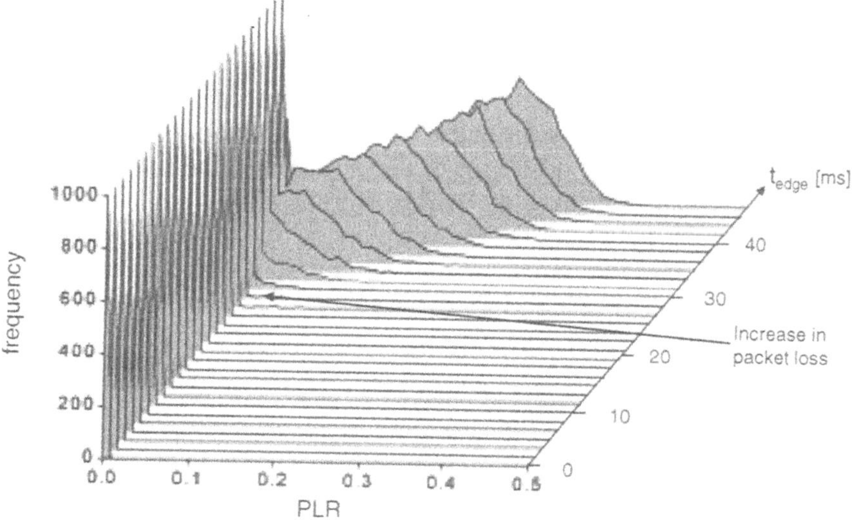

Fig. 5. Graph showing the PLR distribution as a function of edge delay. Significant deviations of the PLR from the mean is observed for Pareto traffic statistics (α = 1.5) at the input for t_{edge} > 30 ms.

Figure 5 shows the further analysis of the simulation results for the PLR, which reveal that significant deviations from the mean PLR can occur, resulting in occasionally excessive packet loss. For the implementation of QoS in OBS networks, therefore, not only the average PLR is to be taken into account, but also occasionally occurring packet loss.

In this analysis, the investigation of the effects of different statistics on the buffer overflow characteristics allows to predict and prevent excessive packet loss by introducing and adjusting threshold values, and, once a pre-defined threshold is exceeded, the burst will be assigned to the next available wavelength. As can be seen from Figure 4, an impairment of this scheme is that the burst aggregation time is limited (e.g. to < 30 ms in Figure 4 for self-similar traffic). Hence buffers are emptied at a higher rate, which increases the number of bursts in the core network at any given time.

Without anticipating results of the analysis in chapter 4, it shall be noted at this place that long burst are beneficial for an efficient use of the network resources, especially to encourage the re-use of given wavelength paths. If, as shown above, bursty traffic entering edge routers reduces the maximum burst size, this will result in a decrease of the network performance.

Therefore, bursty traffic can be represented by the superposition of noise on the mean load, and continuous bit-rate (CBR) traffic represents the limiting case. For the buffer filling process, the decisive metric is the mean filling rate, or the mean flowsize, respectively. This is a valid approach since for identical traffic loads, but different traffic statistics, the mean values of the buffer filling process are identical, but the standard deviations vary significantly. Assuming that the buffer filling process and the mean value are of the main interest, and deviations can be neglected in the first instance, a traffic model independent on the traffic statistics can be applied by using CBR traffic. This allows for the analytical description of the proposed OBS scheme for all possible traffic statistics. In this work a continuous bit-rate (CBR) traffic model was, therefore, assumed. In this case, the flow size L_{flow} increases linearly with the edge delay

$$L_{flow} = t_{edge} \cdot b_{in} \tag{3}$$

The wavelength holding time t_{WHT} denotes the period for which a given wavelength is assigned, typically in milliseconds

$$t_{WHT} = t_{RTT} + (t_{edge}/A) \tag{4}$$

where $A = b_{core}/b_{in}$ is the bit-rate ratio, the significance of this parameter is explained later together with Figs. 7 and 9. t_{RTT} is the time required for the lightpath set-up, including the propagation delay. In these calculations a value of 5 ms was assumed, based on a network with a 1000 km diameter. It is assumed that this represents a propagation delay during which burst assembly takes place. This does not include the wavelength request processing at the control node. The equivalent bandwidth used for a path is defined as bandwidth-per-wavelength

$$B/\lambda = L_{flow}/t_{WHT} \tag{5}$$

For high resource utilisation in the physical layer it is important that a given lightpath is used as efficiently as possible, the bandwidth utilisation U can thus be defined as:

$$U = \frac{B/\lambda}{b_{core}} = \frac{t_{edge}}{A \cdot t_{RTT} + t_{edge}} \tag{6}$$

and is plotted in Figure 6 for t_{RTT} = 5 ms, and b_{core} = 12.5, 40 and 100 Gb/s respectively. These values correspond to bit-rate ratios A = 1.25, 4, and 10 for b_{in} = 10 Gb/s. It turns out that for the lowest core bit-rate, 12.5 Gb/s, the utilisation is maximum. This is due to the fact that the overhead caused by signalling for wavelength path set-up is minimum in this case, resulting in wavelength-holding times t_{WHT} in the range of tens of milliseconds, whereas t_{RTT} = 5 ms was assumed to be constant. The utilisation is very low, however, for high bit-rates such as 100 Gb/s. Even for long edge delays in the range of 50 ms, the utilisation is approximately 50 % only.

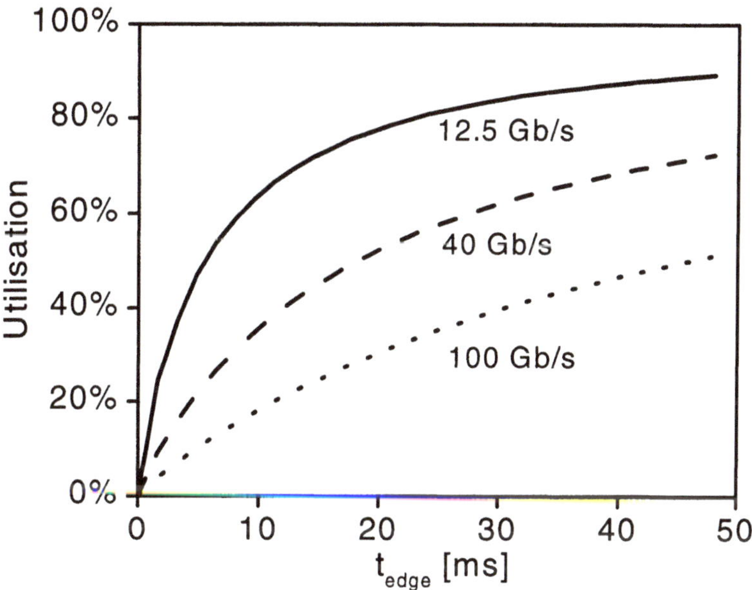

Fig. 6. Utilisation U as a function of t_{edge} for t_{RTT} = 5 ms, b_{in} = 10 Gb/s and b_{core} = 12.5, 40, and 100 Gb/s

To investigate the influence of the edge delay and the bit-rate ratio A on the utilisation for a wider range of parameters, it is plotted in Figure 7 for 0 ms $\leq t_{edge} \leq$ 200 ms, and $0 \leq A \leq 100$. The range of values for t_{edge} and L_{flow} is based on previous simulation results of different packet size and packet inter-arrival times distributions, described in chapter 3.

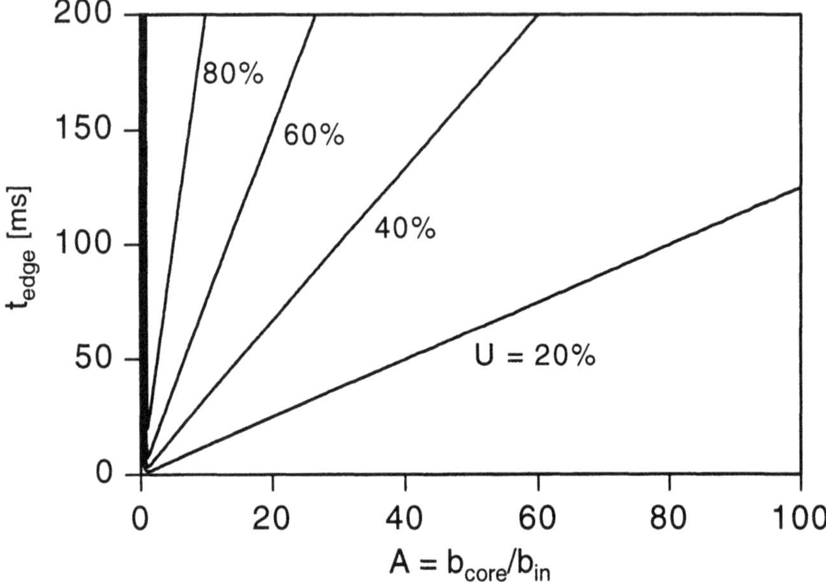

Fig. 7. Bandwidth utilisation, U, as a function of t_{edge} and A for a round-trip time $t_{RTT} = 5$ ms

As $b_{core} \gg b_{in}$ can be assumed for a high-speed optical core, $t_{WHT} \ll t_{edge}$. In this case the time required to transmit a burst and, therefore, the time for which a given wavelength is used is much shorter than the edge delay. In the case of dynamic wavelength allocation an unused wavelength can be assigned to another edge router, and the resultant increase in the wavelength re-use is denoted by a wavelength re-use factor, RUF, defined as:

$$RUF = \frac{t_{edge}}{t_{WHT}} = \frac{A \cdot t_{edge}}{A \cdot t_{RTT} + t_{edge}} = A \cdot U \tag{7}$$

It is plotted in Figure 8 as a function of t_{edge} for $t_{RTT} = 5$ ms, $b_{in} = 10$ Gb/s and $b_{core} = 12.5$, 40, and 100 Gb/s, respectively.

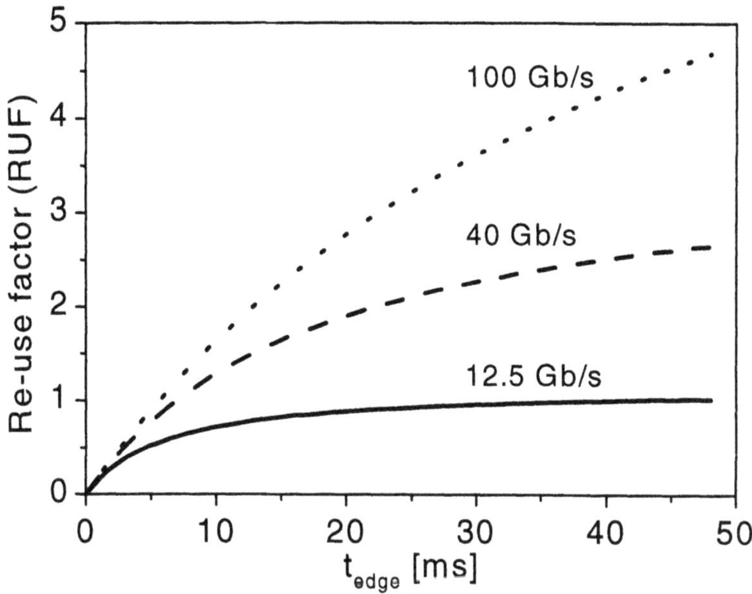

Fig. 8. Wavelength re-use factor RUF as a function of t_{edge} for $t_{RTT} = 5$ ms and $b_{core} = 12.5$, 40, and 100 Gb/s

The wavelength re-use factor is plotted for a wider range of parameters in Figure 9 for 0 ms $\leq t_{edge} \leq$ 200 ms, $0 \leq A \leq 100$, and $t_{RTT} = 5$ ms[1].

From the above analysis it becomes evident that three parameters dominate the performance parameters of OBS networks:

- Edge delay t_{edge}
- Bit-rate ratio A
- Signalling overhead, characterised by the round-trip time t_{RTT}

Of particular importance is the bit-rate ratio A since in core networks the core bit-rate b_{core} will be fixed in most cases. So, for given input bit-rate b_{in}, the bandwidth utilisation U and the wavelength re-use RUF can be directly determined from Figs. 7 and 9 as a function of t_{edge} and t_{RTT}.

[1] It should be noted that the wavelength re-use factor as defined here refers to the efficiency with which a given wavelength channel is used to satisfy other source-destination requests. It does not take into account the wavelength allocation algorithm or its efficiency in wavelength re-use to minimise the wavelength requirements.

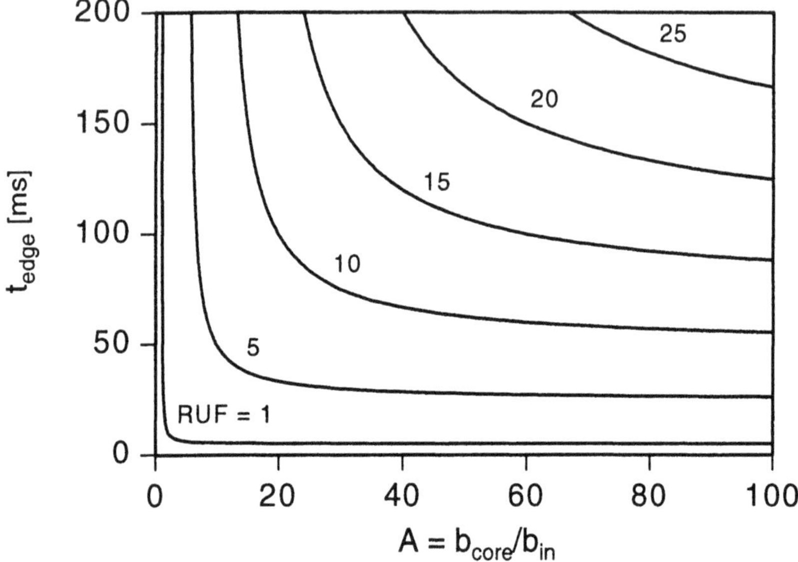

Fig. 9. Wavelength re-use factor, RUF for a round-trip time $t_{RTT} = 5$ ms, as a function of t_{edge} and A

These results clearly show that with this network approach different types of traffic could be accommodated: low delays, required for time-critical types of traffic, are achievable, but with low values of utilisation, U, especially for large bit-rate ratios. It is possible to design a network with U > 80 % for A > 10 and delays > 50 ms. From Figure 9 it can be seen that RUF reaches maximum values with both increasing t_{edge} and A. For comparison, Figure 9 shows also the equivalent to the case of a statically wavelength-routed optical network (WRON) where RUF = 1. For values of RUF < 1 the network would theoretically require more wavelengths than in a static WRON, and this represents the region of network instability where the total input load exceeds the network throughput.

The time to set up a lightpath is t_{RTT}, required for signalling between edge routers and the network control element, either central or distributed. To ensure that RUF > 1 as defined in (6),

$$t_{RTT} < \frac{A-1}{A} \cdot t_{edge} \Leftrightarrow t_{RTT} < t_{edge} \quad \text{for } A >> 1 \tag{8}$$

It is important to note that in order to achieve efficient wavelength re-use, the lightpath set-up time must be as small as possible, and for a fixed t_{edge}, the upper bound in RUF is given by $RUF_{max} = A$.

The proposed architecture can, therefore, accommodate traffic with wide range of delay requirements over the same network. Very delay sensitive traffic, such as voice, would only be queued for 10 - 20 ms before being assigned to a free wavelength with the penalty of a relatively low utilisation (< 50 %) and re-use factor (< 5) as can be seen from Figures 7 and 9. A lower bound for the required edge delay is the round-trip time t_{RTT}. However, less delay sensitive traffic such as data can be routed over the network in parallel, but with longer edge delays allowing for a higher re-use of wavelengths than in the case of latency-sensitive traffic.

4. SUMMARY

New analysis of optical burst-switched networks is described which allows for the quantification of performance parameters key to network design. It was shown that for the limiting case of CBR traffic an analytical model for the edge router can be derived, and that this architecture allows to achieve a range of edge delays to satisfy the latency requirements of different traffic types. Bandwidth utilisation and wavelength re-use were introduced. These allow to quantify accurately the gain with dynamic wavelength allocation, which has the two-fold benefit of reducing wavelength requirements and enabling the network to respond to variable traffic demands. The time required for lightpath path set-up, t_{RTT}, is a lower bound on the achievable edge delays and must be minimised by fast RWA algorithms.

Acknowledgements
The authors would like to express their gratitude to Prof. S. Ferguson (Marconi Communications), Dr. D. Wischik (Cambridge), Dr. R. I. Killey (UCL), E. Kozlovski (UCL), A. Myers (UCL), and I. De Miguel (UCL, on leave from University of Valladolid, Spain) for invaluable comments. Financial support from Marconi Communications, UK EPSRC, and the Royal Society is gratefully acknowledged.

REFERENCES

[1] S. Baroni, P. Bayvel, R.J. Gibbens, S.K. Korotky, "Analysis and design of resilient multifiber wavelength-routed optical transport networks," J. Lightwave Technologies **17**, 743-758 (1999)

[2] S. Baroni, P. Bayvel, "Wavelength requirements in arbitrarily connected wavelength-routed optical networks," J. Lightwave Technology **15**, 242 - 251 (1997)

[3] J.S. Turner, "WDM Burst Switching for Petabit Data Networks," Tech. Dig. OFC 2000, paper WD2-1, 47 - 49 (2000)

[4] J.S. Turner, "Terabit Burst Switching," J. High-Speed Networks **8** (1), 3-16 (1999)

[5] C. Qiao, "Labeled Optical Burst Switching for IP-over-WDM Integration," IEEE Comm. Mag. **38** (9), 104 - 114 (2000)

[6] C. Qiao, M. Yoo, "Choices, Features and Issues in Optical Burst Switching," J. High-Speed Networks **8** (1), 69 - 84

[7] M. Dueser, E. Kozlovski, R.I. Killey, P. Bayvel, "Design Trade-Offs in Optical Burst Switched Networks with Dynamic Wavelength Allocation," Tech. Dig. ECOC 2000, Vol. 2, paper Tu 4.1.4, 23 - 24 (2000)

[8] J. Y. Wei, J. L. Pastor, R. S. Ramamurthy, Y. Tsai, "Just-in-Time Optical Burst Switching for Multiwavelength Networks," IFIP Broadband '99, Hong Kong, Nov. 1999, 339 - 352

[9] S. Verma, H. Chaskar, R. Ravikanth, "Optical Burst Switching: A Viable Solution for Terabit IP Backbone," IEEE Network **14** (6), 48 - 53 (2000)

[10] S.-T. Chuang, A. Goel, N. McKeown, B. Prabhakar, "Matching Output Queueing with a Combined Input/Output-Queued Switch," IEEE J. on Select. Areas in Comms. **17**, 1030 - 1039

[11] N. Likhanov, "Calculating Cell Loss Probabilities for ON-OFF Sources in Large Unbuffered Systems," IEEE Conf. on Communications (ICC'97), Vol. 2, 560 - 564 (1997)

[12] S. Keshav, "An Engineering Approach to Computer Networking: ATM Networks, the Internet, and the Telephone Network," Addison Wesley Longman Inc., Reading (MA), 1997

[13] L. Leland, M. S. Taqqu, W. Willinger, D. V. Wilson, "On the Self-similar Nature of Ethernet Traffic (Extended Version), " IEEE/ACM Trans. on Networking 2(1), 1 - 15 (1994)]

[14] V. Paxson, S. Floyd, "Wide Area Traffic: The Failure of Poisson Modeling," IEEE/ACM Trans. on Networking **3**, 226 - 244 (1995)

[15] M. Parukelar, A. M. Makowski, "Tail probabilities for a multiplexer with self-similar traffic," Fifteenth Annual Joint Conference of the IEEE Computer Societies (INFOCOM'96), Vol. 3, 1452 - 1459 (1996)

Traffic Characterisation using Optical Based Packet Switches with Poisson and Fractal Traffic Sources

David Tarongí[*], Dr. Daniel Rodellar[°], Jaume MasipTorner[+], Josep Solé-Pareta[+], Sebastiano Borgione[+]

*TCOM, EPFL, Telecommunications Laboratory, Ecole Polytechnique Federale de Lausanne, Switzerland.

°SWISSCOM AG, Corporate Technology, Switzerland.

+ CCABA, UPC, Integrated Broadband Communications group, Universitat Politècnica de Catalunya

1 INTRODUCTION

Local area networks traffic and Internet traffic have been analysed for the last 25 years to obtain accurate characteristics that will be used in traffic models. Traditionally, traffic sources were simulated using a Poisson process model, that had the advantage of being easily generated. But as real traffic has been studied, it has been shown that real traffic properties differ from Poisson traffic properties. When designing new communication systems, different alternatives have to be compared with respect to performance, as a first step. In particular when no implementations is available, modelling is often the only way to evaluate quantitatively the alternatives. Besides a model for the system operation, there is a need for traffic models describing the characteristics of the load of the system resource.

Current Internet backbones require high-capacity transport networks as well as high-capacity packet routers in order to properly handle large amounts of traffic fluxes and the optical technology is taking a significant role to achieve this objective. Fiber transport capacities have reached the terabit per second range thanks to wavelength multiplexing techniques. To avoid capacity gaps with the Wavelength Division Multiplexing (WDM) transport network, packet routers need throughput capacities of the same order of magnitude [1]. On one hand, the introduction of optical functions provides the possibility of reaching high throughputs avoiding bottlenecks coming from the high electronic router complexities. On the other hand, all optical routers match the packet routing and WDM network capabilities in a single layer. Optical buffers have special characteristics with respect to conventional electronic buffers, and thus need a special analysis [2], as they take profit of the advances in optical technology to implement all optical routing and buffering functions. The complexity of the optical buffer is independent of the total buffer capacity and its performance highly depends on the internal delay line configuration.

In this paper we aim to characterise the traffic that goes through an optical buffer to find out a delay line configuration that minimises the buffer complexity while maintaining the overall router performance. An optical buffer has been simulated with Poisson and self-similar traffic sources types. The traffic generators are completely characterised in terms of their packet inter-arrivals probability and their packet size distribution. These sources fill the optical buffer queues and the output traffic is then analysed.

2 OPTICAL PACKETS BUFFERING

All optical routers will need optical buffers to solve packet contention. An all optical router with *N* input/output fiber ports is considered and the set of *D* transmission delay lines inside the optical router is called the optical buffer. A possible buffer configuration directly connects each subset of *D*/*N* delay lines to one output. The resulting all-optical router configuration is output buffered.

Figure 1 shows an output optical buffer made up with 5 fiber delay lines. The number of optical delay lines in the output buffer determines the number of incoming packets can be loaded simultaneously towards a required output. The length of the longest delay line determines the total output buffer capacity. Note that the number of parallel accesses determines the complexity of the optical buffer and it is independent of the optical buffer capacity, which is the longest delay line. No optical packet synchronisers are considered and all fibers only contain a single wavelength for shake of simplicity (no WDM).

Each fiber delay line in the subset can have an arbitrary length. Each delay line introduces a different time shift on the scheduled output time. Packets that contend simultaneously for the same output avoid collision using a different time shift path. Incoming packets are always routed to the shortest available delay lines. One delay line can accept an incoming packet when:

- There is no other packet that is being loaded in the same delay line, and,
- Packets loaded on other delay lines are not scheduled with the same output time.

The time shift model in Figure 1 summarises all the available time shifts where incoming packets can be scheduled, one for each delay-line. According to this model, optical buffers with the same complexity (number of delay lines) can have different access distributions, i.e. different time shifting patterns. Figure 2 shows three possible configurations of an optical buffer, one with fixed steps, one with double fixed steps, and one with increasing step lengths.

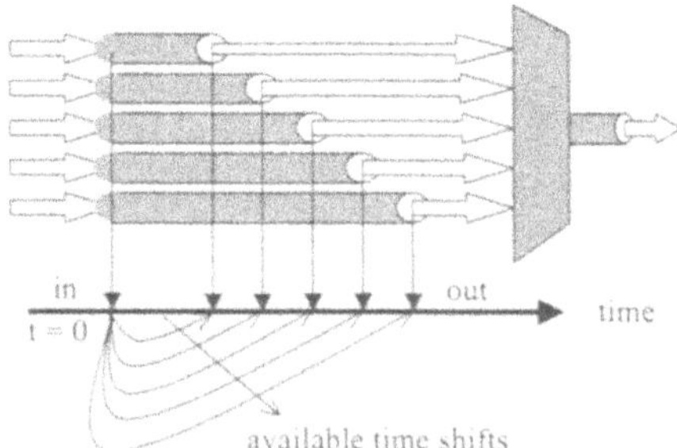

Figure 1: An optical buffer made up with 5 optical packets. The length of the longest delay line determines the total capacity of the optical buffer.

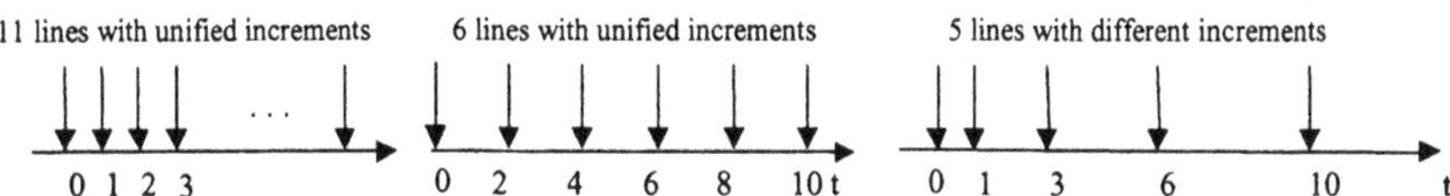

Figure 2: Possible configurations of an optical buffer

3 TRAFFIC-DROP PRIORITY

When the number of incoming packets increases to the congestion point, some of these packets will be lost. To determine which of them are going to be discarded, some traffic-drop priority settings have been defined. The traffic-drop priority scheme works as follows: if the network is not congested there is no difference in how the nodes treat each packet (e.g. no packet discard); and the packets are forwarded in the same order they are received. When congestion occurs best effort packets are discarded first, mitigating the congestion and leaving resources for the assured service packets.

An example of how this strategy can be implemented is given in Figure 3a. When a packet is received, the output port of the OPN node (and the associated queue) is determined. If the corresponding queue depth exceeds a given threshold N1 (congestion is imminent), best effort packets are discarded while assured packets are correctly sent into the queue.

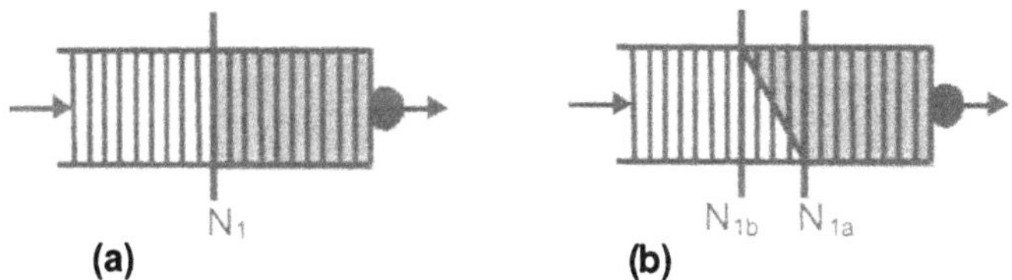

Figure 3: Output buffer with drop priority based on: (a) a threshold, (b) a RED policy.

This threshold mechanism can also be associated with a Random Early Discard (RED) strategy, providing a real advantage in the presence of optical packets carrying TCP/IP datagrams (Figure 3.b). When the buffer occupancy reaches the threshold N1a, the discard probability for a best effort packet begins to be more than zero and its value is increased to 1 when the buffer occupancy reaches the N1b value. In excess of this value only the assured quality packets can be buffered and all the best effort packets are discarded. It is worth noticing that in both cases (plain drop priority and RED) packets are discarded at the input of the OPN node: once packets have been entered the node, they undergo no scheduling modifications. This fits with the particular structure of optical buffers at OPN nodes, in which deleting a packet that has already entered the buffer is difficult and inadvisable.

4 TRAFFIC SOURCES

Three different types of traffic sources have been implemented [3] to feed the optical buffer. The first scheme is a synchronous and periodic source that is absolutely not realistic but can be very useful for the programmers to follow easily the packets inside the optical buffer, and to determine the accuracy of the implemented model. Another scheme implemented is a Poisson model for the packet arrivals, and even if it is not a realistic model for data network traffic, it has been used traditionally for traffic generation. A realistic model, based on the superposition of several Fractal Renewal Processes (FRP) is also implemented and characterised.

Each packet is generated by the sources at a given time, according to the generation distribution selected, and with a given packet size. The packet sizes can be either fixed or with a distribution taken from real Ethernet traces [4].

5 TRAFFIC CHARACTERISATION

Several simulations are carried out with different traffic sources and the traffic is characterised at the input sources and the output of the optical buffer. Results of the simulation using each generation model for the packet inter-arrivals and a packet size distribution from real Ethernet traces are presented in the following paragraphs. The generated traffic is analysed. In the case of a Poisson process the mean and the variance of the traffic are related and correspond to the expectation defined by the number of packets per second generated. In the case of a self-similar (or fractal) process [5][6], a variable X is exactly self-similar with scaling parameter H (Hurst parameter), if for all levels of aggregation m,

$$X^{(m)} = m^{H-1} X$$

Where $X^{(m)}$ is the average of m consecutive values of the variable X. The resulting log-log representation of Var($X^{(m)}$) versus m is called variance-time plot. Intuitively, the critical characteristic of self-similar traffic is that there is no natural length of a "burst". At every time scale ranging from a few milliseconds to minutes and hours, similar-looking traffic bursts are evident. Aggregating streams of such traffic typically intensifies the self-similarity ("burstiness") instead of smoothing it. The words fractal and self-similar must be read in this text as synonymous.

To perform the simulations:

- The buffer was implemented with 5 delay lines, with delays going from 10 ms to 50 ms in steps of 10 ms.
- The discrimination criterion followed in case of congestion is described in the Traffic Drop Priority section, with 1 class of service. The min-max size parameters (N_{1a} and N_{1b}) were 4~5.
- The channel bitrate at the output of the buffer was 10Mbit/s, to be coherent with the packet size histogram data obtained from real Ethernet traces.

- The observed packet size probability distribution function is represented in the Figure 4. From this distribution, the mean packet size is 2.975 Kbit.
- The simulation time was set to allow the generation and treatment of 1000000 packets.
- In a first approach, only Poisson traffic generated by the aggregation of 10 sources was considered. In a most realistic situation, the simulations were performed on self-similar traffic provided by 10 sources. In each one of the input sources, the fractal traffic is obtained by the superposition of several Fractal Renewal Processes (FRP).

For low load scenarios the input and output traffic characteristics were almost identical, since the packet reordering and losses were minimal. Therefore, only the results for high load cases, very close to the congestion point and even with some packet losses are presented.

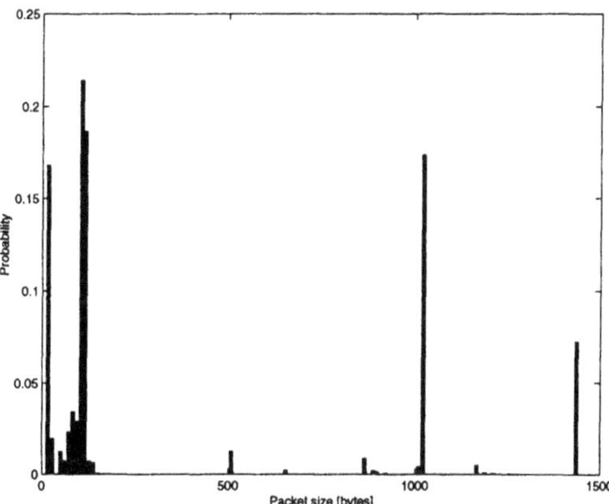

Figure 4: Probability of packet sizes, obtained from real Ethernet traces.

5.1 Input Traffic Traces

In this section the input traffic parameters for the simulations with Poisson and fractal traffic will be described. In both cases the number of sources was set to 10, and the simulation time was about 407 s (1000000 packets in total). The simulated mean incoming bitrate was about 7.3 Mbit/s (7.291 Mbit/s in case of Poisson, and 7.308 Mbit/s for fractal case), which leads to an utilization factor of approx. 73% (input traffic: 0.73 Erlang). The interarrival time probability distribution for each one of the 10 sources is depicted in figures 5.a (Poisson traffic) and 5.b (fractal traffic). We can notice that, in case of Poisson traffic, the curve is well fitted by an exponential function. The mean value for the interarrival time, considering the aggregation of all the sources was 0.4 ms (4 ms for each source taken independently).

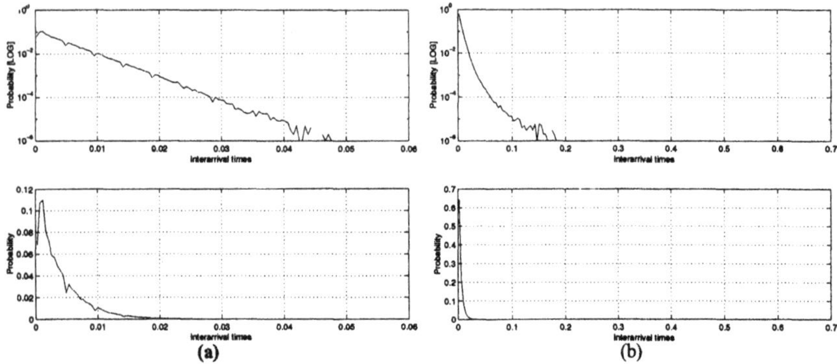

Figure 5: Interarrival time probability distribution, in logarithmic and linear scales, for each: (a) Poisson traffic source, (b) fractal traffic source.

To verify the self-similarity of the traffic provided by each fractal source, the Hurst parameter has been calculated, and the variance-time plot has been obtained from the input traces and represented in the Figure 6.

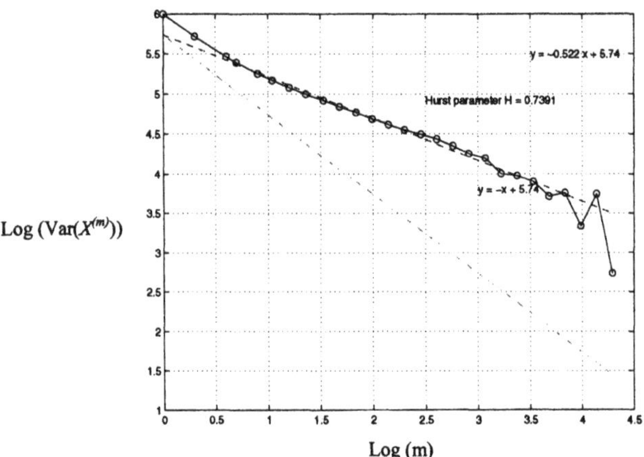

Figure 6: Variance-time plot of the traces. The lower dotted line corresponds to the variance-time plot expected for Poisson traffic.

In the Figure 7 a visualisation of the input traffic (packet generations for all nodes) for both traffic types during the first 10 ms of the simulation is provided. The packet generations are separated by sources on the vertical axis. We can appreciate the conflicts, which appear when some sources transmit their packets simultaneously. As stated in section 5.1, in both cases the traffic parameters were identical. The higher packet density observed in the Figure 7b shows the "burstiness" of this traffic type, taking into account that the interval shown is very short compared to the whole simulation.

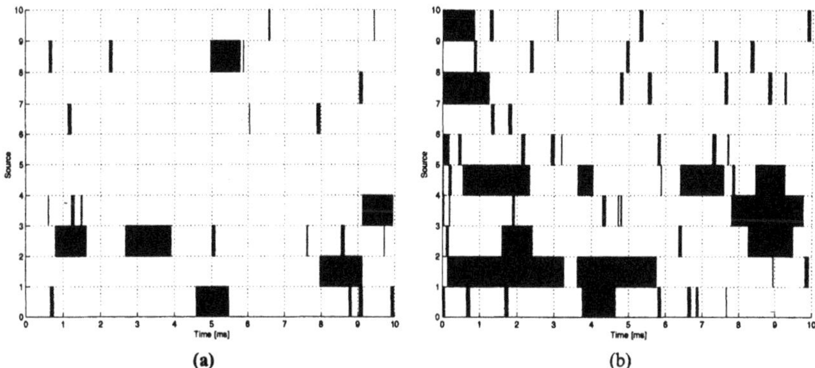

Figure 7: Packets visualisation by sources at the buffer input, considering: (a) Poisson traffic sources, (b) fractal traffic sources.

5.2 Output Traffic Traces

In this section, the results obtained from the simulations with the input traffic parameters specified before are presented. The Figure 8 provides the packets visualisation at the buffer output for the first 25 ms, classified by sources like in Figure 7. As we can notice, there are no conflicts at the buffer output, since the packets have been properly delayed or lost. The most important parameters for the characterisation of the traffic output are presented in the Table 1.

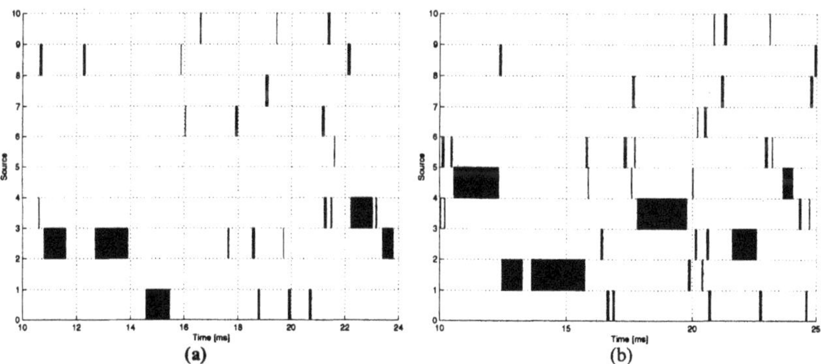

Figure 8: Packets visualisation at the buffer output, considering: (a) Poisson traffic sources, (b) fractal traffic sources.

Parameter	Poisson traffic	Self-similar traffic
Output traffic (Erlang)	0.6103	0.6095
Lost traffic (Erlang)	0.1195	0.1213
Lost packets	92329 (9.2%)	94964 (9.5%)
Mean packet size	2742	2736

Table 1: Output traffic parameters, corresponding to the simulation with the input trace parameters provided in section 5.1

As we can see, the number of packets that have been discarded is slightly higher in the case of fractal traffic, due to the inherent burstiness of this more realistic traffic model. However, we have to notice that the difference is almost insignificant.

Figure 9 shows the interarrival time probability distribution at the buffer output. It is possible to observe the effects of the buffer on the traffic characteristics, more significant in the case of self-similar traffic. In this case, the distribution becomes even similar to a Poison one.

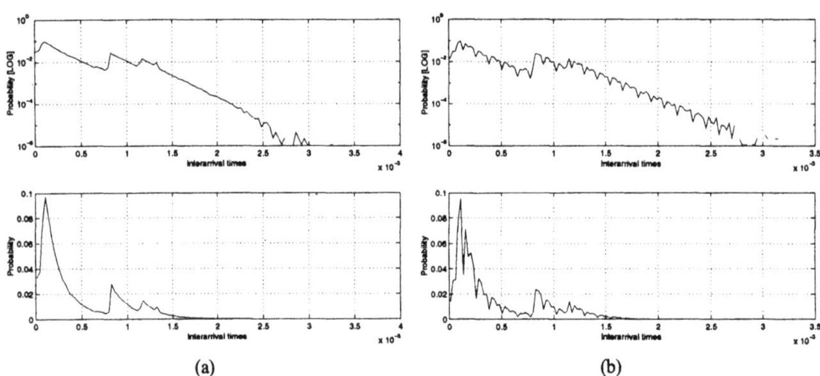

Figure 9: Interarrival time probability distribution, in logarithmic and linear scales, at the buffer output for: (a) Poisson input traffic, (b) fractal input traffic

The packet sizes distributions at the buffer output for both traffic models are represented in Figure 10. Comparing both graphics with the one at the input (Figure 4) it is possible to notice that the biggest packets have been more often discarded than the smallest ones. Therefore, the mean packet size at the output is around 2.740 Kbit (2975 Kbit at the buffer input). Once more, the effect is very similar for both traffic models.

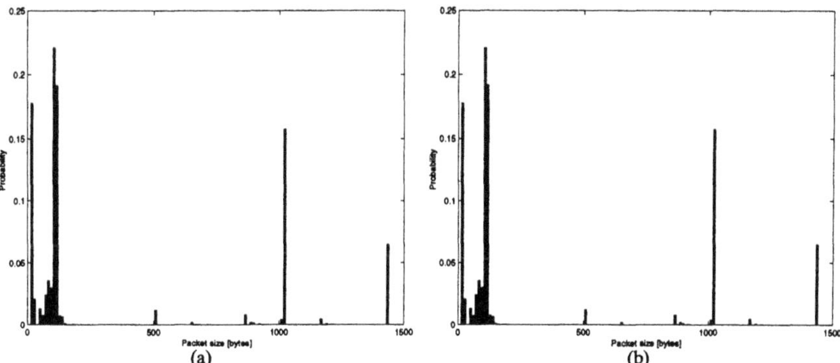

Figure 10: Packet size distribution at the buffer output, for the simulations with: (a) Poisson input traffic, (b) fractal input traffic.

6 CONCLUSIONS & FUTURE WORK

A simple delay line configuration and more sophisticated configurations have been presented in this work. An optical buffer has been simulated with Poisson and fractal source types. The results for both cases have been shown.

The traffic generators are completely characterised in terms of their packet inter-arrivals probability and their packet size distribution. These sources fill the optical buffer queues and the output traffic has been analysed.

The impact of the optical buffer on the traffic characteristics is similar in both Poisson and self-similar input traffic cases. This shows that at high load scenarios the Poisson traffic model is suitable to be used in the characterisation of this kind of devices.

This study has provided a first approach in the characterisation of traffic using optical packets switching devices. Some improvements have to be taken into account in a future work:

- The buffer delay lines have to be designed to be the optimum and a statistical study has to be carried out to find out the optimal values for these delay lines (number of lines, delay distribution...). A first approach for this study could take into account the packet size distribution characteristics.
- A model of the optical for different Class of Services packets has to be created, and then exploited.
- The traffic generator will have to include a new parameter that will select the seeds to generate the packets, to allow several trace generations with different initialisation seeds.
- Different wavelengths must be considered, as well as the possibility of wavelength conversion, to be able to deal with WDM (Wavelength Division Multiplexing) devices.

7 ACKNOWLEDGMENTS

This work has been supported and carried out in the framework of the COST Action Group 266 (Advanced Infrastructure for Photonic Networks) and the CICYT (*Comisión Interministerial de Ciencia y Tecnología*) under contract TIC99-0572-C02-02.

Authors would like to acknowledge Prof. Pierre-Gérard Fontolliet and Dr. Cristian Bungarzeanu from the TCOM laboratory for their continued guidance and support. We would like also to thank the Swiss Federal Office of Education and Science for their financial support.

8 REFERENCES

[1] D.K. Hunter, I. Andonovic, M.C. Chia, "*Optical Buffers for Multi-terabit IP Routers*". 37th Annual Allerton Conference on Communication, Control and Computing. 22-24, September. 1999.

[2] Jaume Masip Torné, Josep Solé Pareta, Sebastiano Borgione, Bruno Bostica, Marco Burzio, "*Providing Differentiated Service Categories in Optical Packet Networks*", International Teletraffic Congress, 7-11 June 1999, IEE, 1999.

[3] D. Rodellar, "*Performance Analysis of Multi-Channel Protocols for Optical Local Area Networks Exploiting Wavelength Division Multiplexing*". Thesis of the Ecole Polytechnique Fédérale de Lausanne, 1999

[4] W. Leland, M. Taqqu, W. Willinger, and D. Wilson, "*On the Self-Similar Nature of Ethernet Traffic (Extended Version)*", IEEE/ACM Transactions on Networking, 2(1), February 1994, pp. 1-15.

[5] W. E. Leland and D. V. Wilson, "*The measurement techniques used in making the traces are described in High time-resolution measurement and analysis of LAN traffic: Implications for LAN interconnection*", Proc. IEEE INFOCOM'91, April 1991, pp. 1360-1366

[6] H. J. Fowler and W. E. Leland, "*Local Area Network Traffic Characteristics, with Implications for Broadband Network Congestion Management*", IEEE JSAC, 9(7), September 1991, pp.1139-1149

Traffic Load Bounds for Optical Burst-Switched Networks with Dynamic Wavelength Allocation

Ignacio de Miguel[1], Michael Düser, Polina Bayvel
Optical Networks Group, Department of Electronic and Electrical Engineering,
University College London, Torrington Place, London WC1E 7JE, UK,
Tel.: +44 20 7679 3843, Fax: +44 20 7388 9325,
E-mail: {imiguel, mdueser, pbayvel}@ee.ucl.ac.uk

Key words: Optical Networks, Optical Burst Switching, Edge delay, Routing and Wavelength Assignment, Traffic load

Abstract: The maximum traffic load that can be supported by a wavelength division multiplexed (WDM) optical burst switched (OBS) network with dynamic wavelength allocation is studied. It is shown that it depends on the requirements of the class of service and on the efficiency of the dynamic routing and wavelength assignment (RWA) algorithm employed. Two methods to build the bursts are presented as well as their influence on the maximum traffic load that can be supported.

1. INTRODUCTION

There are a number of approaches for the design of optical networks. In Wavelength-Routed Optical Networks (WRONs) all-optical channels are established between pairs of nodes by means of *lightpaths* [1]. These networks have been widely investigated (see [2,3] and references there). Although they are relatively easy to design and manage, their main drawback is that the bandwidth which can be provided by established lightpaths is usually much higher than that required to accommodate the

[1] I. de Miguel is also with the Dpt. of Signal Theory, Communications and Telematic Engineering at University of Valladolid (Valladolid, Spain).

average traffic loads between the source-destination nodes, so that the capacity provisioning in the network may be inefficient. Moreover, they have difficulties to adapting to dynamically varying traffic demands, especially if these change on sub-millisecond timescale. A technique that has recently received a lot of attention as a possible solution to that problem is Optical Burst Switching (OBS) [4-12]. With this approach, the network consists of edge routers and core nodes. Packets are buffered in the source edge routers to form a *burst* (also known as *flow*). Then, a control packet is sent and afterwards, the burst is transmitted into the core network. There are several approaches to OBS. In this paper, we study a model called Optical Burst Switching with Dynamic Wavelength Allocation (OBS-DWA) [9-12]. A dynamic routing and wavelength assignment (RWA) algorithm is used to establish a lightpath for the transmission of the burst. Once the lightpath has been established an acknowledgement is sent to the source node, then the burst is transmitted and finally, the lightpath is deleted. This function could be fulfilled by either a centralised or distributed wavelength assignment control. The former option is considered in this paper. The main advantage of this OBS approach is that no buffers are needed in the core nodes and it is possible to ensure a deterministic end-to-end delay for time-critical applications. There are other important parameters, which must be analysed to provide different classes of service, namely the delay jitter and the packet loss rate. It is also a key question to determine the minimum number of wavelengths needed to support a set of quality of service parameters and traffic load.

In this paper, we focus on the analysis of trade-offs between the edge delay (which is related to end-to-end delay), the wavelength requirements and the traffic load of the network in the context of Optical Burst Switching with dynamic wavelength allocation. Two different methods to constructing the bursts are presented and the edge delay is analysed for both of them. Then, we determine the maximum traffic load that the network is able to support due to the imposed edge delay constraints. The wavelength requirements for quasi-static WRONs and OBS-DWA networks are also compared.

2. MODEL AND PARAMETER DEFINITION

In this section, the network model and a set of parameters used throughout the paper are presented. This work is based on the OBS network architecture presented in [9-11], for a network with N edge routers. It is assumed that each edge router has a buffer for every destination and class of service, and that the packet loss rate (PLR) in the edge routers is zero. (More efficient schemes can be applied, although they increase the PLR [12].) When a threshold in the amount of stored data or a timeout (determined by the delay requirements of the class of service) is reached, a request is sent to a centralised control node to establish a lightpath that will be used for the transmission of the burst.

The set of *possible connections* in the network is defined as the set of different {source, destination, class of service} groups so that the source and the destination edge routers are not equal. Only one unidirectional lightpath can be established between a pair of nodes for every class of service. (Throughout this paper all connections referred are unidirectional.) Therefore, considering C classes of service, the number of possible connections is $N(N-1)C$. In this paper, the analysis is limited to a network with one class of service, and so, the number of possible connections is $K = N(N-1)$. Since dynamically varying traffic is considered, a possible connection will alternate between two states. It will be *established* during periods of T_{WHT} time units (which means that a lightpath is joining the source and the destination edge routers), and *idle* (or not established) during periods of T_{idle} time units, where T_{WHT} and T_{idle} are random variables having a mean of $\bar{t}_{WHT}$ and $\bar{t}_{idle}$ respectively.

The idle time represents the elapsed time between two effective connections between a source-destination pair, but the network designer cannot directly control this parameter. The decisions faced by the network designer are the establishment of thresholds and timeouts for the transmission of the requests, and the design of the control node (scheduling of requests, number of processors, choice of the RWA algorithm). Different choices of these will lead to a different value in the average idle time. In this paper, the idle time and its optimum value as a function of the traffic load are considered only. With these assumptions, the maximum traffic load that the OBS-DWA network can support for a desired edge delay can be obtained, and these results are independent of the design options explained above.

For every possible connection the following parameters are defined:

- *Average input bit rate between source-destination nodes* ($\bar{b}_{in}$).
- *Capacity (bit rate) of a lightpath* (b_{core}).
- *Average size of the bursts or flows* ($\bar{L}_{flow}$).
- *Average wavelength holding time* ($\bar{t}_{WHT}$), the average duration of a lightpath. It is defined as the time elapsed from when the lightpath (i.e., the route-wavelength pair) is reserved until the deletion of the lightpath,

$$\bar{t}_{WHT} = \bar{t}_{ack} + \bar{L}_{flow} / b_{core} + \bar{t}_{prop}, \tag{1}$$

where $\bar{t}_{ack}$ is the time for the transmission of the acknowledgement from the control node to the edge router, $\bar{L}_{flow}/b_{core}$ is the time needed for the transmission of the burst, and $\bar{t}_{prop}$ is the propagation time (the lightpath cannot be deleted until the last bit transmitted reaches the destination node, therefore the propagation time is included in the equation).

- *Round trip-time* ($\bar{t}_{RTT}$) is the amount of time that the established lightpath is not used to transmit data. Hence,

$$\bar{t}_{RTT} = \bar{t}_{ack} + \bar{t}_{prop}\,. \tag{2}$$

- *Bandwidth utilisation* (*U*) represents the lightpath utilisation efficiency, that is, the average fraction of time that an established lightpath is used for data transmission.

$$U = \frac{\bar{L}_{flow}/b_{core}}{\bar{t}_{RTT} + \bar{L}_{flow}/b_{core}} = \frac{\bar{L}_{flow}}{b_{core}\,\bar{t}_{WHT}}\,. \tag{3}$$

- *Fraction of time that a possible connection is established* (ρ). As previously stated, a possible connection alternates between two states. It is *idle* during periods of random time with average $\bar{t}_{idle}$, and *established* during periods of average duration $\bar{t}_{WHT}$ (*Figure 2*).

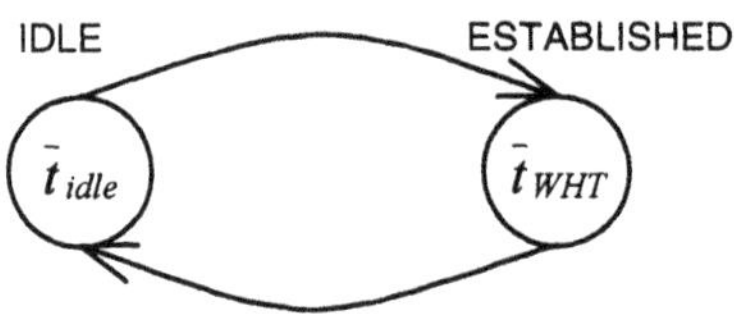

Figure 2. States diagram for the evolution of a possible connection.

Hence, the proportion of time that a possible connection is established is

$$\rho = \frac{\bar{t}_{WHT}}{\bar{t}_{idle} + \bar{t}_{WHT}}\,. \tag{4}$$

- *Traffic load* (ν),

$$\nu = \frac{\bar{b}_{in}}{b_{core}}\,. \tag{5}$$

This parameter is equal to the product of the fraction of time that the possible connection is established and the lightpath utilisation efficiency,

$$\nu = \rho \cdot U . \tag{6}$$

- *Average edge delay* ($\bar{t}_{edge}$), the average time elapsed from the arrival to an edge router of the first bit of the first packet making up the contents of the burst until the transmission of the burst. This parameter determines the end-to-end delay, and it should be minimised, depending on the requirements of the class of service.

The previous parameters refer to individual possible connections, but they can be extended to the complete network averaging the values of all the possible connections. The most important of these network parameters are:

- *Average normalised lightpath load* ($\bar{\rho}$), the average of the ρ parameter for all the connections. It represents the average fraction of lightpaths that are established in the network. Therefore, $0 \le \bar{\rho} \le 1$. Note that the average number of lightpaths established in the network ($\bar{L}$), is $\bar{L} = \bar{\rho} K$ ($0 \le \bar{L} \le K$).
- *Network traffic load* ($\bar{\nu}$),

$$\bar{\nu} = \frac{\bar{\bar{b}}_{in}}{b_{core}} , \tag{7}$$

where the double bar refers to the average for all the possible connections. This parameter is also equal to the product of the fraction of lightpaths established and the average lightpath utilisation efficiency,

$$\bar{\nu} = \bar{\rho} \cdot \bar{U} . \tag{8}$$

3. LIMITS ON THE TRAFFIC LOAD DUE TO THE EDGE DELAY CONSTRAINTS AND THE BURST (FLOW) AGGREGATION METHOD

The edge delay is a key parameter as it determines the end-to-end delay. It is, in turn, determined by the arriving packet statistics and the mechanism of burst/flow aggregation used. We propose two possible methods for the flow aggregation, which have a different effect on the edge delay. While a burst is in the process of transmission, new data arrive to the buffer. In the first method, these new data are not added to the current burst, and therefore, they must wait for another lightpath to be established for their transmission. Hence, when the lightpath is deleted, there may be some data in the buffer. In the second method, the new data arriving to the buffer are considered as part of the current burst, and hence, the lightpath is only deleted when the buffer is completely empty. We call *Limited Size Bursts* (*LS-Bursts*) to the former method and *Not-limited Size Bursts* (*NS-Bursts*) to the latter one.

In this section, it is shown that the requirements on the average edge delay may impose a bound on the maximum traffic load between a source-destination pair, depending on the flow aggregation method used. The results are obtained for every possible connection, but they hold for the entire network when averaging among all the connections.

In a quasi-static WRON where all possible connections between nodes are quasi-permanently established by means of lightpaths, there is no edge delay, as data can be transmitted directly on the lightpath. Therefore, if the network must carry a higher network traffic load than the bounds obtained in this section, the OBS option would not bring any advantages over the quasi-static WRON. If only a few possible connections must carry a higher traffic load than the bound, then it could be advantageous to establish quasi-permanent lightpaths for these.

3.1 Limited Size of the Bursts (LS-Bursts)

For the LS-Bursts method, the length of the burst is known when the data transmission begins. This value is proportional to the time elapsed to build the burst (the edge delay) and to the input bit rate. Therefore, the following relationship applies:

$$\bar{t}_{edge} = \frac{\bar{L}_{flow}}{\bar{b}_{in}}, \tag{9}$$

and using the equations (1–6),

$$\bar{t}_{edge} = \frac{\bar{t}_{WHT}}{\rho} = \bar{t}_{idle} + \bar{t}_{WHT} = \frac{\bar{t}_{idle} + \bar{t}_{RTT}}{1-\nu}. \tag{10}$$

Note that even if the idle time is zero, the edge delay cannot be zero as it depends on the round-trip time and the traffic load.

Figures 3 to *5* show the average edge delay, the bandwidth utilisation factor and the fraction of time that a possible connection is established as a function of the idle time and the traffic load. We assume that $\bar{t}_{RTT} = 5$ ms (which corresponds to a network diameter of 1000 km approximately).

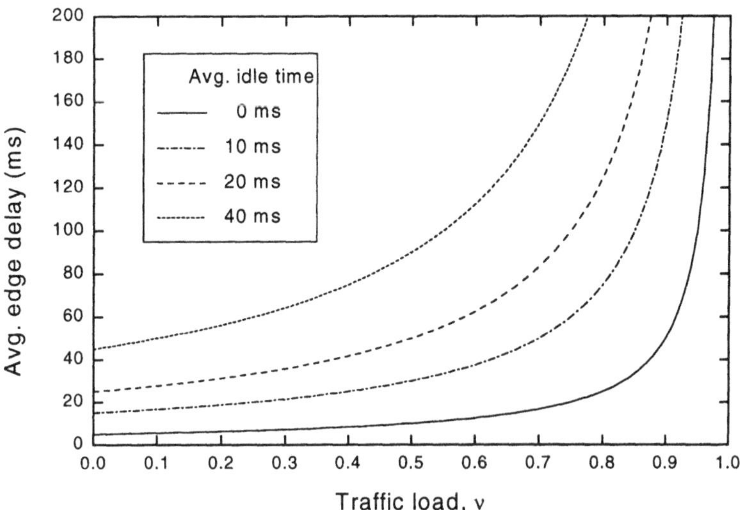

Figure 3. Average edge delay ($\bar{t}_{edge}$) vs. traffic load (ν) for different average idle times.

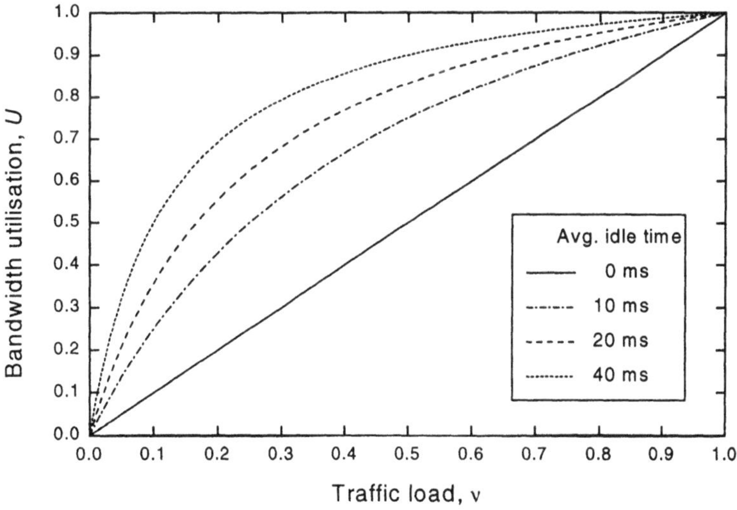

Figure 4. Bandwidth utilisation (or lightpath utilisation efficiency, *U*) vs. traffic load (ν) for different average idle times.

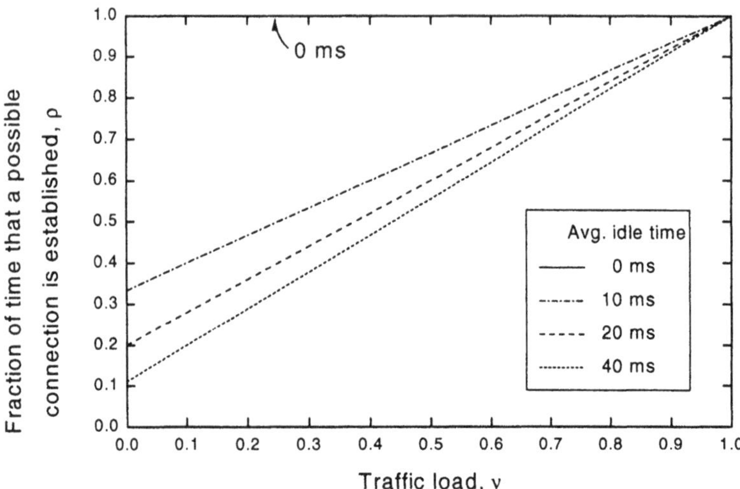

Figure 5. Fraction of time that a possible connection is established (ρ) vs. traffic load (ν) for different average idle times.

As shown in *Figure 3*, the edge delay increases with the increase in the idle time. However, the increase in idle time leads to an increase in the length of the bursts, and since the overhead ($\bar{t}_{RTT}$) remains constant, each lightpath is more efficiently used (*Figure 4*). In turn, this implies that the possible connection remains established a reduced fraction of time (*Figure 5*). Therefore, when considering the entire network[1], fewer lightpaths are needed to transport a given traffic load, and thus a lower number of wavelengths.

Hence, if the network must provide an average edge delay below or equal to a value $\bar{t}_{edge\_required}$, the optimal average idle time ($\bar{t}_{idle\_opt}$) for a given traffic load is that which provides the maximum allowed edge delay by the requirements of the class of service. Hence,

$$\bar{t}_{idle\_opt} = \bar{t}_{edge\_required}(1-\nu) - \bar{t}_{RTT}\,. \tag{11}$$

Since the idle time cannot be negative, the maximum traffic load for a possible connection that the OBS network can support is given by

$$\nu_{max} = 1 - (\bar{t}_{RTT} / \bar{t}_{edge\_required})\,. \tag{12}$$

[1] Note that *Figure 5* also holds for this case, assuming the parameters are the network averages instead of being the values of a single possible connection. Then, the x-axis would represent the network traffic load, and the y-axis the average normalised lightpath load.

Therefore, when this method is used to build the bursts, the edge delay places a limit on the maximum traffic load between pairs of nodes that the network can support.

3.2 Not-limited Size of the Bursts (NS-Bursts)

When using the NS-Bursts method, the duration of the burst is not known when the data transmission begins. Since data arriving to the buffer while the lightpath is still established are also transmitted using the current lightpath, eq. (9) does not hold. As mentioned previously, a lightpath is held until the buffer is completely empty. The time elapsed from the transmission of the last bit of the burst i until the transmission of the first bit of the burst $i+1$ is $\bar{t}_{prop} + \bar{t}_{idle} + \bar{t}_{ack}$. During this time, the burst $i+1$ is built. First of all, there is a finite amount of time until a bit arrives to the empty buffer ($\bar{t}_{silence}$), and then, some time elapses until its transmission, and that is the edge delay ($\bar{t}_{edge}$). Therefore, $\bar{t}_{prop} + \bar{t}_{idle} + \bar{t}_{ack} = \bar{t}_{silence} + \bar{t}_{edge}$, and using eq. (2), $\bar{t}_{idle} = \bar{t}_{silence} + \bar{t}_{edge} - \bar{t}_{RTT}$. In general, $\bar{t}_{silence} << \bar{t}_{edge} - \bar{t}_{RTT}$, except for very low traffic loads (see appendix for details), so

$$\bar{t}_{edge} \approx \bar{t}_{idle} + \bar{t}_{RTT} . \tag{13}$$

Note that *Figures 4* and *5* are also valid for this method (again with $\bar{t}_{RTT} = 5$ ms), as they are plotted as a function of $\bar{t}_{idle}$. When using this method, the minimum edge delay that the network can provide is only bounded by the round-trip time. Regarding the traffic load, the limiting case where the input bit rate between source-destination nodes were equal to the core bit rate, could be handled by the network as the lightpath would be held forever, therefore $\nu_{max} = 1$. Then, with NS-Bursts, the edge delay does not place any limit on the maximum load between pairs of nodes that the network can support.

4. LIMITS ON THE TRAFFIC LOAD DUE TO THE EFFICIENCY OF THE DYNAMIC RWA ALGORITHMS

As mentioned in the introduction, the OBS approach potentially allows a better utilisation of the bandwidth provided by the network. This feature can be exploited to minimise the number of wavelengths used in the network or the number of transmitters and receivers within the nodes. We consider here the first option, and we analyse the limitations on the traffic load imposed by the efficiency in terms of wavelengths of the dynamic RWA algorithm used.

In quasi-static WRONs, the RWA problem is solved through an off-line analysis and the complete set of lightpaths to be established is known a priori. Therefore, efficient algorithms can be applied [3] to obtain a solution that minimises the number of wavelengths. On the other hand, dynamic RWA algorithms must operate in real time so that they must be fast and may not lead to an optimal utilisation of the network resources, specifically the number of wavelengths. Besides, the algorithm must take decisions on a request by request basis. This process involves deciding whether to accept or not a request for a lightpath, and if it is accepted to look for a route and wavelength. If reconfiguration of the routes and wavelengths of the established lightpaths is allowed, then it would be possible to employ always a lower number of wavelengths (or equal) than in a quasi-static WRON, but in this approach, we do not consider reconfiguration due to the short duration of the connections.

For a given topology, RWA algorithm, traffic characteristics and a maximum desired lightpath blocking probability, the number of wavelengths needed is an increasing function of the average normalised lightpath load, $\overline{\rho}$. A new parameter, the *wavelength gain* ($G_W(\overline{\rho})$) can be defined as

$$G_W(\overline{\rho}) = \frac{W_{quasi-static}}{W_{dynamic}(\overline{\rho})}, \tag{14}$$

where $W_{quasi\text{-}static}$ is the number of wavelengths required in a quasi-static WRON (with all the K lightpaths established), and $W_{dynamic}$ is the number of wavelengths required in the dynamic case for a given average normalised lightpath load. The OBS approach will bring advantages when $G_W(\overline{\rho}) > 1$ as a lower number of wavelengths than in the quasi-static case would be used. Therefore, we define the *limiting average normalised lightpath load* ($\overline{\rho}_{lim}$) as the value of $\overline{\rho}$ for which the wavelength gain is equal to one. Hence, for a lightpath utilisation of $U = 1$, the maximum average network traffic load would be limited by $\overline{\rho}_{lim}$,

$$\overline{v}_{max} = \overline{\rho}_{lim}. \tag{15}$$

The maximum network traffic load is therefore limited by the efficiency in terms of wavelengths of the dynamic RWA algorithm. The more efficient the algorithm is, the closer $\overline{\rho}_{lim}$ (and therefore, $\overline{v}_{max}$) will be to one. The main problem is that a more efficient algorithm is usually slower, and therefore, it may not be adequate to fulfil the requirements on the edge delay. Besides, note that for a given RWA algorithm, the requirement of a lower blocking probability implies a lower value of $\overline{\rho}_{lim}$ (and thus of $\overline{v}_{max}$).

5. COMPARISON OF OBS AND QUASI-STATIC WAVELENGTH ROUTED OPTICAL NETWORKS

The aim is to find the range of parameters that make the OBS network with dynamic wavelength allocation a suitable option compared to a quasi-static WRON. When using LS-Bursts, the maximum *network* traffic load is given by:

$$\overline{v}_{max} = \overline{\rho}_{lim} - \frac{\bar{t}_{RTT}}{\bar{\bar{t}}_{edge}}, \tag{16}$$

and when using NS-Bursts, the maximum *network* traffic load is given by:

$$\overline{v}_{max} = \overline{\rho}_{lim} - \left(\frac{\bar{t}_{RTT}}{\bar{\bar{t}}_{edge} - \bar{\bar{t}}_{RTT}} \right)\left(1 - \overline{\rho}_{lim}\right), \tag{17}$$

where $\bar{\bar{t}}_{RTT}$ and $\bar{t}_{edge}$ are the averages of the round-trip times and edge delays (respectively) of all the possible connections.

In *Figure 6* we plot these equations for different values of the edge delay for $\bar{\bar{t}}_{RTT} = 5$ ms.

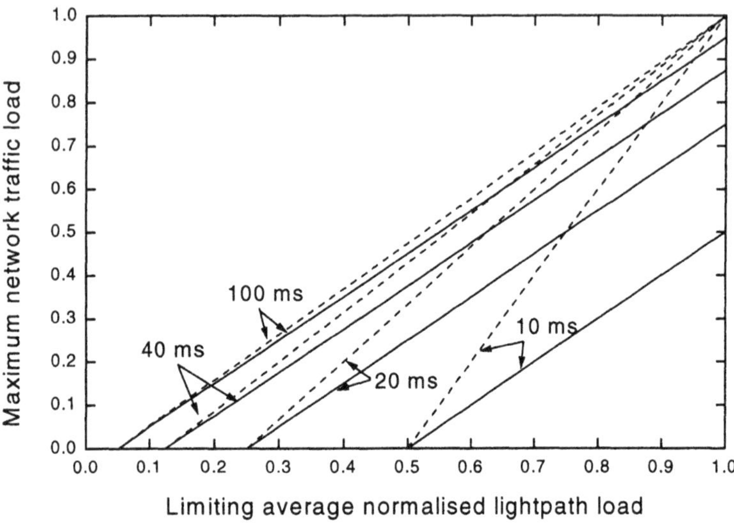

Figure 6. Maximum network traffic load ($\bar{v}_{max}$) vs. limiting average lightpath load ($\bar{\rho}_{lim}$) for different average edge delays ranging from 10 to 100 ms. Solid lines represent the limits for LS-Bursts, and dashed lines those for NS-Bursts.

Therefore, the maximum network traffic load that the network can support is determined by the method used to build the bursts (LS-Bursts or NS-Bursts), the requirements of the class of service (edge delay and PLR, and therefore, the blocking probability of the lightpaths) and the efficiency of the dynamic RWA algorithm. Note that if a wavelength gain higher than one is required, then the network must operate with $\bar{\rho}$ lower than $\bar{\rho}_{lim}$, implying a lower maximum network traffic load.

The comparison between NS-Bursts and LS-Bursts is not as obvious as it may seem. If the same value of $\bar{\rho}_{lim}$ were obtained for both methods, then NS-Bursts would be a better option as the network could support more traffic than with LS-Bursts. This condition approximately holds if the calculation time of the RWA algorithm is zero. But some preliminary simulations have shown that in a not ideal case where the calculation time is not zero, there is a decrease on $\bar{\rho}_{lim}$ for NS-Bursts, which may lead to a lower maximum network traffic load than for LS-Bursts.

By way of an example and comparison with the quasi-static WRON, the NSFNet topology, described in [2], is considered. In the static case, 13 wavelengths are needed to establish all the possible connections (*K* lightpaths) [2]. To study the dynamic case, we have implemented the AUR-EXHAUSTIVE algorithm proposed in [13] and set the maximum desired blocking probability to $p_B = 10^{-4}$. An ideal zero calculation time of the RWA algorithm and a round-trip time of 5 ms have been assumed. The traffic arriving to the edge buffers is modelled as an ON-OFF model (see appendix for details), and we have set a uniform traffic matrix, so that there is the same average traffic load between all pairs of source-destination nodes. When a timer (set to the difference between the desired edge delay and the propagation delay of the request to the control node) is exceeded, the lightpath is immediately established. Since the blocking probability is low, the traffic load is equal for all pairs of edge routers, and there is no interaction between the possible connections (there is no queue in the control node due to the zero calculation time assumption), the number of lightpaths approximately follows a binomial distribution (it would be exact if there were no blocking).

For this example and using LS-Bursts (*Figure 7*), the value of the limiting average normalised lightpath load is $\bar{\rho}_{lim} = 0.68$. Then, if the required average edge delay is 40 ms and $G_W = 1$, the maximum traffic load is $\bar{\nu}_{max} = 0.56$. For a higher wavelength gain, the network must operate with $\bar{\rho}$ lower than $\bar{\rho}_{lim}$, and hence $\bar{\nu}_m$ will decrease too. For instance, for a wavelength gain of 1.18 (such that 11 wavelengths are required rather than 13), the maximum mean normalised lightpath load is 0.51 and $\bar{\nu}_m$ decreases to 0.39. Note that if a lower blocking probability is required, $\bar{\rho}_{lim}$ decreases and thus $\bar{\nu}_m$.

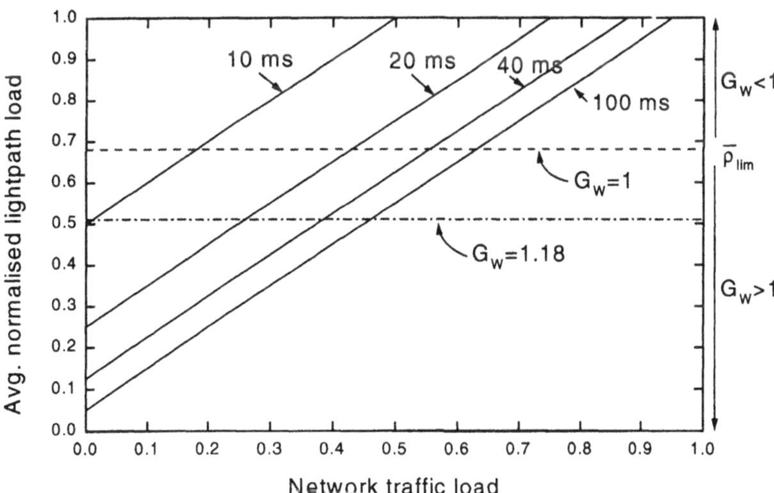

Figure 7. Average normalised lightpath load ($\bar{\rho}$) vs. network traffic load ($\bar{\nu}$) for different average edge delays. $\bar{\rho}_{lim}$ and G_W are represented for the NSFNet, setting a uniform traffic matrix. For clarity of the figure, only edge delays corresponding to LS-Bursts have been shown.

For NS-Bursts (not shown in *Figure 7*), the same value of $\overline{\rho}_{lim}$ is obtained as the calculation time was assumed to be zero, and hence, the maximum network traffic load is higher than for LS-Bursts (0.63 for $G_W = 1$, and 0.44 for $G_W = 1.18$).

6. SUMMARY

The maximum traffic load that an OBS network with dynamic allocation and centralised control can support has been studied. It has been shown that tight constraints in the required edge delay and in the blocking probability, as well as the utilisation of dynamic RWA algorithms with a low efficiency in terms of wavelengths results in a limitation on the maximum traffic load that can be carried. If the traffic load that the network must support is higher than that value, then a quasi-static WRON or semi-permanent lightpaths between selected pairs of edge routers should be used. It has also been shown that there is a trade-off between the edge delay and the number of wavelengths required in the network. Lower edge delays require a higher average number of lightpaths in the network, and therefore a higher number of wavelengths.

Two different approaches to build the bursts have been proposed: LS-Bursts and NS-Bursts. The latter option allows higher traffic loads in the network for the same edge delay assuming an ideal scenario where the calculation time of the dynamic RWA algorithm is zero, but preliminary simulations have shown that in a realistic case, the comparison between both methods is not so obvious. On the other hand, when using LS-Bursts, the length of the burst is known when its transmission starts, and that information could be used by the control node in processing other requests. Therefore, the optimisation of burst-formation requires further analysis.

APPENDIX

First of all the traffic model assumed is presented, and then, it is shown that for NS-Bursts, $\bar{t}_{idle} = \bar{t}_{silence} + \bar{t}_{edge} - \bar{t}_{RTT} \approx \bar{t}_{edge} - \bar{t}_{RTT}$

We have modelled traffic arriving to the edge buffers as an ON-OFF model. The length of the ON (packets arriving) and OFF (interarrival time) periods are modelled according to the Pareto distribution. The length of the ON periods (in bits) is $\lfloor A_{ON}/U^{\alpha} \rfloor$, where U is a random variable uniformly distributed on [0,1], $1 < \alpha \leq 2$ so that the length of the periods has finite mean and infinite variance, and A_{ON} is the minimum size of the packets. The length of the OFF periods is set using the same distribution and value of α, but the minimum size of the periods is adjusted in order to achieve the desired average input bit rate (and hence the desired traffic load). During the ON period data is arriving at a bit rate of b_{ON}, with $b_{ON} \leq b_{core}$. Obviously, $b_{ON} \geq \bar{b}_{in}$. The relationship between these parameters is

$$\bar{b}_{in} = \frac{b_{ON}\bar{t}_{ON}}{\bar{t}_{ON} + \bar{t}_{OFF}},$$

where $\bar{t}_{ON}$ is the average duration of the ON periods:

$$\bar{t}_{ON} = \left(\frac{A_{ON}\alpha}{\alpha - 1}\right) \Big/ b_{ON},$$

and $\bar{t}_{ON}$ is the average duration of the OFF periods,

$$\bar{t}_{OFF} = \left(\frac{A_{OFF}\alpha}{\alpha - 1}\right) \Big/ b_{ON}$$

In order to achieve a certain traffic load ν, the minimum length of the OFF periods is set to:

$$A_{OFF} = A_{ON}\left(\frac{\nu_{ON}}{\nu} - 1\right), \text{ where } \nu_{ON} = \frac{b_{ON}}{b_{core}}.$$

$\bar{t}_{idle} \approx \bar{t}_{edge} - \bar{t}_{RTT} \Leftrightarrow \bar{t}_{edge} - \bar{t}_{RTT} >> \bar{t}_{silence}$. The parameter $\bar{t}_{silence}$ is the average time elapsed from when the buffer is completely empty until the first bit arrives, so $\bar{t}_{silence} < \max\{\bar{t}_{OFF}, 1/b_{ON}\}$. Hence, if $\bar{t}_{edge} - \bar{t}_{RTT} >> \max\{\bar{t}_{OFF}, 1/b_{ON}\}$, then $\bar{t}_{idle} \approx \bar{t}_{edge} - \bar{t}_{RTT}$. Therefore the approximation holds if

$$\bar{t}_{edge} - \bar{t}_{RTT} >> \frac{1}{b_{ON}} \text{ and } \nu >> \frac{A_{ON}(b_{ON}/b_{core})}{A_{ON} + (\bar{t}_{edge} - \bar{t}_{RTT})b_{ON}\left(\frac{\alpha - 1}{\alpha}\right)}.$$

For a desired $\bar{t}_{edge}$=10 ms and $\bar{t}_{RTT}$=5 ms the first inequality holds if $b_{ON} >> 200$ bps, which is clearly satisfied. The worst case for the second inequality is obtained with $b_{ON} = b_{core}$. If we set A_{ON}=3200 bits (400 bytes), α=1.5, and $b_{ON} = b_{core}$ = 1 Gbps, the second part holds if $\nu >> 1.9 \cdot 10^{-3}$. Therefore, $\bar{t}_{idle} \approx \bar{t}_{edge} - \bar{t}_{RTT}$ except for very low traffic loads.

In the simulation presented in section 5, the model and parameters described have been used.

ACKNOWLEDGEMENTS

The authors would like to express their gratitude to Dr. Damon Wischik (Statistical Laboratory, University of Cambridge), Eugene Kozlovski, Giancarlo Gavioli and Dr. Robert Killey (UCL), and to the reviewers for valuable comments. The financial support from the Royal Society, Marconi Communications and University of Valladolid is gratefully acknowledged. Ignacio de Miguel is also very grateful to Dr. Evaristo J. Abril, and the other members of the Optical Communications Group at University of Valladolid for their help during his stay at UCL.

REFERENCES

[1] I. Chlamtac, A. Ganz, G. Karmi, "Lightpath communications: An approach to high bandwidth optical WAN's," *IEEE Transactions on Communications*, Vol. 40, No. 7 , Jul. 1992, pp. 1171-1182

[2] S. Baroni, P. Bayvel, "Wavelength requirements in arbitrarily connected wavelength-routed optical networks," *Journal of Lightwave Technology*, Vol. 15, No. 2, Feb. 1997, pp. 242-251

[3] S. Baroni, P. Bayvel, R.J. Gibbens, S.K. Korotky, "Analysis and design of resilient multifiber wavelength-routed optical transport networks," *Journal of Lightwave Technology*, Vol. 17, No. 5, May 1999, pp. 743-758

[4] J.S. Turner, "Terabit burst switching," *Journal of High-Speed Networks*, Vol. 8, 1999, pp. 3-16

[5] C. Qiao, M. Yoo, "Optical bust switching (OBS) – A new paradigm for an optical Internet," *Journal of High-Speed Networks*, Vol. 8, 1999, pp. 69-84

[6] J.S. Turner, "WDM burst switching for petabit data networks," *Technical Digest OFC 2000*, paper WD2-1, 2000, pp. 47-49

[7] C. Qiao, M. Yoo, "Choices, features and issues in optical burst switching," *Optical Networks Magazine*, Apr. 2000, pp. 36-44

[8] S. Verma, H. Chaskar, R. Ravikanth, "Optical burst switching: A viable solution for terabit IP backbone," *IEEE Network*, Vol. 14, No. 6, Nov./Dec. 2000, pp. 48-53

[9] M. Düser, E. Kozlovski, R.I. Killey, P. Bayvel, "Distributed router architecture for packet-routed optical networks," *Proc. Fourth Working Conference on Optical Network Design and Modelling (ONDM 2000)*, Feb. 2000, pp. 202-221

[10] M. Düser, E. Kozlovski, R.I. Killey, P. Bayvel, "Design trade-offs in optical burst switched networks with dynamic wavelength allocation," *Technical Digest ECOC 2000*, paper Tu 4.1.4, 2000, pp. 23-24

[11] M. Düser, P. Bayvel, "Bandwidth utilisation and wavelength re-use in WDM Optical Burst-Switched Packet Networks," *Proc. Fifth Working Conference on Optical Network Design and Modelling (ONDM 2001)*, 2001

[12] E. Kozlovski, M. Düser, R.I. Killey, P. Bayvel, "The design and performance analysis of QoS-aware edge-router for high-speed IP optical networks," *Proceedings of the London Communications Symposium 2000*, Sep. 2000, pp. 185-190

[13] A. Mokhtar, M. Azizoglu, "Adaptive wavelength routing in all-optical networks," *IEEE/ACM Transactions on Networking*, Vol. 6, No. 2, Apr. 1998, pp. 197-206

Part Four

Advances in optical network technologies

SKEW COMPENSATION IN ALL OPTICAL BIT PARALLEL WDM SYSTEMS

M. E. Vieira Segatto* , F. N. Timofeev, R. Wyatt, R. Kashyap†
Corning Research Centre
Adastral Park, Martlesham Heath, Ipswich
IP5 3RE, United Kingdom.
segatto@ele.ufes.br

J. R. Taylor
Femtosecond Optics Group, The Blackett Labs. Imperial College, London
SW7 2BZ, United Kingdom.
jr.taylor@ic.ac.uk

Abstract A bit skew and dispersion compensation scheme using fibre gratings for a 10 Gb/s-channel, 2-bit, Bit Parallel WDM system over 100 km of standard single-mode fibre is presented. The combination of channelised dispersion, and additional fibre delay, is sufficient to give full compensation.

Keywords: Optical Parallel Transmission, Bit Skew Compensation

Introduction

High-speed digital transmission between two points can be implemented basically in three different ways as illustrated in figure 1. On the serial system (figure 1(a)), a Serialiser is used to convert the bits of a digital word in a stream of bits. This stream is then used to modulate an optical beam and sent through the fibre to the receiver. At the receiver side, the optical signal is converted to electrical pulses and a deserialiser generates the original binary word. As the bits are transmitted serially, to form a word the time equivalent of a number of bits in a word at

*Federal University of Espírito Santo, Brazil. Also on leave of absence at Imperial College.
This work was partially supported by CAPES/Brazil.
†Raman Kashyap is now with Corvis Canada.

the bit period plus the time for serial-to-parallel and parallel-to-serial conversions is required. Figure 1(b) shows an alternative to the serial transmission system. In order to avoid the serial-parallel-serial conversions, each bit is sent using a different fibre. The Bit Parallel WDM (BP-WDM) shown in figure 1(c) uses WDM technology to transmit information. Each bit is used to modulate the optical source at a different wavelength. The wavelengths are then multiplexed and launched into the fibre. After the transmission, the wavelengths are separated and the digital word is recovered. Besides the well-known impairments in the serial system such as attenuation, dispersion and non-linearities, the parallel system is also quite sensitive to the propagation delay differences among the bits of a digital word. For both parallel systems, Figures 1(b) and 1(c), each bit travels in different optical paths. While in the ribbon fibre system the refractive index and length differences are responsible for most of the bit-skew, for the BP-WDM system it is caused by the fibre dispersion characteristic. Practical systems using ribbon fibres are restricted to short distances and low-speed due to high values of bit-skew having its use restricted to boards and inter-frame connections and to short-distance links [1, 2, 3]. For high performance computer interconnections, for instance, the BP-WDM system can be a good option since it is able to cover higher bit-rates and distances by using just one fibre [4, 5, 6].

In this paper we present results on a two-channel (two-bit) BP-WDM system operating at 10Gb/s per channel over 100-km of Standard Single Mode (SSM) fibre. The bit skew and dispersion were compensated using two apodised step Chirped Fibre Gratings (CFG). Section 1 presents the basic ideas of BP-WDM systems. Bit skew and bit skew compensation are discussed in sections 2 and 3. Experimental results for the bit skew compensator is presented in section 4.

1. BP-WDM TRANSMISSION

In 1988, Loeb and Stilwell [7] proposed the fist long-distance parallel system using WDM technology. In his system, each bit of a digital word is transmitted on a separated wavelength, as illustrated in figure 2. This Bit Parallel WDM system uses a single fibre, eliminating the limitations imposed by multiple optical path lengths. Bit-skew remains a problem in a long-distance optical link operating in the first and third transmission windows where a standard fibre exhibits large group delay dispersion.

Three major questions must be answered when designing a BP-WDM system. The first question is in which transmission window the system should work. VCSEL and photodiode arrays operating in 850 nm

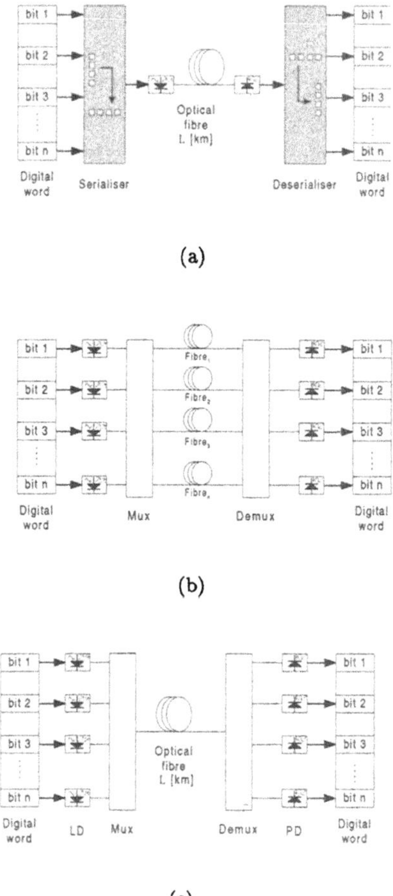

Figure 1 High-speed digital systems. (a) Serial system, (b) ribbon fibre system, and (c) parallel WDM system. LD:laser diode, PD: Photodetector.

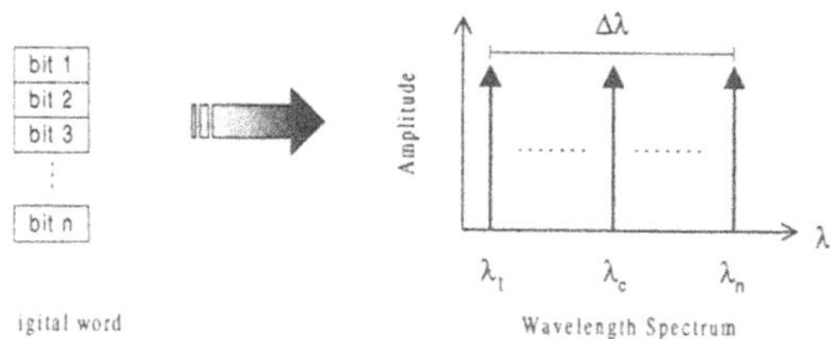

Figure 2 Wavelength allocation in BP-WDM systems.

wavelength window have low manufacturing costs but intramodal and intermodal dispersion are very high for fibres at this wavelength window and link length would be severely limited. Single mode fibres operating in the second window have low dispersion but relatively higher losses if compared with the losses in the third window. On the other hand, the low losses and the availability of optical amplifiers in the third window favour long-distance transmission. Another factor is that WDM systems are available commercially mostly in the third window [8, 9].

The second question is the rate at which each bit-channel should operate. The main aim of a serial WDM transmission is to fully utilise the potential throughput of an optical fibre, by keeping the data bit rate for individual wavelength channels as high as possible. Instead, a BP-WDM system endeavours to reduce the speed per bit-channel by the number of optical channels, thus simplifying implementation of the required signal processing in optical networks. Although paralleling the data path increases the number of electronic circuits by a factor equal to the number of channels, lowered speed translates into lower prices for the electronic systems. For the experiments shown in this paper, the rate per bit-channel chosen was 10 Gbit/s for two reasons. From the electronic point of view the performance of computers and the associated integrated circuits increases by a factor of two every 18 months [10] which means that in few years computer clocks at 5 or 10 GHz will be part of our reality. From the optical point of view, the effects of dispersion in the system performance are apparent for a few kilometres of fibre and can also be studied. The third question is link length. Bit skew and dispersion are the import factors to answer this question. The maximum link length is mainly defined by the fibre dispersion and any attempt to increase it will depend on bit skew compensation techniques.

Bit skew compensation techniques can be divided into two categories, electronic compensation [11] and all optical compensation [12]. Electronic compensation implies more complicated transmitter and receiver systems and consequently in more expensive systems. However, all optical compensation can be used without increasing the complexity of the system.

2. BIT SKEW

The bit skew is the major problem in BP-WDM systems. It can be caused by several factors such as turn-on delay in semiconductor lasers, dispersion in optical devices such as filters and (de)multiplexer, and mainly by the group delay dispersion of the optical fibre [12].

As illustrated in figure 2, a digital word with n bits is mapped into the optical domain using one wavelength per bit. As each bit is represented by a different wavelength, they suffer different group delay due to the dispersion characteristics of the fibre and arrive at the end of the fibre at different times.

The bit skew in the fibre can be written as a function of the dispersion as follow: The total dispersion in a single mode fibre, expressed as function of the dispersion slope S_0, the zero-dispersion wavelength λ_0 and wavelength λ, is given by the Sellmeier expression [13]

$$D(\lambda) = \frac{S_0 \lambda}{4}\left[1 - \left(\frac{\lambda_0}{\lambda}\right)^4\right]. \tag{1}$$

The maximum bit skew can be determined by the difference in the group delay between the fastest and the slowest channel. For the sake of simplicity, we assume that all channels are on the same side of the zero-dispersion wavelength. Then, the delay between the first channel and the last channel gives the maximum bit skew.

If we define a system with n channels (i.e. a digital word with n bits) equally spaced and total bandwidth $\Delta\lambda$, as shown in figure 2, the wavelength of the first channel and of the last channel can be expressed as a function of the central wavelength λ_c as

$$\begin{aligned} \lambda_1 &= \lambda_c - \Delta\lambda/2, \\ \lambda_n &= \lambda_c + \Delta\lambda/2. \end{aligned} \tag{2}$$

The group delay τ_g in a fibre of length L is determined by integrating the total dispersion over the wavelength, and

$$\frac{\tau_g}{L} = \int D(\lambda')\, d\lambda'. \tag{3}$$

The maximum bit skew per unit of length is given by

$$\frac{\Delta\tau_g}{L} = \int_{\lambda_1}^{\lambda_n} D(\lambda') d\lambda'. \tag{4}$$

Substituting Eq. (1) in (4) and using (2), we have

$$\frac{\Delta\tau_g}{L} = \Delta\lambda \frac{\lambda_c S_0}{4}\left[1 - \frac{\lambda_0^4}{\left(\lambda_c^2 - \Delta\lambda^2/4\right)^2}\right]. \tag{5}$$

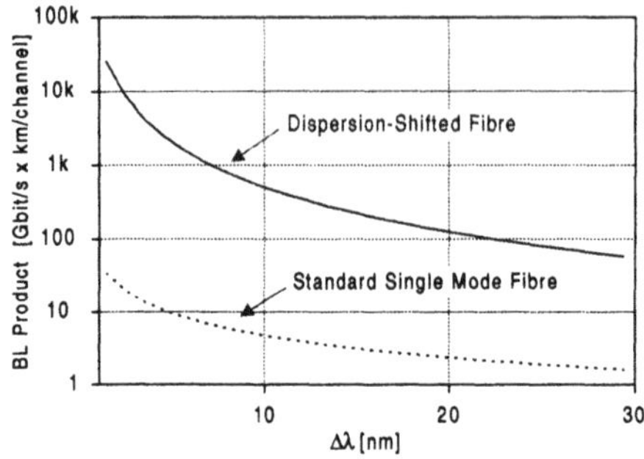

Figure 3 The bit rate-distance product as a function of the total bandwidth for standard single mode and dispersion-shifted fibres when the maximum bit skew allowed is equal to one bit period.

Assuming that $\lambda_c \gg \Delta\lambda$, then

$$\frac{\Delta\tau_g}{L} \cong \Delta\lambda \frac{\lambda_c S_0}{4}\left[1 - \left(\frac{\lambda_0}{\lambda_c}\right)^4\right] = \Delta\lambda D\left(\lambda_c\right). \tag{6}$$

Equation (6) shows that the maximum bit skew is a function of the total bandwidth and of the dispersion of the fibre at the central wavelength. The only assumptions made here are that the central wavelength must be much greater than the total bandwidth and all wavelengths are in the same side of the fibre dispersion curve. This expression holds even if different expressions are used for the dispersion equation (1).

The bit rate-distance (BL) product can be easily computed from (6) if a maximum bit skew is defined. Figure 3 shows the BL product per channel for standard single mode fibre and for dispersion-shifted (DS) fibre[1] when the central wavelength is 1538 nm and the maximum bit skew allowed is equal to one bit period. High products can be obtained for DS fibres [4] if the spacing between wavelengths is kept small.

3. BIT SKEW COMPENSATION

A number of techniques have been proposed to deal with bit skew. These techniques include the use of dispersion-shifted fibre [4], electronic bit realignment in the receiver [11], and a shepherd pulse technique [14] based on nonlinear optics. Drawbacks of these techniques include lack of resynchronisation accuracy ($\sim$ 100 ps) or high pulse energy to induce fibre nonlinearities. In addition, the shepherd pulse technique requires that the passive component skew be within the capture range of the

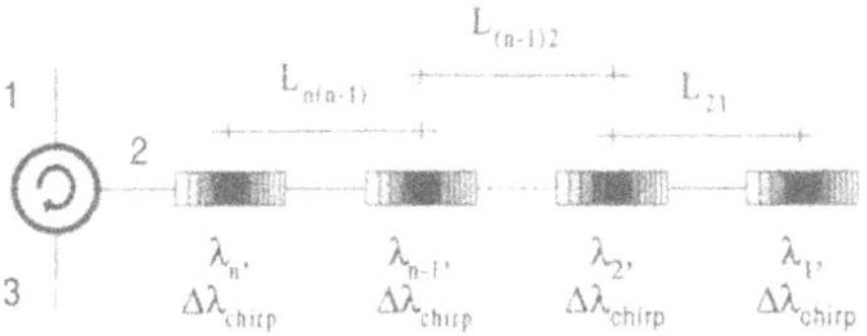

Figure 4 The fibre grating configuration used for bit skew and dispersion compensation.

shepherd pulse ($\sim$ 5 ps), which complicates the design of optical couplers and demultiplexers [9].

In this paper, we propose the use of of step Chirped Fibre Gratings (CFG) to simultaneously compensate bit skew and dispersion in 10 Gbit/s-channel parallel WDM systems. The system works as follows: The group delay in Standard Single Mode fibres (SSM) increases with the wavelength in the anomalous-dispersion regime. Bits travelling in lower wavelengths will arrive at the receiver end before those in higher wavelengths, so additional delay must be added to the faster bits.

Figure 4 shows the Fibre Grating Bit Skew Compensator (FGBSC). It consists of a three port circulator and a set of fibre gratings. Light coming from the fibre enters into the circulator and is reflected by the set of gratings. Each wavelength is reflected in a different position of the set of gratings, so the delay imposed by the FGBSC can be used to compensate to the fibre skew.

Chirped gratings have been extensively used as dispersion compensation devices [15] due their compactness and dispersion characteristics. Each grating acts as a selective mirror, reflecting one wavelength. The gratings are sliced in such way that long wavelengths are reflected first and the device delay is defined by the separation of the gratings. Some degree of tuning is possible by mechanically stressing the gratings. The two mechanisms, the separation of the gratings and the chirp are responsible to compensate to the fibre dispersion and completely eliminate the bit skew.

4. EXPERIMENTAL RESULTS

The experimental setup used to measure bit skew and dispersion is shown in Figure 5. The optical transmitter includes two semiconductor lasers with external fibre grating [16], directly modulated at 10 Gb/s at wavelengths 1554 nm and 1556.8 nm. After combining, the two wavelengths optical power $\sim$ 3 dBm per channel was launched into 50-km of

SSM fibre, optically amplified and launched again into a further 50-km of SSM fibre. Two 10-cm long step-chirped gratings [15, 17] with reflection bandwidth 0.42 nm and 0.48 nm, respectively, separated by a short length of SSM are used to compensate the chromatic dispersion in each channel, and the overall bit skew. The per-channel dispersion of around 1200 ps/nm, corresponding to about 72 km of SSM fibre is compensated by the grating chirp, while the overall bit skew is compensated by the optical separation between the centres of the gratings (Δl). Figure 6 shows the transmitted and received optical spectra. A fibre grating operating in reflection in conjunction with a circulator provides $\sim$ 1 dB insertion loss for both channels. Very small grating reflectivity outside the reflection band provides good filtering of optical ASE noise from in-line optical amplifier. Both gratings were fine-tuned using strain to the wavelength of the corresponding channel. Eye diagrams measured for channel 1 both back to back and after fibre + compensator are shown in Figure 7(a) and 7(b), confirming good reshaping of signal from directly modulated laser sources. The transmitted and received waveforms for both channels are shown in Figure 8, demonstrating total bit skew compensation. The bit skew after 100 km of fibre was around 2.5 ns. The propagation time delay compensation has been done using $\Delta l \sim 9$ cm of fibre spliced between the two gratings.

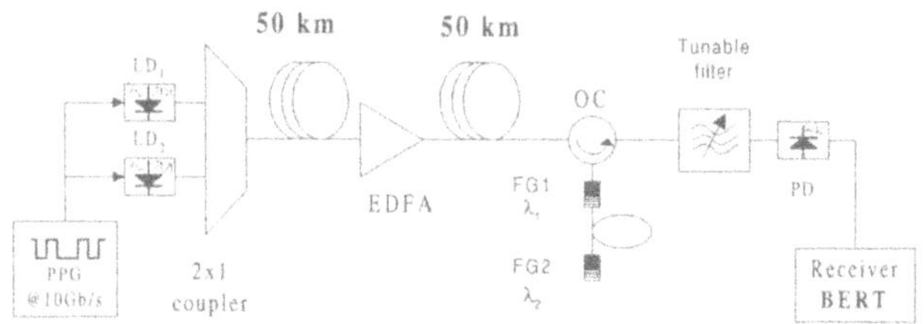

Figure 5 Experimental setup.

5. CONCLUSIONS

In this paper, we have proposed a new scheme for all optical bit skew compensation in bit parallel WDM systems. We have demonstrated 2 bits BP-WDM transmission, 10 Gb/s per channel, over 100 km of SSM fibre using directly modulated lasers with external fibre gratings. Simultaneous chromatic dispersion (-1200 ps/nm) and bit skew (2.5 ns) compensation were achieved. We believe that systems using BP-WDM will be useful for computer interconnects in high-speed parallel systems and ring networks.

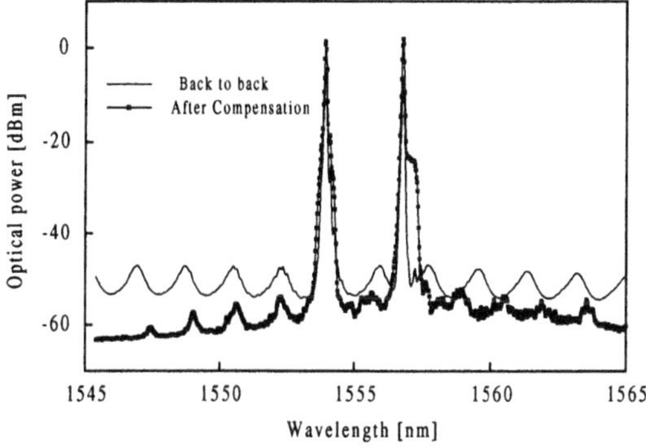

Figure 6 Transmitted and received optical spectra.

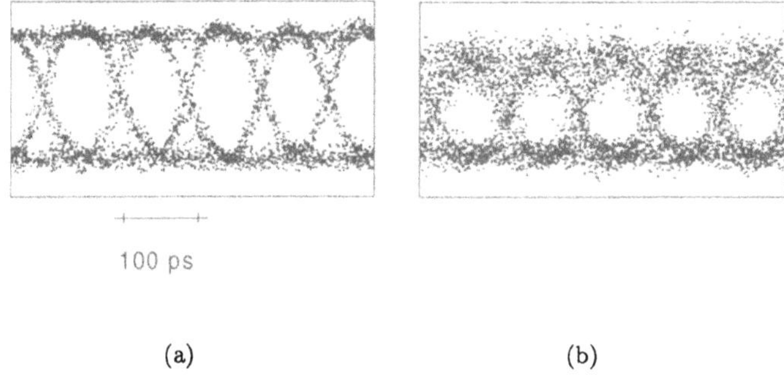

(a) (b)

Figure 7 (a) Transmitted and (b) received eye diagrams.

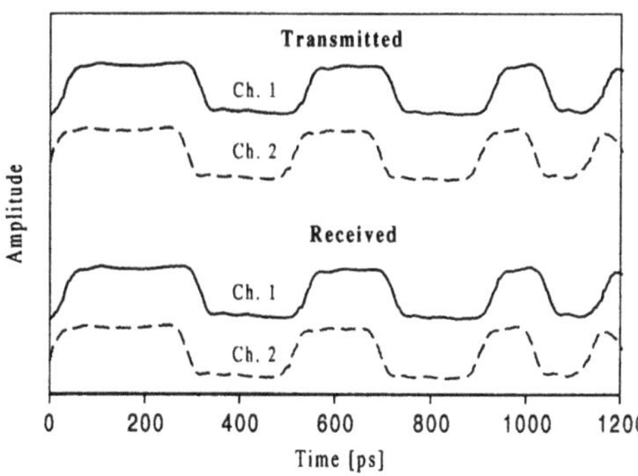

Figure 8 Transmitted and received waveforms for both channels.

References

[1] N. Fujimoto, K. Wakao, and M. Yano. Intelligent parallel optical link - parallel optical transmission system and LSIs. *FUJITSU Sci. Tech. J.*, 30(2):214–23, December 1994.

[2] A. Takai, T. Kato, S. Yamashita, S. Hanatani, Y. Motegi, K. Ito, H. Abe, and H. Kodera. 200-Mb/s/ch 100-m optical subsystem interconnections using 8-channel 1.3-μm laser diode arrays and single-mode fiber arrays. *IEEE J. Lightwave Tech.*, 12(2):260–270, February 1994.

[3] H. Westphal and J. M. Spaus. Interfacing parallel computers to fiber-optic communication links for plant automation. In *Proc. IEEE Electronic Components and Technology Conference*, pages 805–810, 1994.

[4] L. Bergman, J. Morookian, and C. Yeh. An all-optical long-distance multi-Gbytes/s bit-parallel WDM single-mode fiber link. *IEEE J. Lightwave Technol.*, 16:1577–1582, September 1998.

[5] K. M. Sivalingam. A comparison of bit-parallel and bit-serial architectures for WDM networks. *Journal on Photonic Network Communications*, 1:1–27, June 1999.

[6] L. Kleinrock *et. al.* The supercomputer supernet testbed: A WDM-based supercomputer interconnect. *IEEE J. Lightwave Technol.*, 14(6):1388–1399, June 1996.

[7] M. L. Loeb and G. R. Stilwell Jr. High-speed data transmission on an optical fiber using a byte-wide WDM system. *IEEE J. Lightwave Technol.*, 6:1306–1311, August 1988.

[8] L. Bergman, J. Morookian, and C. Yeh. WDM component requirements for bit-parallel fiber optic computer networks. In *Proc. SPIE*, pages 2–13, 1998.

[9] A. J. Mendez. Design of constant optical path couplers for bit-parallel WDM systems. In *Proc. SPIE*, volume 3234, pages 94–99, 1998.

[10] A. F. J. Levi. Optical interconnects in systems. *Proceedings of IEEE*, 88(6):750–757, June 2000.

[11] M. L. Loeb and G. R. Stilwell Jr. An algorithm for bit-skew correction in byte-wide WDM optical fiber systems. *IEEE J. Lightwave Technol*, 8(2):239–242, February 1990.

[12] M. E. Vieira Segatto, R. Kashyap, G. D. Maxwell, J. R. Taylor, V.A. Bhagavatula, G.E. Berkey, and A.F. Evans. Multi Gb/s bit parallel WDM transmission using dispersion managed fibers. *IEEE Photonics Tech. Letters*, 12(8):925–927, August 2000.

[13] J. M. Senior. *Optical Fiber Communications: Principles and Practice.* Prentice Hall, London, 1992.

[14] L. Bergman and C. Yeh. Experimental verification of the pulse shcpherding concept in dispersion-shifted single mode fiber for bit-parallel wavelength links. In *Proc. Fourth International Conference on Massively Parallel Processing Using Optical Interconnections (MPPOI '97)*, pages 25–29, Montreal-Canada, June 1997.

[15] R. Kashyap. *Fibre Bragg Gratings.* Academic Press, San Diego, 1999.

[16] F. N. Timofeev, I. A. Kostko, P. Bayvel, O. Berger, R. Wyatt, R. Kashyap, I. F. Lealman, and G. D. Maxwell. 10 Gbit/s directly modulated, high temperature-stability external fibre grating laser for dense WDM networks. *Electron. Letters*, 35(20):1737–1739, September 1999.

[17] I. Riant, S. Gurib, J. Gourhant, P. Sansonetti, C. Bungarzeanu, and R. Kashyap. Chirped fiber gratings for WDM chromatic dispersion compensation in multispan 10-Gb/s transmission. *IEEE Journal of Selected Topics in Quantum Electronics*, 5(5):1312–1324, September 1999.

Effect of EDFA Cross-Gain Saturation on the Transmission of Packetized Burst-Mode Data over WDM

Miroslav Karásek* and Mourad Mennif+

**Institute of Radio Engineering and Electronics, Academy of Sciences of the Czech Republic, Chaberska 57, 182 51 Prague, +Centre for Optics, Photonics and Lasers (COPL), Laval University, Department of Electrical and Computer Engineering, Québec, (Québec), Canada G1K 7P4*

Abstract. In this paper, we demonstrate the importance of the number of WDM channels and the network traffic variability on the dynamics of a cascade of EDFAs fed by packet traffic in burst mode. The dynamics are determined using a numerical model incorporating time variation effects in the EDFA. Calculations are based on the solution of a transcendental equation describing the time evolution of the *reservoir*, i.e. the total number of excited ions, for each EDFA. Traffic on WDM channels is modeled as statistically independent ON/OFF time-slotted sources. We find that the cross-gain saturation effect depends on the number of WDM channels and also on network traffic variability. As the number of WDM channels increases, and even though each one is highly variable in time, the dynamics of the total input power exhibit less fluctuation, allowing the reservoir to have a more suitable behavior. In addition, the swings of the reservoir and the total output power decrease when we increase the network utilization factor. This positive effect is not felt on the output power excursion of an individual WDM channel. This is due to the fact that the cumulative effect of the reservoir fluctuations along a cascade will lead to further broadening of any individual channel power PDF. Finally, we investigate the effect of gain clamping of the first amplifier in the cascade – by implementing a ring laser and propagating the lasing power through the cascade – on the statistics of several measurable entities.

Keywords: Modeling, optical communication, optical fiber amplifiers, wavelength-division multiplexing, transient analysis

1. Introduction

Among the most important characteristics of erbium-doped fiber amplifiers (EDFAs) for wavelength division multiplexing (WDM) applications is a gain ripple over the required wavelength range. Therefore, factors such as the number of WDM channels, their input powers, or the network utilization factor, influence significantly the behavior of a cascade of several EDFAs. As for any design procedure, we must try to ensure optimal functioning for a cascade of EDFAs over a point-to-point link. For this to happen, it is important to verify the consequence of the above external parameters on the dynamics of the cascade.

An EDFA can be modeled as a dynamic non-linear system with only one state variable: the total number of excited ions in the amplifier, known as the *reservoir* (Bononi and Rusch, 1998). The reservoir is defined by $r(t) \triangleq \rho A \int_0^\ell N_2(z,t)\, dz$, where $N_2(z,t)$ is the fraction of population in the $^4I_{13/2}$ metastable level, ℓ is the length of erbium-doped fiber (EDF), and ρ is the Er^{3+}-ion density in the doped fiber core of effective area A.

The response of EDFAs to non-periodic changes in the power of input channels (due to channel add/drop, network reconfiguration, fiber cuts, or packetized traffic) has been the subject of much recent research (Srivastava and Sun, 1997; Ziskind and Srivastav, 1996; Bononi and Tancevski, 1998). Due to changes in the input power, the gain excursion in an EDFA leads to the occurrence of wide swings in the output power and output optical signal-to-noise ratio (OSNR) (Karasek and Vallés, 1998; Tancevski and Bononi, 1999). Moreover, the gain of an EDFA is in general wavelength dependent, which leads to different gains among WDM channels. Hence, signals traversing a cascade of several amplifiers will experience an increasing output power spread among individual channels.

Any modification of the total input power of WDM signals is reflected in reservoir dynamics. If some channels are not present for a certain time, the total EDFA input power decreases. From a system perspective, the removal of input fluxes under constant pump conditions will lead to a rising reservoir level. Conversely, if some channels are added, the total input power increases, and the reservoir falls. EDFA gain fluctuations are a direct consequence of such reservoir variations (Bononi and Tancevski, 1998). The variations of one channel, hence, affects the gain of all the other channels, leading to excursions in individual channel output power, even though their input power is maintained constant. Such interaction is defined as *cross-gain saturation.*

In this work, we investigate the effect of traffic burstiness and of the number of WDM channels on the dynamics of an EDFA cascade.

2. Simulation parameters

The numerical model used for the simulations is based on the dynamic model of EDFA described in detail in (Karasek and Bononi, 2000) and assumes homogeneously broadened gain medium and absence of excited-state absorption. The effect of amplified spontaneous emission (*ASE*) is taken into account. We consider a cascade of six identical single-stage amplifiers with typical *Lucent Technology* EDF pumped at 1480 nm. The pump power and the length of each EDFA are identical and are respectively $P_p = 25\,\mathrm{mW}$ and $l = 25\,\mathrm{m}$. All emission and absorption cross-sections are spectrally resolved in the region from 1450 to 1650 nm, subdivided into $N_{ASE} = 1000$ bins of $\delta\lambda = 0.2$ nm. The inter-amplifier loss is assumed to be 20 dB. For a typical transmission fiber loss of 0.25 dB/km, that represents a distance between neighboring amplifiers of 80 km.

In our analysis, we consider burst-mode packet data transmission, such as native mode ATM, Ethernet, or IP over WDM. Such traffic has been shown to be self-similar (Taqqu and Willinger, 1997; Leland and Taqqu, 1994), i.e., there exists a very high variability in the duration of the ON and OFF bursts on each channel. A self-similar traffic entails a wide variation of the total input power level over time scales comparable to the EDFA chain reaction time (Tancevski and Bononi, 1999). This can result in large variations in the EDFA gain level, and output power of the channels, due to the effect of EDFA cross-gain saturation.

The time varying bursty signals representing traffic of the WDM channels were randomly generated as independent ON/OFF sources, i.e., each as a succession of ON and OFF periods. The duration of each ON/OFF period is assumed to be a random variable with a rounded Pareto distribution (Tancevski and Bononi, 1999). Using a random number U uniformly distributed on [0, 1], statistically independent ON and OFF intervals (in units of slots) were generated for each WDM channel via

$$T_{i,\mathrm{ON}} = T_{slot}\left\lfloor \frac{1}{U^{1/\alpha_{\mathrm{ON}}}} \right\rfloor, \qquad T_{i,\mathrm{OFF}} = T_{slot}\left\lfloor \frac{1}{U^{1/\alpha_{\mathrm{OFF}}}} \right\rfloor \tag{1}$$

where α_{OFF} and α_{ON} are parameters regulating the burstiness of the traffic, and $\lfloor x \rfloor$ is the floor function. The elementary size of one ON/OFF period is 53 bytes, corresponding to one ATM cell, $T_{slot} = 2.83\,\mu s$. The average utilization of the aggregate WDM traffic is given by the utilization factor u

$$u = \frac{\sum_{i=1}^{NCH} E[T_{\mathrm{i,ON}}]}{\sum_{i=1}^{NCH} E[T_{\mathrm{i,\,ON}}] + \sum_{i=1}^{NCH} E[T_{\mathrm{i,OFF}}]} \tag{2}$$

where NCH is the number of WDM channels, $E[T_{\mathrm{i,ON}}]$, $E[T_{\mathrm{i,OFF}}]$ are the mean values of the ON and OFF bursts (in slot units) of individual WDM channels. Different network utilization factors u have been implemented in order to evaluate the behavior of a cascade with various traffic types and densities. In order to obtain different utilization factor, we must change the value of α_{ON} and α_{OFF}, which are indicators of the degree of variability of the traffic, as indicated in (Tancevski and Bononi, 1999).

In our simulations, we maintain the total input power to the cascade constant while the number of WDM channels varies. Indeed, we consider three cases: 2, 8 and 32 WDM channels centred at 1550 nm with a spacing of 0.4 nm. In these three cases, we assume that the average input power for an ON period is respectively $P_{ch}^{in} = -13, -19, -25$ dBm for each WDM channel. In addition, we assume that power of one ON period is not constant, as suggested in previous studies (Bononi and Tancevski, 1998; Tancevski and Bononi, 1999), where the ratio of "0" and "1" in the ON slots is assumed to be exactly 50% for each packet. On the contrary, we model the ON packet power as a binomial random variable due to two main reasons. The reservoir time response has not the same order as the packet time, and we suppose that the power in one ON slot is related to the number of "1" which can be supposed as a bernoulli random variable. Also, we have considered that during the OFF period, there is a fixed input value $P_{in}^{ch} = -50$ dBm. This value is comparable with a thermal noise generated by the transmitter.

Alongside the WDM channels, we have added a continus wave (CW) placed before the WDM comb having a constant input power $P_{sur}^{in} = -25$ dBm, corresponding to a single channel power in the case of 32 WDM channels. With such a low input power level, the effect of the CW channel on the behavior of the cascade is negligible. Besides the fact that it is an easily measurable quantity, the CW channel is an indicator of the gain excursion.

The effect of the bursty packetized traffic is evaluated statistically. The probability density function (PDF) histograms of the reservoir, the total output power, the CW output power, and also the output power of the reference wavelength are determined along the cascade.

To suppress eventual large gain and output power variations, we examine the use of all optical gain-clamping (Dai and Pan, 1997). The first EDFA in the chain is clamped by a ring laser configuration, with two 10:90 couplers in the input and in the output. In the feedback loop of the AOGC EDFA, an optical band-pass filter centered at the lasing wavelength of $\lambda_l = 1543$ nm, and having a Gaussian transmittivity was simulated. The ring configuration at the first EDFA creates a lasing signal that counteracts the fluctuations in the input channel power level, and tends to stabilize the doped fiber inversion level to the steady-state value. Lasing power from the first gain-clamped EDFA is allowed to propagate down the cascade, thus stabilizing to some extent the successive EDFAs.

3. results and Discussion

This section is divided into three parts. First, a transient analysis gives considerable insight into the effects of cross-gain saturation and their dependence on the number of channels. In the following part, the implications of cross-gain saturation are analyzed in detail from a statistical point of view. Clamping results are depicted in the last subsection.

3.1. Transient Analysis

We start with the time evolution of the CW output power and the total output power at the end of the first EDFA for a time period of 340 μs with a network utilization factor $u = 0.5$ ($\alpha_{ON} = \alpha_{OFF} = 1.2$). Figure 1a shows the high variability of the CW channel in case of 2 and 32 WDM channels with the same power budget.

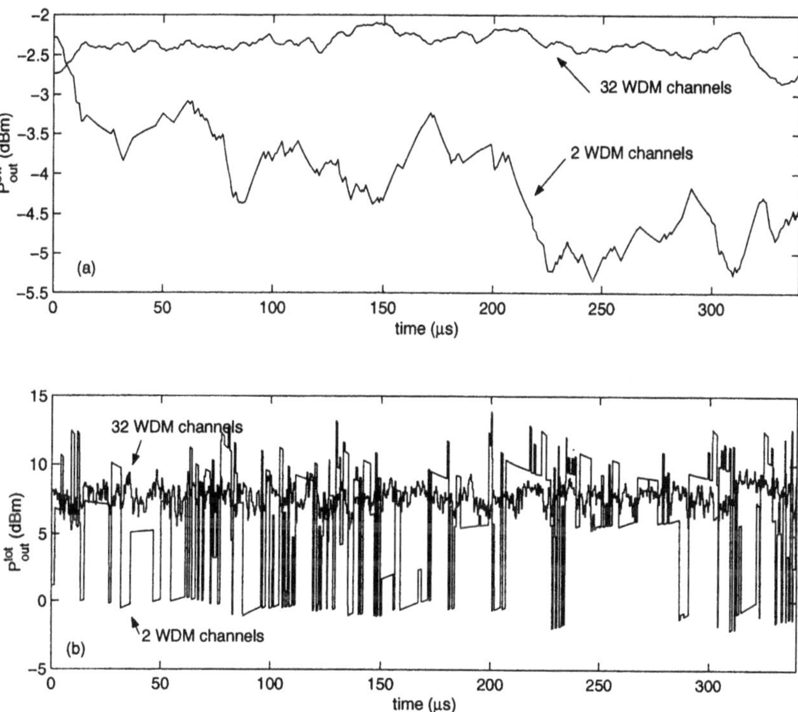

Figure 1. (a) Time evolution of continuous wave at the output of an EDFA with 2 and 32 WDM channels. (b) Time evolution of total output power at the output of an EDFA with 2 and 32 WDM channels.

The significant CW power swings in the case of only 2 WDM channels are caused by large input power variations. If in one of the two channels a long OFF period occurs, input power to the amplifier drops by 50 %. This gives the EDFA enough time to acquire gain greatly exceeding the average value. With increasing number of channels the probability of concurrent occurrence of long OFF periods in several channels decreases which results in the decrease of gain fluctuations. Figure 1b plots time evolution of total output power when 2 and 8 WDM channels are transmitted.

3.2. Statistical Analysis

The results of the simulation in a cascade of six EDFAs are shown in Fig. 2. In this figure, the rows correspond to the number of WDM channels in the system. Three different configurations were considered: 2, 8, and 32 WDM channels. The three columns represent the three network utilization factors implemented: $u = 0.25$ ($\alpha_{ON} = 1.91, \alpha_{OFF} = 1.2$), $u = 0.5$ ($\alpha_{ON} = 1.2, \alpha_{OFF} = 1.2$) and $u = 0.75$ ($\alpha_{ON} = 1.2, \alpha_{OFF} = 1.91$). For each number of WDM channels and network utilization factor, the PDFs of the reservoir of the 1st, 3rd, and 6th EDFAs are plotted. The first general observation is that the fluctuations of the

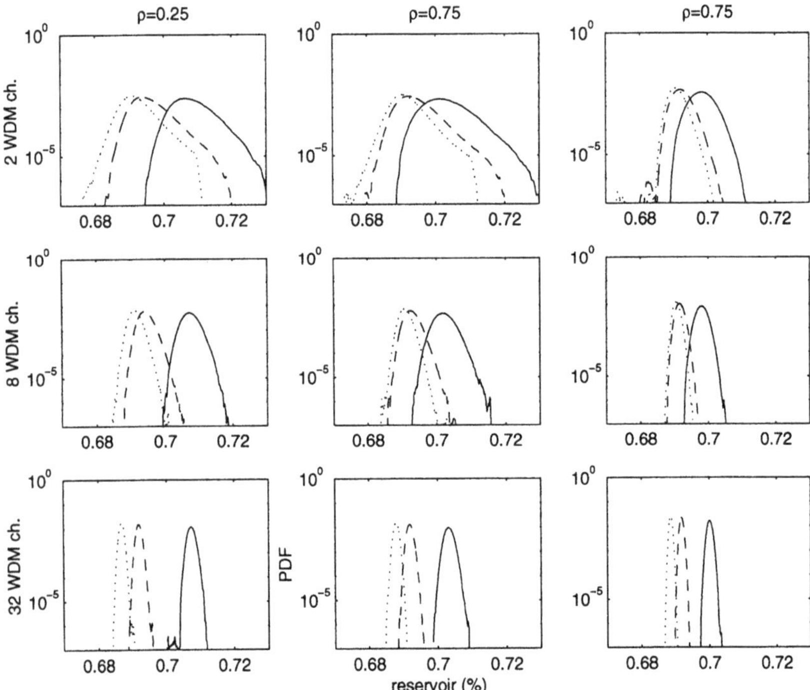

Figure 2. Probability density function of the reservoir for an unclamped cascade for different numbers of WDM channels and different utilization factors. Amplifier 1, 3 and 6 are represented by the solid, dashed and dot line, respectively.

reservoir decrease along the cascade in all the cases. We also observe that the mean value of the reservoir decreases along the cascade due to the propagation of ASE. We remark also that the excursion of the reservoir decreases when we increase the utilization factor or the number of WDM channels. However, the narrowing is more significant with the increase of the number of channels. In both cases, the decrease of reservoir fluctuations is best explained with total power arguments, as we will see below.

In order to realize the importance of cross-gain saturation on the behavior of a cascade of several EDFAs, we evaluate statistically the PDF of the total output power along the cascade as shown in Fig. 3. The three distinct maxima of the two-channel case are due to the

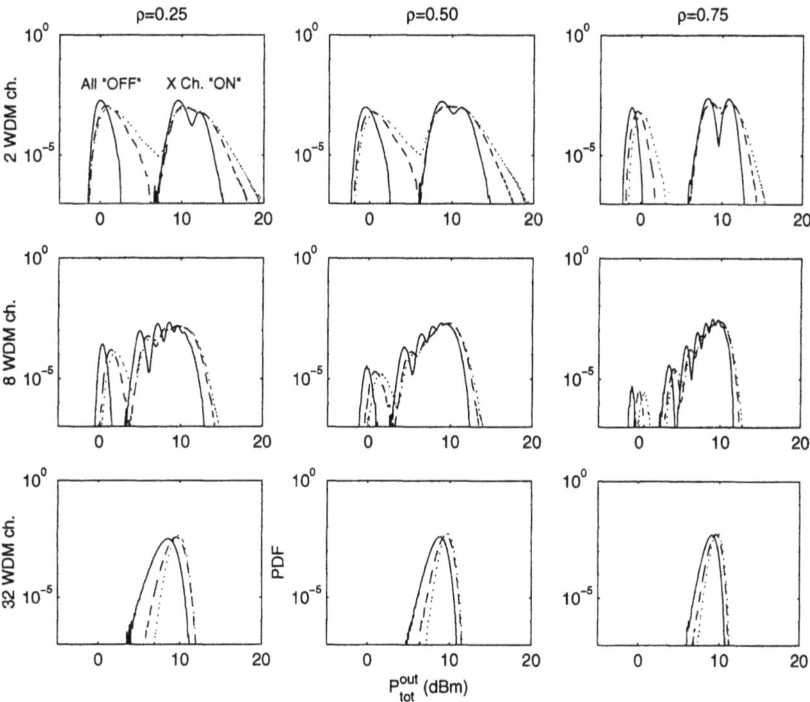

Figure 3. Probability density function of the total output power for an unclamped cascade for different numbers of WDM channels and different utilization factors. Amplifier 1, 3 and 6 are represented by the solid, dashed and dot line, respectively.

frequent occurrence of the three extreme cases: "all OFF", "1 ON" or "all ON". As the number of WDM channels increases, the intermediate states become more frequent, and the PDFs approach a gaussian shape. In addition, it is interesting to mention that the probability of having all channels in the OFF state (represented by the left-hand PDF peak) decreases as the network utilization factor increases, as can be expected.

We note that the general behavior of the total output power is the same as that observed for the reservoir histograms: First, the excursion of the total output power decreases along the cascade; second, we notice that as the utilization factor increases, the fluctuation of the total output power decreases. In contrast to the reservoir statistics, we observe that the mean value of the total output power increases along the cascade due to growing ASE.

A comparison between Fig. 2 and Fig. 3 shows clearly the dependence of reservoir excursions on the variability of total input power. Indeed, the reservoir dynamics are directly related to the total input flux. As the number of WDM channels or the network utilization factor increase, the variability of the total input power becomes less pronounced which allows a more stable reservoir behavior.

Let us now verify whether a high number of WDM channels and a high utilization factor will affect the dynamics of a single WDM channel. We evaluate, in Fig. 4, the output power PDF of the reference channel at $\lambda_{ref} = 1550\,\text{nm}$.

It is clear from Fig. 4 that the excursion of output power of the reference channel increases along the cascade, for any number of WDM channels and network utilization factor. We note that the excursion of each PDF decreases, but not significantly, as the utilization factor increases. It is important to notice that, in contract with the total power PDFs, the reduction in fluctuation of individual output powers due to the increasing number of channels or having a denser traffic were not so substantial. This can be explained by two main reasons. First, the input power of each channel is highly variable along time. The second reason lies in the relation between the output power and the reservoir shown below:

$$\begin{aligned} P_{\lambda_i}^{(k,out)}(t) &= P_{\lambda_i}^{(k,in)}(t)\,.\exp\left(B_i r^{(k)}(t) - A_i\right) \\ &= P_{\lambda_i}^{(k,in)}(t)\,.G_i^{ss}.\exp\left(B_i \Delta r^{(k)}(t)\right) \end{aligned} \tag{3}$$

with G_i^{ss} is the gain value at λ_i at the steady-state and $\Delta r^{(k)}(t) = r^{(k)}(t) - r^{(k,ss)}$ as the difference between the level of the the actual value and steady-state.

From eq.(3), it is clear that any perturbation in the reservoir level leads to a broadening of the output power PDF following the exponentiel relation. Recall that an ON power for each packet is modeled as a uniform variable, hence generating a broad PDF. Any variability, even slight, in the reservoir will lead to further broadening of that PDF. That explains the observation that although reservoir PDFs get nar-

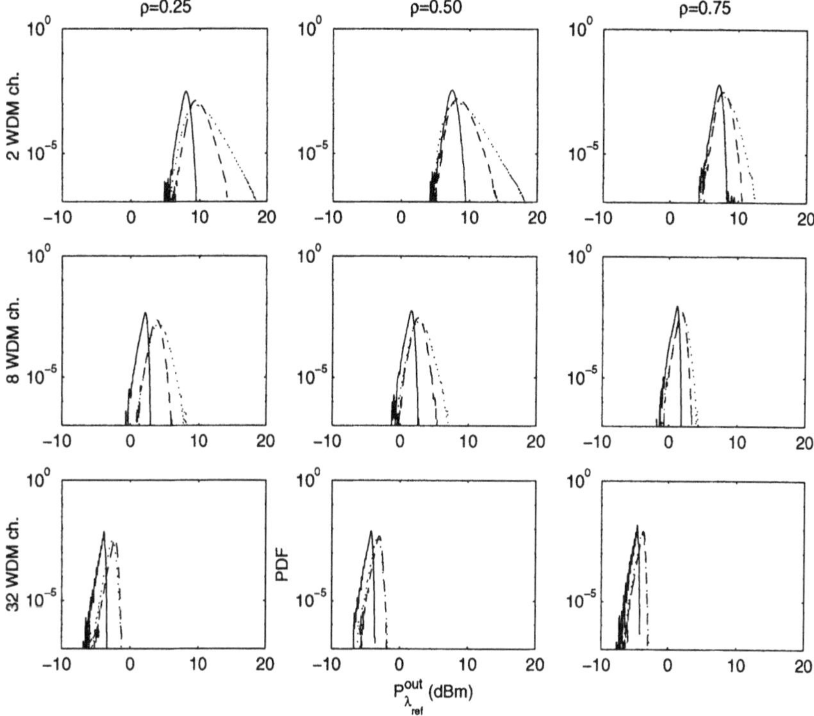

Figure 4. Probability density function of the output power of the reference channel for an unclamped cascade for different numbers of WDM channels and different utilization factors. Amplifier 1, 3 and 6 are represented by the solid, dashed and dot line, respectively.

rower along the cascade (due to cross-gain saturation), the cumulative effect of the reservoir fluctuations after several EDFAs leads to the broadening of the individual channel power PDFs. The growth of the output power fluctuations along the cascade is better visualized in the CW channel as illustrated in Fig. 5. Besides, the CW channel can be used experimentally to measure the gain and the reservoir fluctuations. Fig. 5 shows the PDF of the output power of the CW channel for a cascade of 8 WDM channels and a utilization factor $u = 0.5$.

The broadening of individual channel power PDFs and the narrowing of reservoir and total power PDFs along the cascade are not contradictory. The fact that individual channel power fluctuations do not necessarily imply similar variations on total output power is due to cross-gain saturation. Indeed, the switching of a single channel to the OFF state, for instance, induces the increase in gain, hence output power, of all other channels. The output power decrease in the dropped

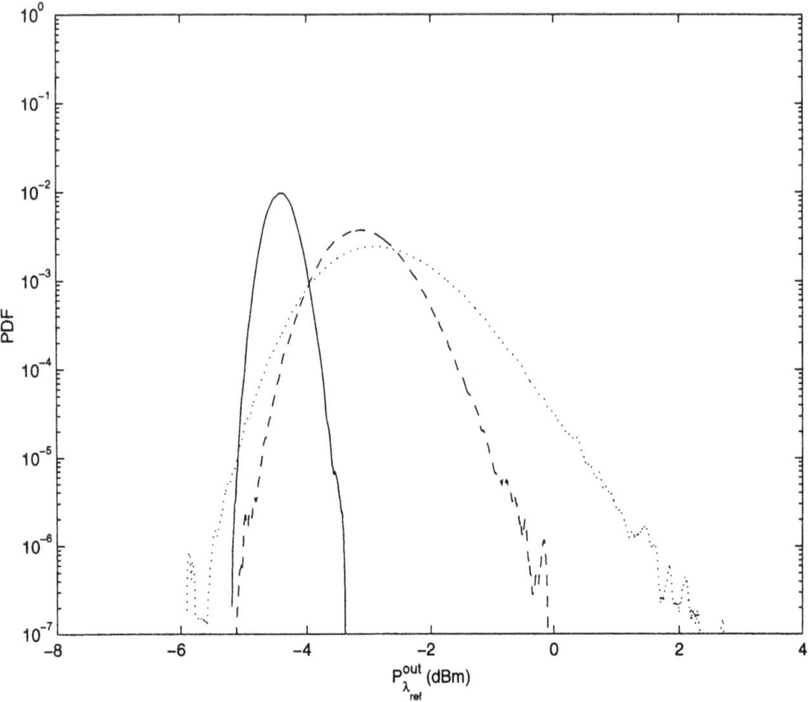

Figure 5. Probability density function of the continuous wave for an unclamped cascade with 8 WDM channels and $u = 0.5$. The outputs of amplifiers 1, 3 and 6 are represented with solid, dashed and dot lines, respectively.

channel is, in a sense, balanced by the increase in the output power of the other channels. Those variations are therefore reflected on the dynamics of individual channels rather than on the total output power (or the reservoir), as is clear from the PDFs of Fig. 2 and Fig. 4.

The fluctuations in the reservoir level have been identified as the direct cause of gain instability in EDFAs and EDFA cascades. The effect of cross-gain saturation depicted in Fig. 2 is therefore beneficial for it results in the reduction of reservoir excursions. However, it has no effect on individual output power swings (see Fig. 4 and Fig. 5). In order to stabilize them, any reservoir fluctuations must be suppressed, especially at the beginning of the cascade, where the largest reservoir fluctuations occur. All-optical gain clamping has been proposed and demonstrated for that purpose (Dai and Pan, 1997; Zirngibl, 1991). In the following, we examine the effect of using AOGC on the the statistics of the reservoir and reference channel output power.

3.3. Clamping Effect

The advantages of clamping the first EDFA are shown in Fig. 6. This figure shows the PDF of the reservoir of the 1st, 3rd, and 6th EDFAs for a unclamped (Fig. 6a) and clamped (Fig. 6b) cascade with 8 WDM channels and an utilization factor $u = 0.5$. First, we note that the

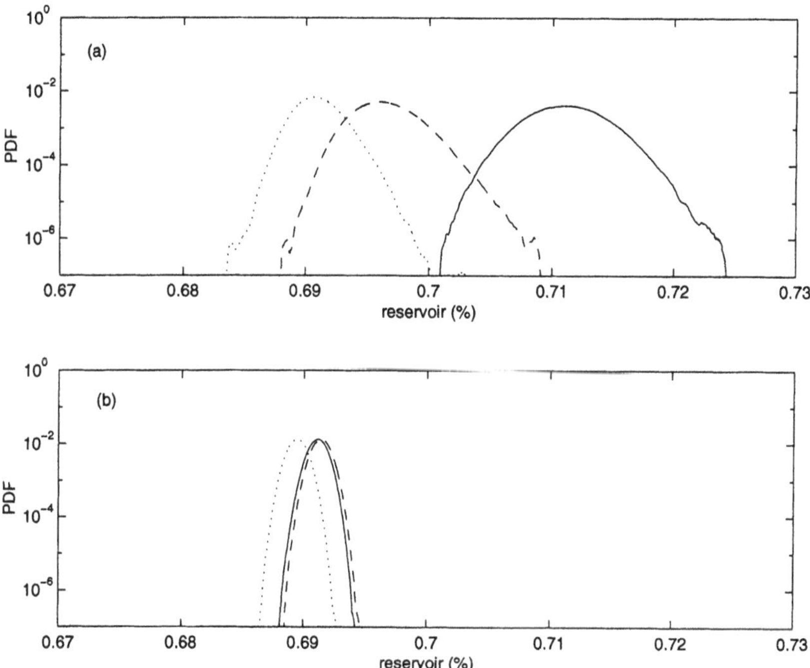

Figure 6. Probability density function of the reservoir for an unclamped (a) and clamped (b) cascade with 8 WDM channels and $u = 0.5$ at the amplifier 1, 3 and 6 with solid, dashed and dot line, respectively.

reservoir mean values are slightly lower in the clamped case due to the decrease in gain caused by the propagation of lasing power. More importantly, the largest spread of the reservoir probability at the level of 10^{-6} is less than 0.5% along the cascade, in contrast with 3% in the unclamped cascade, corresponding to the first EDFA. Nevertheless, the effect of cross-gain saturation is still visible in the clamped cascade, for reservoir excursions decrease along the cascade.

In Fig. 7a, b, the histograms of signal power of reference channel at the 1st, 3rd, and 6th EDFAs are plotted for an unclamped and clamped cascade, respectively. The PDFs are plotted for a 0.5 utilization factor, a cascade of 6 EDFAs and 8 WDM channels, an ON average power of -19 dBm, and a fixed OFF power of -50 dBm.

We notice that also for the clamped cascade, the output power fluctuations grow along the cascade. However, a significant improvement is perceptible. Reduction of at least 4 dB is achieved with respect to the unclamped cascade. The width of the reference channel output power PDF at the end of the unclamped cascade, at the level of 10^{-6}, is 6.75 dB. For the clamped cascade, we obtain only 3.75 dB.

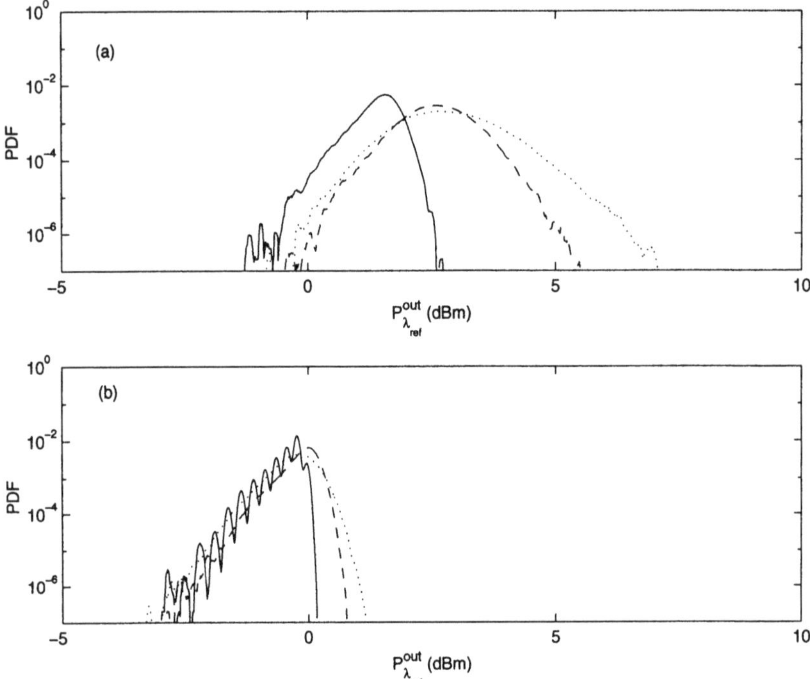

Figure 7. Probability density function of the output power of the reference WDM channel for an unclamped (a) and clamped (b) cascade representing ON periods with 8 WDM channels and $u = 0.5$ at the amplifier 1, 3 and 6 with solid, dashed and dot line, respectively.

4. Conclusion

We have shown that the effect of cross-gain saturation depends on the number of WDM channels, and also on network traffic variability. Indeed, the swing of the reservoir and the total output power decreases when we increase the number of WDM channels or the network utilization factor. As the number of WDM channels increases, the dynamics of the total input power no longer reflects the individual channel power fluctuations, thus allowing the reservoir to have less excursions. This advantage does not affect individual output power excursions. In fact, the cumulative effect of the reservoir fluctuations along a cascade will lead to further broadening of all individual channel power PDFs.

References

Bononi, A. and L. A. Rusch, Doped-Fiber Amplifier Dynamics: A System Perspective. *Journal of Ligthware Technology*, 16:945–956, May 1998.

Srivastava, A.K., Y. Sun, J.L. Zyskind, and J.W. Sulhoff, EDFA Transient Response to Channel Loss in WDM Transmission System. *IEEE Photon. Technol. Lett.*, 9:386–388, 1997.

Zyskind, J.L., A.K. Srivastava, J.C. Ellson Y. Sun, G.W. Newsome, R.W. Tkach, A.R. Chraplyvy, J.W. Sulhoff, T.A. Strasser, J.R. Pedrazzani, and C. Wolf, Fast Link Control Protection for Surviving Channels in Multiwavelength Optical Networks. in *Proc. European Conf. Opt. Commun. Oslo*, June 1996.

Bononi, A., L. Tancevski, and L. A. Rusch, Fast Dynamics and Power Swings in Doped-Fiber Amplifiers Fed by Highly Variable Multimedia Traffic. *in Optical Fiber Communication conference, OSA Technical Digest Series (Optical Society of America, Washington)*, 2:213–214, Feb. 1998.

Karasek, M., and J. A. Vallés, Analysis of Channel Addition/Removal Response in All-Optical Gain-Controlled Cascade of Erbium-Doped Fiber Amplifiers. *IEEE Journal of Lightwave Technology*, 16(10):1795–1803, Oct. 1998.

Tancevski, L., A. Bononi, and L. A. Rusch, Output Power and SNR Swings in Cascades of EDFAs for Circuit Packet-Switched Optical Networks. *IEEE Journal of Lightwave Technology*, 17(5):733–742, May 1999.

Karasek, M., A. Bononi, L.A. Rusch, and M. Menif, Effectiveness of Gain Control in EDFAs against Traffic with Different Levels of Bursty Behaviour. *IEE Proc. Optoelectronics*, 147(5):394–411, 2000.

Taqqu, M., W. Willinger, and R. Sherman, Proof of a Fundamental Result in Self-Similar Traffic Modeling. *ACM-SIGCOMM Computer Communications Rewiew*, 27(1):5–23, Feb. 1997.

Leland, W. E., M. S. Taqqu, W. Willinger, and D. V. Wilson, On the Self-Similar Nature of Ethernet Traffic (Extended Version). *IEEE/ACM Transactions on Networking*, 2(1):1–15, Feb. 1994.

Dai, H., J.Y Pan, and Ch. Lin, All-Optical Gain Control of In-Line Erbium-Doped Fiber Amplifiers for Hybrid Analog/Digital WDM Systems. *IEEE Photonics Technology Lettres*, 9:737–739, 1997.

Zirngibl, M. Gain Control in Erbium-Doped Fiber Amplifiers by All-Optical Feedback Loop. *Electron. Lett.*, 27:560–561, 1991.

Influence of intensity noise in spectrum-sliced WDM systems

Mingshan Zhao, Geert Morthier, Johan Dekoster*, Bart Moeyersoon and Roel Baets

Department of Information Technology (INTEC), University of Gent-IMEC, Sint-Pietersnieuwstraat 41, B-9000 Gent, Belgium.

**Siemens Atea-ICN OT DH, Atealaan 34, B-2200 Herentals.*

E-mail: mszhao@intec.rug.ac.be

Key words: wavelength division multiplexing, spectral slicing, intensity noise, semiconductor optical amplifier

Abstract: We discuss the influence of intensity noise on the achievable bitrates and transmission distances of spectrum-sliced WDM systems and the reduction of the intensity noise using a saturated SOA. Also the influence of the gain-saturation levels and of the bias currents of the SOA on the noise reduction was studied. For the optimum operation of the saturated SOA, a noise reduction of 17dB at low frequencies and 10 dB at high frequencies was obtained experimentally.

1. INTRODUCTION

Spectral-slicing is a WDM technique in which optical filters are used to obtain a spectral slice of light from a broadband source and modulators encode data onto the slice. Spectrum-sliced WDM systems utilising incoherent broadband sources such as LEDs, superluminescent LEDs (SLEDs) and Erbium-doped fibre amplifiers (EDFAs) have advantages of low cost, high wavelength selectivity and temperature stability as compared to conventional DWDM systems. Such systems are strong candidates for future fibre-to-the-home access networks and have potential for cost-sensitive local area network applications. Spectrum-sliced incoherent light however, exhibits a large intensity noise that places limits on the achievable system performances.

In this paper, we present results about the influence of the intensity noise on the achievable bitrates and transmission distances of spectrum-sliced WDM systems and about the reduction of the noise using a saturated semiconductor optical amplifier (SOA). A noise reduction of 17dB at low frequencies and 10 dB at high frequencies was obtained, which is much better than what has so far been published in the literature (10dB at low frequencies) [1].

2. INFLUENCE OF INTENSITY NOISE ON THE TRANSMISSION PERFORMANCE OF SPECTRUM-SLICED WDM

The setup used for the transmission experiment using a single spectrum sliced WDM channel is shown in Fig. 1. The amplified spontaneous emission (ASE) from a SLED (SLED 1550MHFD, Opto speed) was spectrally sliced using a fibre Bragg grating (FBG) filter. The slice of light was NRZ-modulated using a $LiNbO_3$ external modulator with a 2^{23}-1 pseudorandom-bit-sequence (PRBS) pattern, and then detected using a receiver with electrical low pass filters and an oscilloscope (Tektronix 11801B digital sampling oscilloscope).

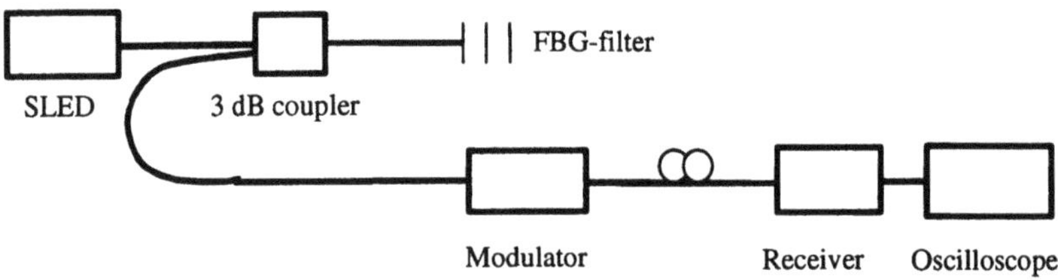

Figure 1. Setup for the fibre link experiment with spectrum-sliced WDM

Fig. 2 shows the eye diagrams of the spectrum-sliced WDM channel for different bitrates and for back to back operation. There is a considerable contribution of intensity noise to the high level of the eye. At low bitrates, e.g. several hundreds of Mb/s, the eyes are open well, but at high bitrates, e. g. 2.488Gb/s, the intensity noise becomes significant and the eyes are almost closed. The eye diagrams for different slice bandwidths are shown in Fig. 3. Larger slicewidth results in a higher signal to noise ratio and thus a better eye diagram. These results are consistent with theoretical analysis. With the assumption of a Gaussian probability density function, the signal to noise ratio (SNR) of a spectral slice is given by

$$SNR = \sqrt{\frac{B_o}{2B_e}} = \sqrt{\frac{2B_o}{3B}} \qquad (1)$$

when intensity noise dominates (that means, thermal noise and shot noise can be neglected), where B_e (= $0.75B$) is the electrical bandwidth of the receiver, B_o the optical bandwidth of the spectral slice, and B the data bitrate.

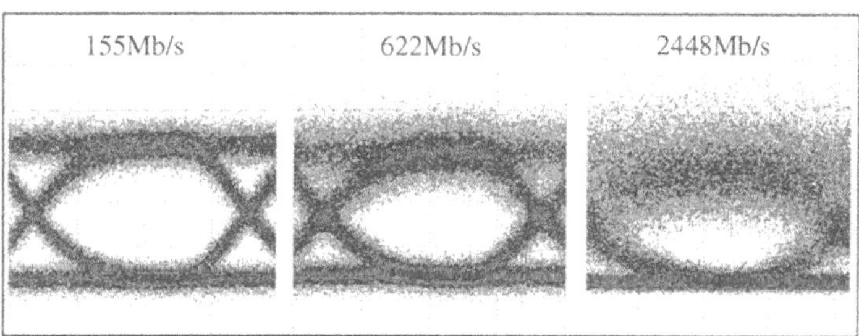

Figure 2. Eye diagrams for different bitrates at back to back operation. Slicewidth: 0.5nm

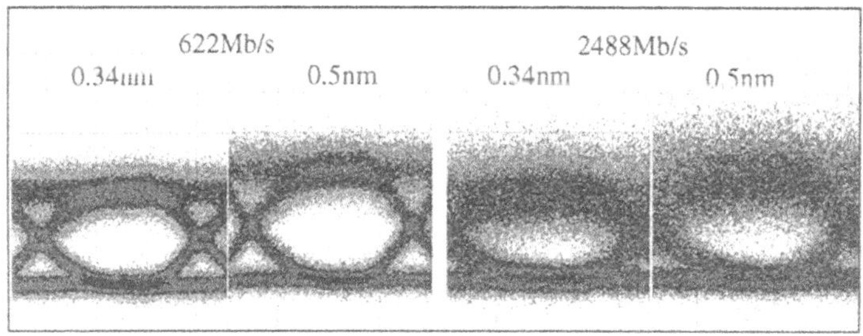

Figure 3. Eye diagrams for different slicewidths and different bitrates. Back to back operation

Equation 1 shows that the signal to intensity noise ratio of a spectral slice is proportional to the square root of the slicewidth B_o, but inversely proportional to the square root of the data bitrate B. For a given slicewidth, higher bitrate results in a lower SNR. A lower bitrate results in a higher SNR and thus a larger eye height. For a given bitrate B, the SNR can be increased by increasing the slicewidth. However, a larger slicewidth also leads to larger dispersion induced inter-symbol interference, which seriously degrades the system performance. So, intensity noise places limits to the bitrate and transmission distance in spectrum-sliced WDM. With PTDS (Photonic Transmission Design Suite) from Virtual Photonics Incorporated [2], simulations of the transmission of a single spectrum-sliced WDM channel were performed. The results show that, as can be seen in Fig. 4 and 5, for a bitrate of 625 Mb/s, transmission over 100km with a BER of 10^{-9} can be achieved, for a slicewidth of 0.8nm. But for a bitrate of 2.5Gb/s, less than 10km transmission with a BER of 10^{-9} can be achieved, for a slicewidth of 1.4nm.

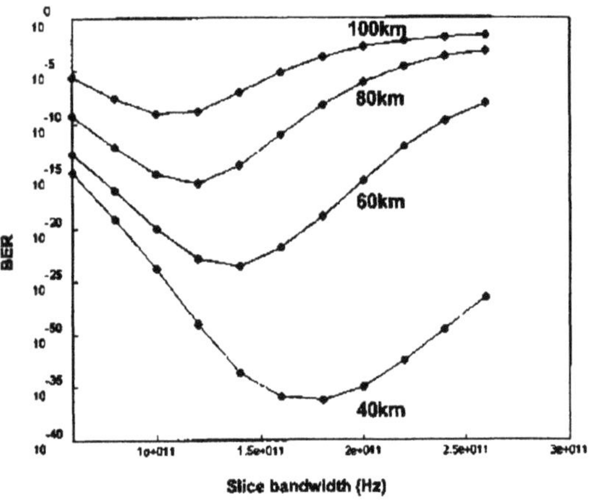

Figure 4. BER vs. slicewidth for different lengths of fibre links at a bitrate of 625Mb/s.

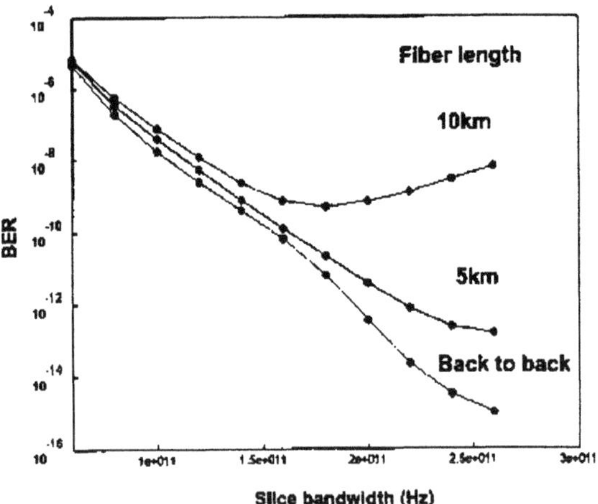

Figure 5. BER vs. slicewidth for different lengths of fibre links at a bitrate of 2.5Gb/s.

3. REDUCTION OF INTENSITY NOISE IN SPECTRUM-SLICED WDM USING A SATURATED SOA

Both the experimental and the simulation results given in the above section have shown that large intensity noise in spectrum-sliced WDM strongly limits transmission bitrate and distance. In order to break the limitations and improve the performance of spectrum-sliced WDM, the noise property of the spectral slice must be modified. A saturated SOA exhibits a suppressed intensity noise for narrow band detection[3]. Techniques for reducing the intensity noise using a saturated SOA have recently been proposed and experimentally demonstrated [1,4]. We have further investigated the possibility of intensity noise reduction using a saturated SOA. The influence of the gain-saturation level and the bias current of the SOA on the noise reduction has been investigated experimentally, and the optimum condition for a high noise suppression ratio has been derived. In contrast to previous experiments we have used a Gain-Clamped SOA (GC-SOA).

Our experimental setup for intensity noise reduction is shown in Fig. 6. The ASE from a SLED was spectrally sliced using a 0.5nm-bandwidth FBG filter. The spectrum-sliced incoherent light centred at 1550.8nm was amplified using an EDFA and then injected into a gain saturated GC-SOA. The output of the GC-SOA was modulated using a $LiNbO_3$ external modulator. To remove the ASE from the amplifiers and the lasing-mode from the GC-SOA, a 1.2-nm bandwidth filter was used before the modulator. With a lightwave signal analyser (HP71400), a digital sampling oscilloscope (Tektronix) and an Error Detector (HP70842A), relative intensity noise, eye diagrams and BERs with and without GC-SOA were measured, respectively.

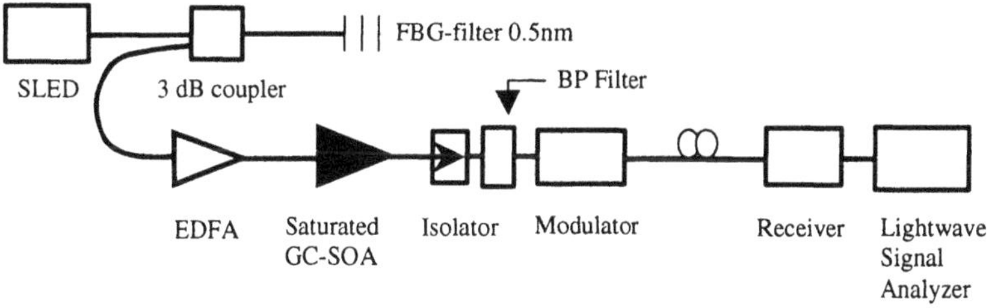

Figure 6. Experimental setup for intensity noise reduction

Figs. 7 and 8 show the measured relative intensity noise spectra of spectrum-sliced incoherent light and the noise suppression ratios, respectively. The noise level of the input was significantly suppressed over a bandwidth of several GHz. The amount of the noise reduction is proportional to the gain compression (difference between a large signal gain and a small signal gain) and bias current of the SOA but depending on the frequency. At low frequency, a noise reduction of 17dB was obtained with a high bias current of the GC-SOA. At high frequency, e. g. 2.5GHz, 10dB reduction of the intensity noise can be reached.

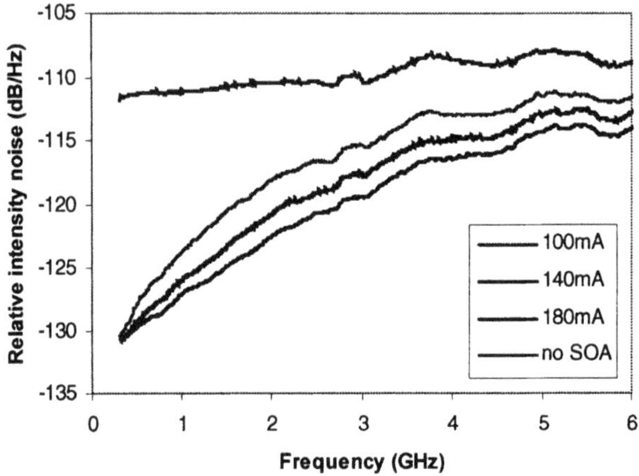

Figure 7. Measured relative intensity noise without and with a gain-saturated GC-SOA at different bias currents

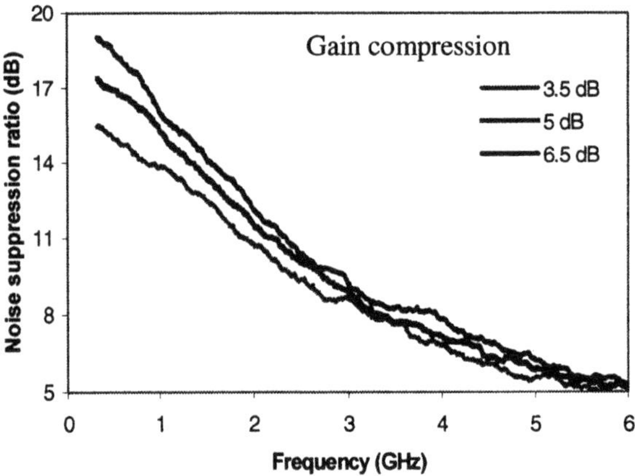

Figure 8. Noise suppression ratio for different gain-saturation levels

These points are shown more clearly in Fig. 9. At low frequency, e. g. 622MHz, the noise suppression ratio increases both with increasing bias current and with increasing gain-saturation level of the GC-SOA, but the

dependence on bias current is less strong, particularly in the high bias current region. At high frequency, e. g. 2.5GHz, the noise suppression ratio increases with increasing bias current, but it is not effective to improve the noise suppression by increasing the gain-saturation level. These experimental results above are consistent with the theoretical analyses on the noise characteristics of a saturated SOA[3] , which show that the 3dB bandwidth of noise suppression for a given input power is proportional to the intensity of the output and thus the bias current of the saturated SOA. So, in practical applications we can improve the noise suppression by increasing the gain-saturation level of the SOA at low bitrate and the bias current at high bitrate.

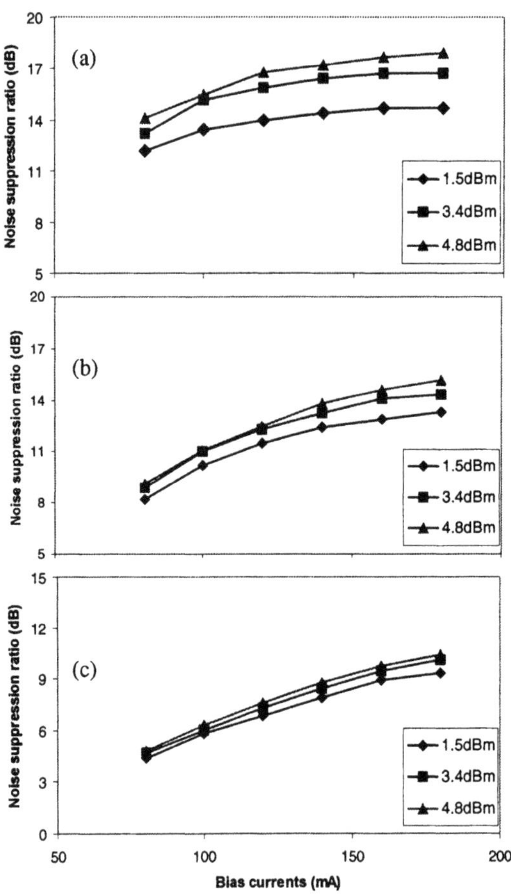

Figure 9. Noise suppression ratios vs. bias currents of the GC-SOA at different frequencies and input powers, (a) 622MHz, (b) 1.25GHz, and (c) 2.5GHz

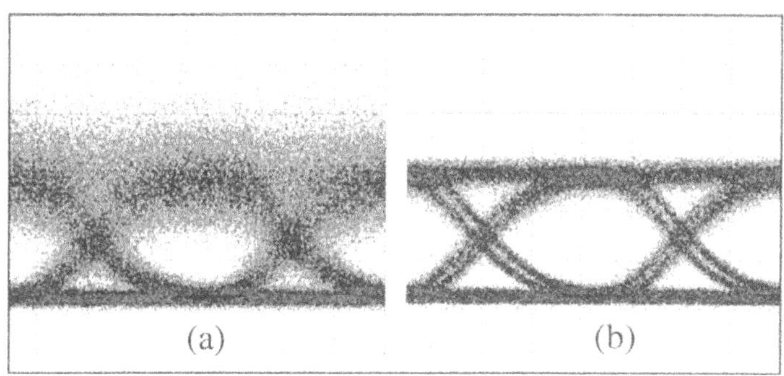

Figure 10. Eye diagrams for back to back operation for a data bitrate of 2.488Gb/s (2^{23}-1 PRBS): (a) without and (b) with GC-SOA

Fig. 10 shows the eye diagrams for 2.488Gb/s back to back operation. With the GC-SOA, the intensity noise is suppressed dramatically. The bias current of the GC-SOA is 180mA, and the gain compression is about 6.5dB.

We also measured the BER for 622Mb/s back to back operation. The results show that without GC-SOA the BER is rather high and there is a floor at around 1×10^{-6}. However, with GC-SOA the bit error rate is improved by several orders of magnitude and the error floor is removed. When the received power is larger than –18dBm, the transmission is error free. More experiments about the influence of gain saturation level and bias current on the BER are on going.

4. CONCLUSION

We have presented the influence of the intensity noise on the achievable bitrates and transmission distances of spectrum-sliced WDM systems and the reduction of the intensity noise using a saturated SOA. The influence of the gain-saturation level and the bias current of the SOA on the noise reduction has also been studied. For the optimum operation of the SOA, a noise reduction of 17dB at low frequency and 10 dB at high frequency was obtained experimentally.

ACKNOWLEDGEMENTS

This work has been supported by the Flemish Government-IWT.

References:

[1] T. Yamatoya, F. Koyama and K. Iga, "Noise suppression and intensity modulation using gain-saturated semiconductor optical amplifier", *Technical Digest, Integrated Photonics Research 2000,* Canada, July, 2000. pp. 72-74.

[2] PTDS was supplied by Virtual Photonics Incorporated: www.virtualphotonics.com.

[3] M. Shtaif and G. Eisenstein, "Noise characteristics of nonlinear semiconductor optical amplifiers in the Gaussian limit", *IEEE J. Quantum Electron.,* Vol. 32, pp. 1801-1809, 1996.

[4] S. J. Kim, J. H. Han, J. S. Lee and C. S. Park, "Intensity noise suppression in spectrum-sliced incoherent light communication systems using a gain-saturated SOA", *Photonics Technology Letters*, Vol. 11, No. 8, 1999, P 1042-1044.

OPC-TDM network performance improvement by the use of full-scalable optical packet compression/ decompression units

Slaviša Aleksić, Kemal Bengi, Vjeko Krajinović
Vienna University of Technology, Institute of Communication Networks
Favoritenstrasse 9/388, A-1040, Vienna, Austria
e-mail: Slavisa.Aleksic@tuwien.ac.at

Key words: Optical packet compression/decompression, OTDM, performance study

Abstract: An optical packet compression/decompression scheme applicable in ultrahigh-speed OTDM networks is proposed. This full-extensible scheme allows high compression rates and large packet sizes by relaxing the medium access constrains. The improvement of the Optical Packet Compression-TDM (OPC-TDM) network performance achieved by reducing the rate-conversion latency is shown.

1. INTRODUCTION

Optical Packet Compression-Time Division Multiplexing (OPC-TDM) networks provide ultrahigh-speed access to the optical medium employing the optical packet compression/decompression technique [1,2]. Fig. 1 shows an OPC-TDM access node architecture. Depending on the packet destination address, incoming high-speed packets can be either dropped or forwarded by an optical switch located at the node. An optical delay line is employed in the optical path to compensate the header processing latency. The essential part of an OPC-TDM access node is the Optical Packet Compression/ Decompression Unit, responsible for rate-conversion of ultra high-speed optical data streams to lover rate data streams that can be detected, saved and processed electronically. On the other hand, the low bit-rate packet acquired from the transmitting queue must be at first up-converted to the high medium data rate and after that inserted in the network by replacing of a dropped or an empty packet.

Recently, several methods for optical rate conversion (compression and decompression) have been proposed [3,4,5]. Most of these optical rate conversion schemes are based on an optical buffer (optical storage loop) and a sampling technique. However, there are some restrictions concerning bit-rate and packet size if this technique is used. Those restrictions are caused by difficulties in storage of ultrahigh-speed large optical packets for a larger period because of the optical signal impairing caused by dispersion and ASE (Amplified Spontaneous Emission) noise accumulation during amplification. A further technique for optical packet compression is a feed-forward delay-line structure consisting of $q = log_2(n)$ stages reported in [5]. This technique allows simultaneous compression and decompression of n-bit large optical packets using the same device. It's simple to build by using only passive components, but for larger packet lengths the splitting losses (3 dB per stage) must be compensated with an optical amplifier. However, the maximal packet size $L_{p,max}$ is limited by the compression rate K as follows:

$$L_{p,\max} \leq K-1 \quad ; \quad K=\frac{T_0}{\tau_0} \tag{1}$$

Consequently, for an 100:1 compression ($K = 100$) the $L_{p,max}$ is limited to 99 bits. Such short packets are usually impractical in many applications.

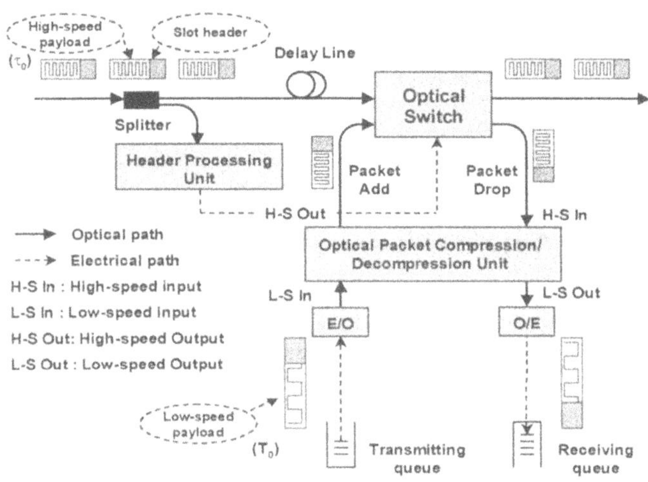

Fig. 1 OPC-TDM access node architecture.

This paper is organized as follows. In Section 2 of this paper, we propose a full scalable optical packet compression/decompression scheme using a parallel arrangement of Optical Delay Line Lattices (ODLLs). This scheme allows simultaneously compression and decompression of optical packets not limited in size and compression rate. In Section 3, a network performance analysis is performed in order to investigate the impact of this scheme on the network performance, especially on the average packet delay.

2. OPTICAL PACKET COMPRESSION/DECOMPRESSION UNIT

A figure-of-merit for OPC-TDM networks is the packet rate-conversion time relative to the packet length ($\alpha = T_{prc}/T_p$). In the case of a packet compressor consisting of a parallel arrangement of M ODLLs the α is given by:

$$\alpha = \frac{T_{prc}}{T_p} = \frac{K-1}{M} \tag{2}$$

The α signifies the number of packets in the network not accessible by a receiving node caused by down-conversion latency. In other words, the receiver must first down-convert the received packet, before it is ready for receiving another packet. For a conversion rate $K = 100$ and by the use of only one ODLL for the rate conversion ($M = 1$), the receiver can access only each 100th packet in the network. This hard restriction leads to an inefficient bandwidth utilization.

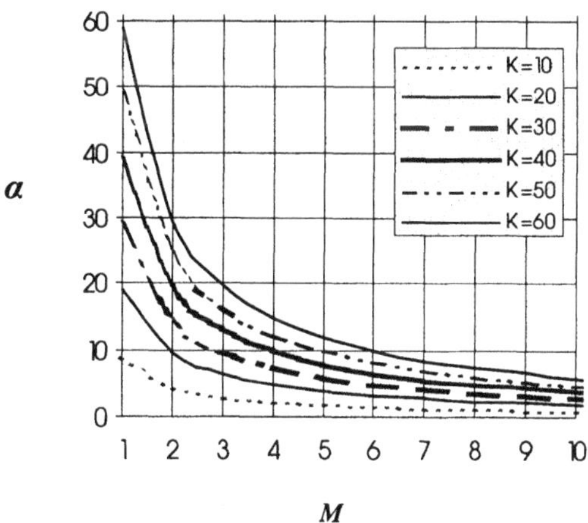

Fig. 2 Packet down-conversion time relative to packet length.

The impact of compression rate, packet size, and the number of deployed parallel transceivers on the OPC-TDM ring network performance has been partly investigated [2,6]. The results show that decreasing the α significantly improves network performance. Fig. 2 plots α over the number of deployed parallel ODLLs related to the different compression rates (K). The α shows a gradual decrease if M increases. For example, an additional ODLL causes a decreasing of α value from 49 to 24.5 for a compression rate K = 50. Therefore, we propose and investigate a packet compression/decompression scheme using a parallel arrangement of delay lines, which provides lower α values for high compression rates and large packet sizes.

The proposed scheme depicted in Fig. 3 consists of M parallel compression/decompression units, each of them responsible for rate-conversion of a part of the optical packet. The operation of the device can be described as follows: high-speed input packet (H-S In) is divided into M successive n-bit sequences using a splitter and M bi-directional optical gates. Those sequences are then copied n times in a n-bit ODLL. Each copy is delayed with respect to the next copy of the packet by $T_0 - \tau_0$ as shown in Fig. 4(a). An ultrafast OTDM demultiplexer selects within very narrow switching window bits spaced at the bit period T_0, whereby the down-conversion of the high-speed sequence is done. Finally, all M sequences are delayed by an appropriate optical delay line $DL_{2,m}$ and combined by a M x 1 coupler. Thereby, the whole high-speed input packet is down-converted.

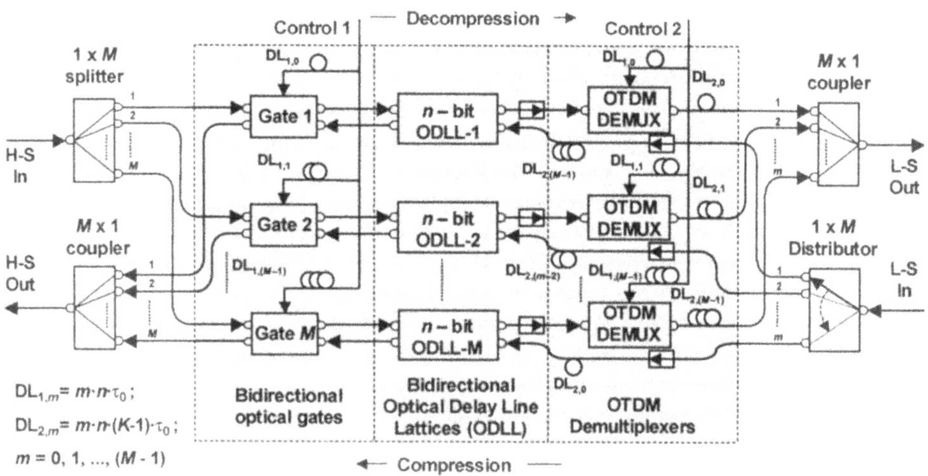

Fig. 3 Optical packet compression/decompression unit.

Moreover, using the same device in the reverse direction, a low-speed packet (bit period T_0) can be compressed as shown in Fig. 4(b). The low-speed optical packet (L-S In) is first divided into M successive sequences by an 1 x M distributor. Each sequence is then delayed by an appropriate delay line $DL_{2,m}$ and compressed by an n-bit ODLL. The fully compressed sequences are finally selected by the gates and combined by a M x 1 coupler.

The result is the high-speed output signal (bit period τ_0) that has the same bit pattern sequence as the low-speed packet. Note that both compression and decompression can occur simultaneously in the same device.

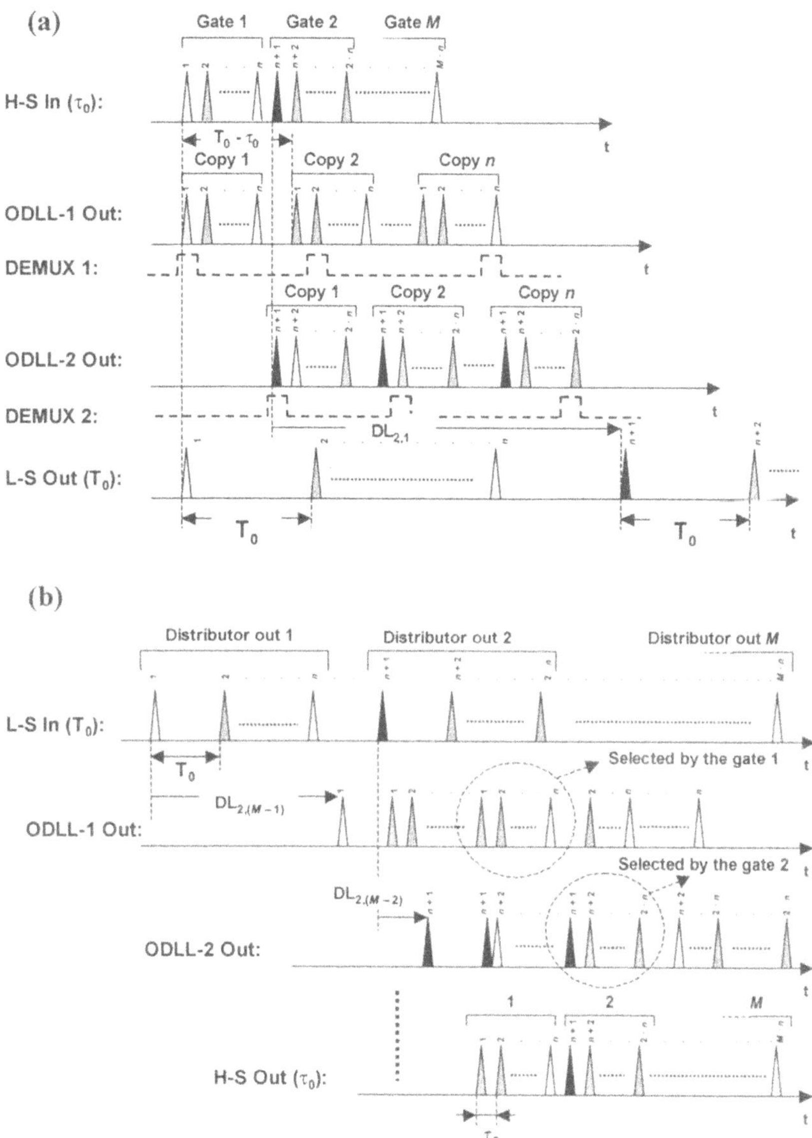

Fig. 4 Timing chart of the optical packet (a) decompression and (b) compression operation

Simulations are carried out in order to investigate the feasibility of the proposed optical packet compression/decompression scheme. The simulation setup is shown in Fig. 5(a).

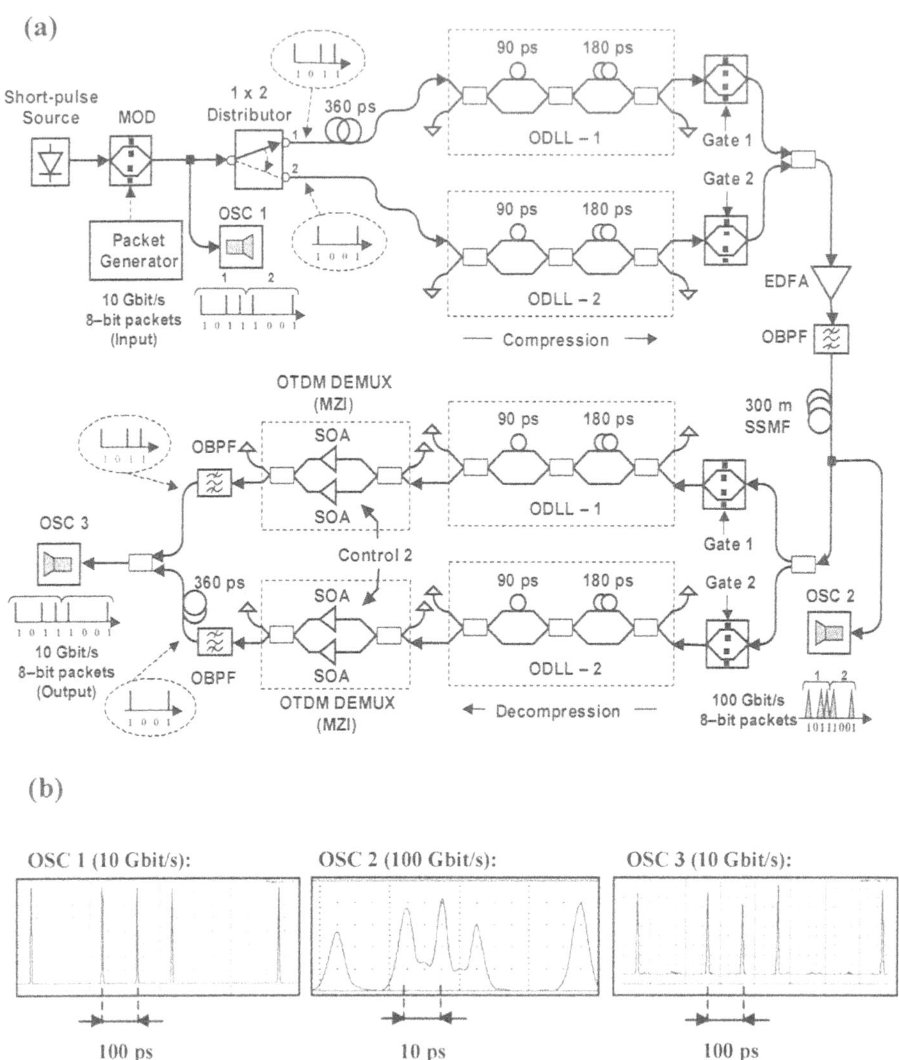

Fig. 5 Simulation setup and signal traces

8-bit large packets at 10 Gbit/s (4 ps FWHM pulses at 1.55 μm wavelength) were generated and compressed by a compressor composed of two parallel 4-bit ODLLs. The compressed packets were then transmitted over 300 m of standard single mode fiber (SSMF) and finally decompressed by the receiver using the proposed scheme. In the transmission path an EDFA (Erbium Doped Fiber Amplifier) and an OBPF (Optical Band Pass Filter) were deployed for fiber loss compensation and amplifier ASE removing respectively. A symmetric Mach-Zehnder Interferometer (MZI) with SOAs in its arms was used as an OTDM demultiplexer. An 168 to 10.5 Gbit/s demultiplexing experiment has already been reported using a similar structure [7]. We used a $LiNbO_3$ modulator for implementing the gating functionality. A fast all-optical switch (e.g. MZI) can be also deployed if ultra-fast gates are required. A 10 Gbit/s input packet (OSC 1), a 100 Gbit/s optical signal after SSMF (OSC 2), and decompressed 10 Gbit/s packet by receiver (OSC 3) are shown in Fig. 5(b). The signal traces show the general feasibility of the proposed scheme, where in our case an extinction ratio better than 13 dB was measured at the low-speed output (OSC 3). Note that an important advantage of this scheme is its scalability. It is possible to choose an optimal configuration (optimal M and n values) for achieving a lower α value along with large packet sizes and large compression rates. In comparison with a compression/decompression unit consisting of only one n-bit ODLL (requiring $q = log_2(n)$ stages), the compression/decompression unit consisting of M ODLLs can be realized using ODLLs with only $q = log_2(n/M)$ stages for compression of n-bit large packets. In this case a small number of stages implies less splitting losses. Thereby, optical signal amplification needs not to be applied.

3. OPC-TDM NETWORK PERFORMANCE STUDY

In this Section, a mathematical performance study is performed in order to investigate the impact of the proposed optical packet compression/ decompression scheme on the network performance. In this performance study, particular attention is paid to the comparison of the proposed scheme with a single feed-forward delay-line structure considering average packet transfer delay.

3.1 Protocol Description

We assume a system based on the slotted synchronized single-ring architecture as shown in Fig. 6. The MAC-protocol used in this system is simple and based on the Slotted Ring scheme with destination release and spatial slot reuse. Since only every $(\alpha + 1)$-th slot is accessible by the receiving node (due to the strict time constraints associated with the packet rate-conversion time), the slots are grouped into frames. Each slot within a frame is dedicated to a receiving station on the ring. The frame size is then $N = (\alpha + 1)$ slots, where N denotes the number of nodes in the system. The length of the time slots is a design issue and in our studies we set the slot length to the average message length in the system.

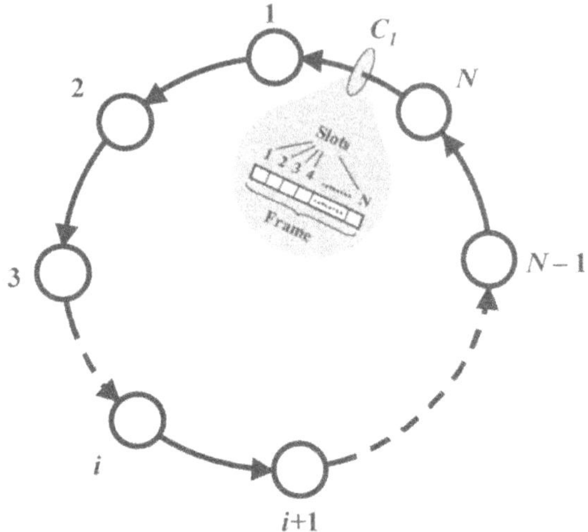

Fig. 6 Considered OPC-TDM single-ring network

3.2 Performance Study

In this section, a preliminary performance study of the considered slotted ring network concerning different packet compression/decompression unit realizations is performed.

3.2.1 Average Packet Transfer Delay

Let us assume that the packets arrive at each transmitting queue with a Poisson arrival process. Further, each of N slots in a frame is uniquely assigned to a destination station forming a receiver-oriented TDMA system. As we know, in TDMA the transmission capacity of the medium C is divided into N portions (channels), each with the capacity equal to C/N. Moreover, we assume a uniform traffic, i.e. a packet is equally likely to be transmitted to any of the $N - 1$ other stations. The average packet transfer delay is defined as the expected delay in transmitting a message from any source node to any destination node including packet transmission time (T_p), waiting time in the queue (T_{queue}), and end-to-end transmission time (τ) of the network. For a slotted-ring OPC-TDM network it will also include the packet rate-conversion time (T_{prc}), and the time a node must wait for a free slot assigned to corresponding destination node (T_{access}). Therefore, the average packet transfer delay in our case is given by:

$$D = T_p + T_{prc} + T_{queue} + T_{access} + \tau \qquad (3)$$

The packet transmission time (T_p) corresponds to the time it takes to transmit a data packet of X bits at R bit/s data transmission rate. The packet rate-conversion time (T_{prc}) is given above by Equation (2). The time a packet must wait in the transmitting queue including the waiting time for the i-th slot assigned to the destination node i is given by [8]. Considering the frame synchronisation period along with the mean waiting time of the M/D/1 queue model, we obtain:

$$T_{queue} = \frac{N}{2}T_p + \frac{N}{2}T_p\frac{\rho}{1-\rho} = \frac{N}{2(1-\rho)}T_p \quad (4)$$

If the *i-th* slot assigned to the destination node i is already busy, the transmitting node must wait for the next frame. To find the time a packet must wait for a free slot assigned to the destination node i, we determine a probability P_i, which denotes the probability that the slot i within the frame is busy. For overloaded network with uniform traffic and destination release the probability P_i is given by $P_i = (N-2)/(2\cdot(N-1))$. Thus T_{access} can be expressed as:

$$T_{access} = P_i \cdot T_{frame} = \frac{N-2}{2(N-1)}N \cdot T_p \quad (5)$$

Using the expression (3), combining $T_{prc} = (N-1)\cdot T_p$, and the results of (4), and (5) we obtain the average packet transfer delay as:

$$D = \frac{T_p N}{2}\left(2 + \frac{1}{1-\rho} + \frac{N-2}{N-1}\right) + \tau \quad (6)$$

Finally, N in (6) can be replaced by K and M using $N = (\alpha + 1)$ and (2) then gives companion expression to (6) for D, namely:

$$D = T_p \frac{M+K-1}{2M}\left(3 + \frac{1}{1-\rho} - \frac{M}{K-1}\right) + \tau \quad (7)$$

Curves of D versus offered load ρ related to different number of the parallel ODLLs (M) are plotted in Fig. 7. Fig. 8 shows the average packet transfer delay in dependence on the number of slots in the frame related to different network loads. For all plotted curves we assumed a compression rate of $K = 100$, a packet length of $T_p = 10$ kbits · 10 ps = 0.1 μs (at 100 Gbit/s), and a 20 km fiber-ring network with $\tau \approx 97$ μs.

From the plots it can be observed that adding additional ODLLs can reduce the average packet delay. If the offered load increases, the differences between the curves increase as well. The significant improvement can be seen for $M = 2$. Another advantage of the proposed packet compression/ decompression scheme is decreasing the frame length. Particularly, for higher load (> 0.8) the average packet delay increases significantly if the frame length increases.

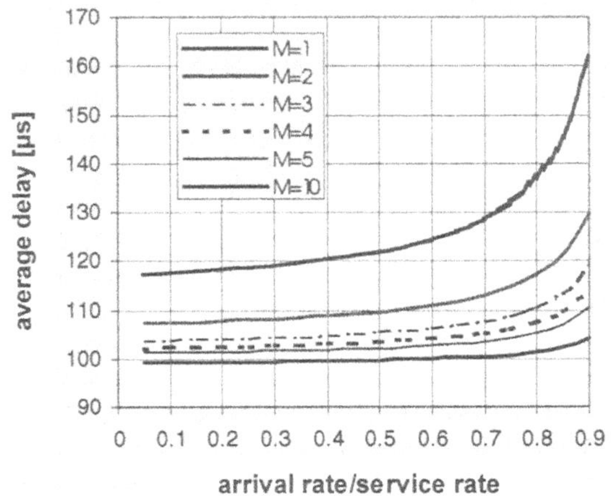

Fig. 7 Average delay vs. offered load related to M

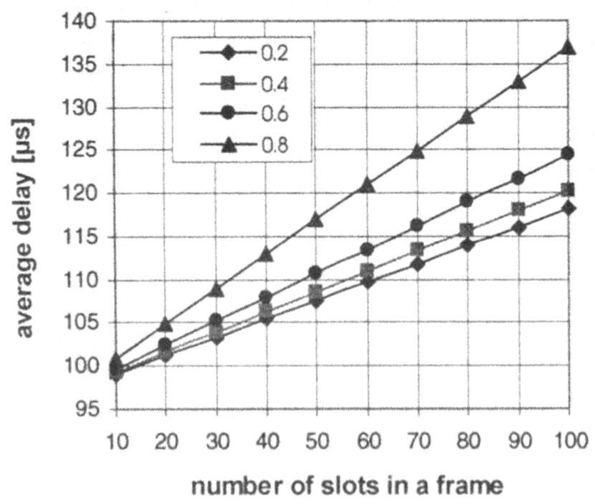

Fig. 8 Average delay vs. number of slots in the frame related to offered load

4. CONCLUSION

In conclusion, an optical packet compression/decompression scheme using a parallel arrangement of optical delay line structures is proposed and investigated in this paper. The proposed scheme allows high compression rates and large packet sizes by reducing the medium access constrains. Our simulations show feasibility of the proposed scheme. The impact of the proposed scheme on the network performance is theoretically investigated. Particularly, the average packet transfer delay in dependence on the number of deployed optical delay line lattices (ODLLs) is shown. A significant performance improvement in comparison with a single feed-forward delay-line structure can be achieved by adding additional ODLLs.

ACKNOWLEDGEMENTS

This work is supported by the Austrian Science Fund FWF (*Fonds zur Förderung der wissenschaftlichen Forschung*) under contract P13144-INF. The tool PTDS was employed for the simulations.

REFERENCES

[1] S. W. Seo, K. Bergman, P. R. Prucnal, "Transparent Optical Networks with Time Division Multiplexing", IEEE JSAC, Vol. **14**, No. 5, (1996), pp. 1039 – 1051.

[2] K.Bengi, S.Aleksic, "Design Considerations for a Slotted OTDM Ring-LAN", in NOC2000 (Stuttgart, 2000), pp. 191 – 198.

[3]H. Toda, F. Nakada, M.Suzuki, A. Hasegawa, „An Optical Packet Compressor Using a Fiber Loop for a feasible all optical TDM Network", in ECOC'99, (Nice, France,1999), pp. I-256 – I-257.

[4] N. S. Patel, K. L. Hall, K. A. Rauschenbach, "Optical Rate Conversion for High-Speed TDM Networks", IEEE Photon. Techn. Letters, Vol. **9**, (1997), pp.1277 – 1279.

[5] P.Toliver, K.L. Deng, I. Glesk, P.R. Prucnal, "Simultaneous Optical Compression and Decompression of 100 Gb/s OTDM Packets Using a Single Bidirectional Optical Delay Line Lattice", IEEE Photon. Techn. Letters, Vol. **11**, (1999), pp. 1183 – 1185.

[6] K. Gokyu, K. Baba, M. Murata, "Path Setup Algirithms for Uniform/Nonuniform Traffic in Unidirectional Optical Compression TDM Rings", in OptiComm 2000 (Dallas, Texas, October 22 – 26, 2000), pp. 98 – 109.

[7] S. Nakamura, et al, "168 Gbps error-free demultiplexing with hybrid-integrated symmetric Mach-Zehnder all-optical switch", in OFC2000 (Baltimore, Maryland, 2000), pp. ThF3-1 – ThF3-3.

[8] J.Spragins, J. Hammond, and D. Pawlikowski, "Telecommunications: Protocols and Design", Addison-Wesley Publishing Co., Reading, MA, 1991.

A TRANSPONDER FOR GIGABIT ETHERNET OVER WDM

D.A. Schupke[1]
and S. Weigert[1]
and O. Jahreis[2]
[1] *Munich University of Technology*
Institute of Communication Networks
80290 Munich, Germany
Tel.: +49 89 289-23511, Fax: +49 89 289-63511
Dominic.Schupke@lkn.ei.tum.de
Stefan.Weigert@t-online.de

[2] *Optisphere Networks Inc., A Siemens Company*
Reston, VA, U.S.A.
Tel.: +1 703-262-2507
oliver.jahreis@optisphere.com

Abstract The deployment of Gigabit Ethernet in corporate and ISP networks makes the transmission over WDM networks necessary to provide for the interconnection over metropolitan and wide area WDM networks. As the specifications of the optical signal between Gigabit Ethernet and the WDM-network mismatch, a transponder is needed. We propose an architecture for a Gigabit Ethernet over WDM transponder and report on a transponder prototype implementation. We further show how to use the transponder for Gigabit Ethernet networks over WDM rings.

1. INTRODUCTION

The interconnection of decentralized corporate networks to form a virtual private network (VPN) gains more and more importance today. At the same time wavelength-division multiplex (WDM) networks are emerging supporting multiple protocol-transparent client links and are therefore best suited to provide for broadband VPNs. One important WDM network architecture is the ring-based metropolitan area network (MAN).

As currently most corporate local area networks (LAN) are Ethernet based, one efficient method is to interconnect these LANs by Gigabit Ethernet over WDM-links [1], if they are demanding high bandwidth among each other. By this the coexistence of multiple protocols and protocol conversion at the LAN/MAN-edge is avoided.

Gigabit Ethernet has also gained importance as "routed Gigabit Ethernet," where this protocol is used as a point-to-point data link only [2]. Gigabit Ethernet performs no switching and is only present at the ports of an IP router. Thus routed Gigabit Ethernet can form the backbone of an Internet service provider (ISP).

In both cases inexpensive Ethernet equipment can be used, which is popular and known among network users.

Gigabit Ethernet, however, cannot be directly carried over a transport WDM-network. The standard port signal has to be adjusted to the specifications of the WDM-network which is accomplished by a transponder.

In section 2 we propose an architecture for a Gigabit Ethernet transponder and describe a prototype implementation based on this in section 3. We discuss in section 4 the issue of the transponder location and show in section 5 how the transponder fits in favorite network structures. In section 6 we conclude this paper.

2. TRANSPONDER ARCHITECTURE

An optical Gigabit Ethernet connection is operated in full-duplex mode. The 1 Gbit/s client signal is encoded by a 8B10B line-code to achieve DC balance (which is important for the setting of the receiver's threshold) and to cancel long 0/1-sequences (which is needed for clock recovery at the receiver) [3].

The encoding results in a symbol rate of 1.25 Gbaud. The main function of the transponder is to convert this Gigabit Ethernet signal present in 1000BASE-LX or -SX (as defined in [4]) to a WDM signal. Moreover the transmission power at the sender and the sensitivity at the receiver has to be adjusted.

We propose the basic structure depicted in the block diagram of figure 1. On the left side is the optical in/output to connect to the end system in normal operation (working).

To achieve network redundancy as described in section 5 the end system signal can be switched by S1 to a protection in/output. There it is possible to connect to a second end system (optically) or to another transponder (electrically). At S1 the input signal is chosen based on

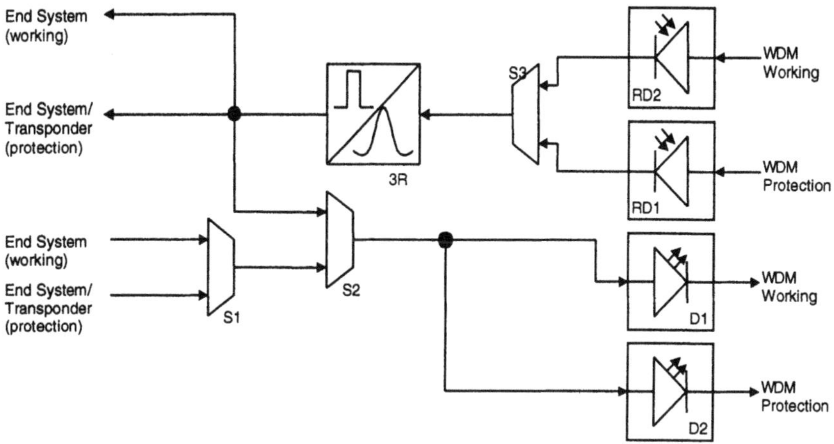

Figure 1: Transponder block diagram

the Loss-of-Signal (LOS) indication of the electrical preamplifier (not shown).

On the right side are working and protection interfaces for the WDM signal to allow for 1+1 protection on the WDM network. With 1+1 protection the signal is fed into two laser sources and thus transmitted redundantly. The receiver obtains both signals and chooses with S3 the forward signal based on the received power.

We propose a modular transmit-receive-unit for both working and protection instances. Therefore the protection unit can be omitted if 1+1 protection is unnecessary. This reduces the cost of the transponder, since expensive laser optic can be saved.

As we do o/e/o-conversion at the transponder, in the receiver part we can easily incorporate electrical 3R-regeneration (resizing, retiming, reshaping) of the signal chosen by S3. This is introduced, since the signal carried over the WDM network can have already low quality. The regenerated signal then goes to the two outputs for the end systems simultaneously.

For fault localization or testing a loopback can be performed at S2. In normal operation S2 switches the input signal to the sending unit(s). In loopback state S2 switches the regenerator signal to the sending unit(s). Thus the WDM signal is received, regenerated and sent back.

To obtain a stable WDM signal of both senders, a separate controler for the lasers is needed, too (not shown).

3. PROTOTYPE IMPLEMENTATION

The transponder prototype was developed based on the structure shown in figure 1. It is further equipped with a controler which is responsible for evaluation of measured data, supervising the hardware and performing some additional functions for Operation and Maintenance (OAM).

The controler can be connected to the Gigabit Ethernet network to use the error database of the equipment. By this the transponder is able to get, e.g., information about the frame error rate which is provided by the simple network management protocol (SNMP). This is important for performance monitoring, because Ethernet does not prove bit error rate monitoring.

The SNMP interface can also be used for the management of the transponder from a remote network management system. Because of the utilization of the Gigabit Ethernet infrastructure no additional management and control links between the transponder and the network management system are necessary.

For radiation protection an automatic laser safety mechanism was also implemented in the transponder. This mechanism works without a signaling system, but is realized on the optical layer.

The selection of the receive paths is made by the measurement of the optical input power

The system measures for each of the two redundant lines the analog optical input power, which is digitized by the controler. If the input power is too low to regenerate the data stream, traffic will be routed over the redundant line and the corresponding laser diode will be switched off. At the far end of the optical link between the two transponders the same mechanism is used. When the bi-directional link is down, the transmitters at each end of the system probe the link periodically with short flashes of light. Only if one of these flashes can be received at the system's far end, the lasers are activated again.

The prototype is realized as a set of boards (euro format). One of these boards (the controler board) contains all the functionality for switching, signal regeneration and measurement. On the controler board the analog signals are switched by 2 to 1 multiplexers. The electrical path between the optical front ends is controled and initialized by the controler of the transponder.

The redundant transmitters and receivers are made as two identical boards, so the transponder can be equipped with either one or two optical WDM-ports. These transmit-receive-units contain the optical front ends (laser and receiver diode) with the transmission stages (an amplifier in

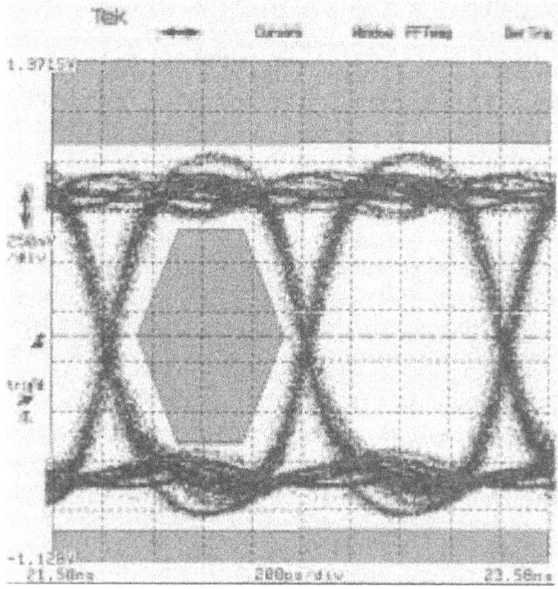

Figure 2: Eye diagram of the received signal

the receive path and a driver in the transmission path). Furthermore each of the transmit-receive-units uses an external laser diode controler in the prototype, in order to control the wavelength of the transmission diode.

The 3R-regenerator was provided as a single chip solution. The regenerator can be switched to different bit rates up to 1.25 Gbit/s. For this reason the transponder can also be used for rates under 1.25 Gbit/s.

The WDM-ports of the prototype maintain a sensitivity of at least -25 dBm and a maximum output power of +3 dBm. The wavelength divergence is at most ± 0.001 nm. The eye diagram of a received and regenerated Gigabit Ethernet signal is depicted in figure 2 (approx. 15-20 dB attenuation). The signal is in the white area and thus fulfills the requirements of [4].

4. TRANSPONDER LOCATION

For the employment in a network we have to determine where to install the transponder appropriately: at the end system location or at the WDM access node.

One decision parameter is whether the network operator of the Gigabit Ethernet equipment or the WDM network operator is responsible for operation and management of the transponder. We assume the latter case in the following.

If the transponder is located at the WDM node, the optical power of the WDM channel has to be adjusted for the WDM network only. The WDM operator has an easy access to the transponder and can perform out-band management of the transponder. But using standard Gigabit Ethernet ports for transmission, the access link is restricted to span a short distance (5 km).

If the transponder is located at the Gigabit Ethernet equipment, we obtain a definite network termination. Since the distance from the end system to the transponder can be short (in-house, some 10 meters), we can also use an inexpensive copper connection (1000BASE-CX or -T) between the two elements. As the optical power of the WDM channel at the WDM access node, however, is dependent on the distance of the access link, more adjustment is necessary and high power levels may be reached. In absence of parallel access links, in-band signaling has to be used for managing the transponder.

Another criterion is the protection of the access link. The transponder can perform this protection only if located at the Gigabit Ethernet equipment. If located at the WDM node, the protection from the end system to the transponder has to be modified.

5. NETWORK STRUCTURES

In figure 3 we show a realization of a Gigabit Ethernet line network A-B-C over a WDM ring, e.g. a transparent MAN ring. The line network consists basically of two point-to-point links between A-B and B-C, each requiring transponders at the endpoints. Since node B can fail and a communication between A and C should still be possible, B can be bypassed by the two adjacent transponders. For this the protection in/outputs of the transponders are connected by a bypass link. If node B fails, S1 in both transponders switch to the protection connection.

Figure 4 depicts a star network over a WDM ring, where the nodes A, C and D are connected to the star node B. Node B has a backup node B' for redundancy. Here the protection in/outputs of the transponders are used to connect the nodes A, C and D to node B' if node B failed.

Usually, Gigabit Ethernet also provides by itself survivability mechanisms like load-balancing, 1+1-protection and reconfiguration by the spanning tree algorithm.

1+1-protection requires two and load-balancing requires at least two ports and two connections with transponders between the Gigabit Ethernet end systems. However, to not overprovide redundancy the transpon-

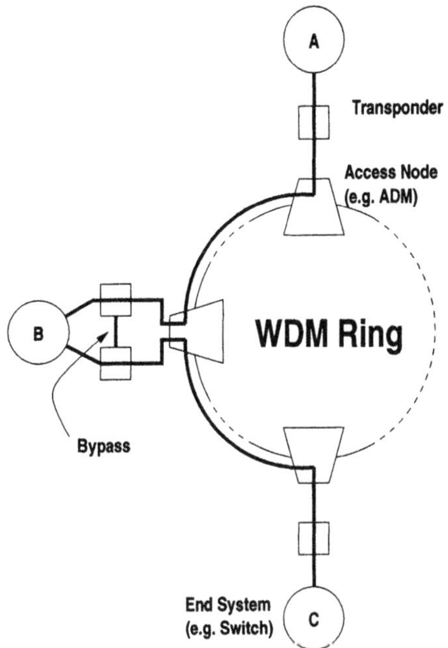

Figure 3: A line network over a WDM ring

ders only need to be equipped with one transmit-receive-unit (working). By this, the link from the Gigabit Ethernet equipment to the WDM ring can be made redundant, too. For instance, two disjoint access links and WDM nodes can be used (dual homing).

6. CONCLUSIONS

We presented an architecture for a Gigabit Ethernet over WDM transponder with modular transmit-receive-units for working and protection. The transponder also does 3R-regeneration, automatic laser safety switching and is manageable over Gigabit Ethernet. The functionality of the transponder was demonstrated by a prototype. We have further shown how to locate the transponder (geographically and administratively). We presented possible configurations of Gigabit Ethernet line and star networks over WDM rings.

From the technical point of view the bandwidth can be increased easily to 2.5 Gbit/s. This becomes an issue for transmissions of future 10 Gigabit Ethernet rates which are composed of four 2.5 Gbit/s signals.

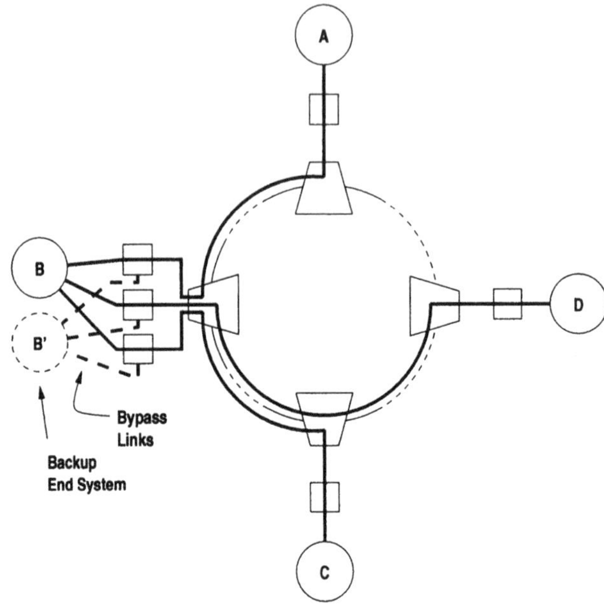

Figure 4: A star network over a WDM ring

7. ACKNOWLEDGMENTS

This work was supported by the German Federal Ministry of Education and Research (BMBF) and by Siemens AG, Munich in the KomNet project. The content of this paper is solely the responsibility of the authors.

References

[1] D.A. Schupke. Broadband Internet Access using Gigabit Ethernet over Wavelength-Division Multiplexing (WDM) Networks. In *Proceedings of the Fifth EUNICE Open European Summer School, Barcelona, Spain*, 1999.

[2] M. Jäger, H.-M. Foisel, H.-M. Westphal, J. Chawki, K. Ovsthus, and J.-C. Bischoff. Evaluation of Network Architectures for the Integration of IP over Optical Networks. In *2nd Int. Workshop on the Design of Reliable Communication Networks (DRCN)*, April 2000.

[3] R. Ramaswami and K.N. Sivarajan. *Optical Networks: A Practical Perspective.* Morgan Kaufmann Publishers, 1998.

[4] Institute of Electrical and Electronics Engineers (IEEE). *802.3z: Media Access Control (MAC) Parameters, Physical Layer, Repeater and Management Parameters for 1000 Mb/s Operation*, 1998.

Part Five

IP and WDM-based network architectures

LINK AND PATH ASYMMETRY ISSUES IN IP OVER WDM TRANSPORT NETWORKS

D.A. Schupke
Munich University of Technology
Institute of Communication Networks
80290 Munich, Germany
Tel.: +49 89 289-23511, Fax: +49 89 289-63511
Dominic.Schupke@lkn.ei.tum.de

Abstract In this paper we consider link and path asymmetry in IP over WDM transport networks. While transport networks are traditionally symmetric the carried Internet traffic is very asymmetric involving a potential for major network cost reductions. We determine where asymmetry can be introduced and indicate necessary protocol adaptations. We raise investigation issues in asymmetry which need to be pursued in future.

1. INTRODUCTION

Measurements in the Internet show a highly asymmetric characteristic of the traffic flow [1, 2]. One of the reasons for this is that a large portion originates from servers connected to the Internet, which tend to involve much more generated traffic (information supply) than received traffic (requests, etc.) With the currently emerging Internet applications, it is expected that this characteristic will remain for the future.

Today in transport networks traffic is carried over symmetric links with protocols like SDH/SONET or point-to-point Gigabit Ethernet which in turn are operated over WDM networks. Therefore one transmission direction is underutilized by the asymmetric Internet traffic. Since optical line port equipment (in particular the lasers) is still a big cost factor, cost reductions are expected when introducing asymmetric links in the network.

Furthermore in transport networks routing is done symmetrically, e.g. a bidirectional path traverses the same nodes for both directions. This can result in inefficient usage of the network resources.

In the following we classify the different possibilities for link and path asymmetry. We discuss realizations of asymmetry in IP over WDM networks and indicate necessary protocol adaptations. In conclusion we raise the investigation issues in asymmetry which need to be pursued in future.

2. ASYMMETRY IN IP OVER WDM NETWORKS

We consider links and paths on both the IP layer and the WDM layer. Paths on the WDM layer represent (virtual) links seen by the IP layer. Dealing with future high traffic volumes in transport networks we do not consider multiplexing of layers between IP and WDM (which can provide links and paths by themselves).

There are both unidirectional (simplex) or bidirectional (in this context full-duplex) links and paths possible. On the WDM layer paths are also called lightpaths [3]. Link (path) symmetry means that a bidirectional link (path) has the same characteristic in each direction. Then asymmetry can occur in different fashions:

- IP and WDM layer: Two nodes communicate in one direction only (unidirectionality).

- IP and WDM layer: Different nodes are traversed by the two unidirectional paths for each direction which together provide a bidirectional connectivity between two nodes (path asymmetry).

- WDM layer: On a segment the wavelengths of a bidirectional lightpath are different for each direction (wavelength asymmetry).

- IP layer: The bandwidth capacity is different between each direction of a bidirectional link (bandwidth asymmetry). Thus unidirectionality is a special case of bandwidth asymmetry.

- WDM layer: The number of lightpaths provided to an endpoint pair on the IP layer differs between both directions. This translates to bandwidth asymmetry seen by the IP layer.

Asymmetry also occurs with multicasting and protection schemes (e.g. one sided protection switching), which is not covered here.

From the network management point of view asymmetry introduces more effort in network configuration and in the administration of network data. Therefore for each of these different possibilities the efficiency gain has to be weighted against the increase of network management complexity.

3. UNIDIRECTIONALITY

In IP networks to make two routers adjacent a bidirectional link connecting them is required. The support for unidirectional links has been motivated by IP over broadcast satellite networks and is equally applicable in high bandwidth IP over WDM networks. A mechanism to emulate bidirectional connectivity over a unidirectional link at the link layer is proposed in [4], and can be used as illustrated in figure 1.

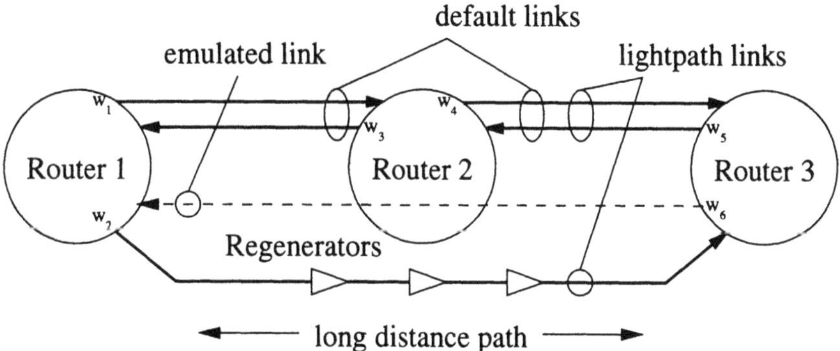

Figure 1: A unidirectional link with a tunnel as proposed in [4].

The bidirectional default connections are provided from router 2 to routers 1 and 3. If there is a high traffic demand from router 1 to 3 but not from router 3 to 1, it becomes desirable that a unidirectional (virtual) link is established between them (short-cut). As the IP layer needs to see a bidirectional link, by the extension in [4] a unidirectional tunnel is established from router 3 to 1 operated over the default links and router 2.

In comparison to a bidirectional link, a sending unit at router 3 was saved, e.g. a Packet over SDH/SONET (POS) interface not equipped with a laser.

The physical distance between routers 1 and 3 can become very long requiring expensive regenerators to refresh the signal. A bidirectional link needs regenerators for each direction. In comparison, the unidirectional constellation needs only half of the regenerators of a bidirectional link, since the return link is provided over an existing path.

In this sense unidirectionality generally requires less resources in the WDM network than bidirectionality. This also holds for networks with nodes having partial wavelength conversion. Reducing the number of lightpaths in effect reduces the number of potentially needed converters.

Therefore with unidirectional links the costs introduced by regenerators and wavelength converters can be kept at minimum.

One important question is the allocation of the IP routing weights. Referring again to figure 1, we can provide an asymmetric setting of $w_2 \leq w_1 + w_4$ to let the traffic flow over the short-cut and $w_6 > w_3 + w_5$ to reduce the tunneling effort. As these weights can have an impact on network-wide routing, this technique has to be considered when allocating the weights in the whole network (see also section 5).

We note here that establishing undirectional lightpaths should be possible even in WDM networks operating with automated lightpath set-up, e.g. provided by a MPLS control plane (so-called MPλS [5]). Paths in MPLS are by definition unidirectional. There are proposals for the extension of MPLS to directly set up bidirectional paths in optical networks [6], mainly to avoid contention problems. These extensions can coexist with unidirectional path set-up.

4. PATH AND WAVELENGTH ASYMMETRY

Firstly we consider wavelength asymmetry in bidirectional paths. Figure 2 depicts the situation of a four node star network [3]. Assume that the nodes are optical cross-connects without wavelength converters interconnected by a bidirectional fiber pair and sources can choose from two wavelengths in the system.

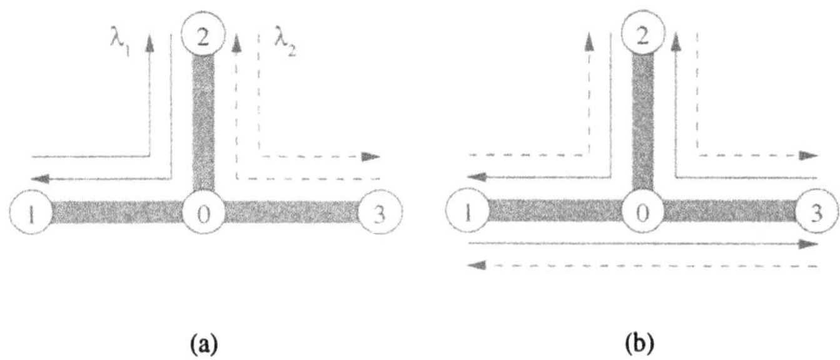

Figure 2: A star network with (a) wavelength symmetry and (b) with wavelength asymmetry of paths.

A full mesh demand of one bidirectional path between the nodes 1, 2 and 3 should be provided. With wavelength symmetry in figure 2 (a) only 2 out of 3 bidirectional paths can be provided, whereas the asymmetric assignment in figure 2 (b) can provide for the full demand.

Since wavelength selection seems not to be a crucial point for the routing process [7], wavelength asymmetry can become feasible for WDM networks. For example in MPλS the wavelength assignment on a link is likely to be done by the signaling protocol for each direction independently.

Another example is the converterless WDM bidirectional line-switched ring using two fibers (BLSR/2) [3] which also requires wavelength asymmetry. Such a BLSR/2 consists of two counterrotating (and unidirectional) rings. One half of the wavelengths is used for working traffic, and the other half for protection. To avoid wavelength converters, two complementary sets of wavelengths are assigned to both rings. The working wavelengths of one ring are the protection wavelengths of the other. As bidirectional paths are routed symmetrically on the ring, different wavelengths are needed for each direction.

Path asymmetry, however, is highly controversial, because it makes routing information more unmanageable. Path asymmetry in WDM networks comes to front when introducing partial wavelength conversion in the network. Figure 3 (a) depicts the situation of a grid-like network with four nodes and existing lightpaths. Assume that the nodes are optical cross-connects each having a single wavelength converter available. The cross-connects are interconnected again by bidirectional fiber pairs and sources can select one of two wavelengths in the system.

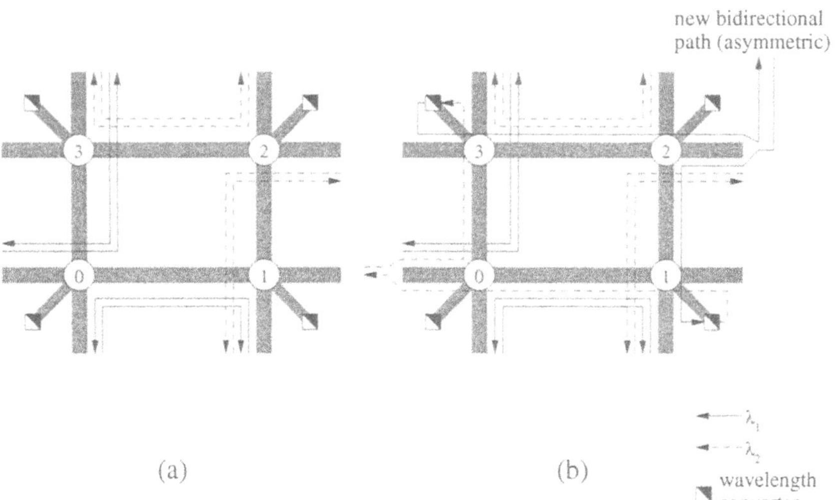

Figure 3: (a) A network with existing lightpaths. (b) A new bidirectional lightpath demand routed asymmetrically.

Consider the following scenario. A new bidirectional lightpath between nodes 0 and 2 has to be provided without changing the existing lightpaths. This can only be done by an asymmetric path as depicted in figure 3 (b).

For WDM networks with partial wavelength conversion asymmetric paths introduce a further degree of freedom which results in efficient network utilization. Thus it becomes an investigation issue what can be gained from path asymmetry in WDM networks.

On the IP layer nodes are operated with distributed routing protocols like OSPF, which relies on routing weights assigned to links. On one hand different weight assignments for both directions of each link in a network do not necessarily result in path asymmetry.

On the other hand even when assigning equal weights for both directions of every link in the network we can obtain path asymmetry when there is more than one shortest path between two nodes. Depending on the implementation of the routing protocol, the nodes can choose different paths for the two directions of a connectivity, since the paths are evaluated as equally "shortest." This last issue has been avoided so far by dimensioning the routing weights such that unique shortest paths exist between the node pairs [8, 9].

In networks with IP as a connectionless protocol packets can take an arbitrary route, but in IP backbones the network service operators have dimensioned the routing weights in such a way to obtain predictability of the traffic flow. Thus it becomes an issue how to assign the weights such that IP path symmetry including unidirectional and parallel links (see below) is achieved and traffic flow predictability is still assured.

5. BANDWIDTH ASYMMETRY

Firstly we consider the case where the bandwidths are different in the two directions of a bidirectional lightpath. We assume that line ports become more expensive with growing bitrate. For POS point-to-point systems there is no reason against having an asymmetric bandwidth allocation as long as bidirectional communication for IP is possible. The POS protocol specification, however, has to be modified slightly for this.

Consider a point-to-point transmission with traffic of 10 Gb/s in one and 2.5 Gb/s in the other direction which should be operated with POS as link layer (asymmetry ratio of 4:1). With today's equipment two STM-64 ports are needed where one direction is utilized by 25% only. Assume half of the cost of the two ports to provide for each direction of the link. Assume furthermore that the cost of a STM-N system doubles as N quadruplicates. Then an asymmetric 10 Gb/s$\uparrow$-2.5 Gb/s$\downarrow$ system would be 75% of the costs of the symmetric STM-64 system. Thus again we obtain cost reductions for these links.

An important bandwidth asymmetry also occurs when two routers are connected bidirectionally by lightpaths, but with a different number of lightpaths for each direction [2], see figure 4. The advantage of this configuration is again that we can economize on sending and regenerating equipment.

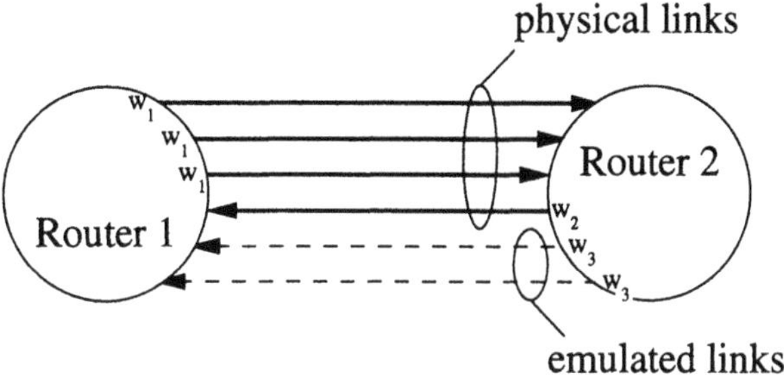

Figure 4: Bandwidth asymmetry by different lightpath allocations for each direction and a tunnel as in [4].

There are three possibilities to handle this case at the interface WDM/IP:

1 The link layer provides a virtual interface (with asymmetric bandwidth) and performs inverse multiplexing to split the data on the lightpaths. A major problem becomes chromatic dispersion causing that the signals on the lightpaths experience different delays. Thus synchronization on the receiver side becomes too difficult.

2 The mechanism in [4] again provides a solution. Referring to figure 4 there is one "physical" and two emulated bidirectional links between router 1 and 2. Thus the IP layer sees the parallel links and uses them simultaneously if at router 1 the weights are equal (at router 2: $w_2 < w_3$.)

3 Another possibility is that the IP layer sees the parallel links provided by the mechanism in [4] and distributes the traffic depending on some classification, e.g. flow classification or QoS class. Thus a router with traffic engineering capabilities is required.

For the possibility 2 again the setting of the weights in the whole network becomes an issue. Moreover as chromatic dispersion is present, packets on one lightpath can overtake packets on other lightpaths. If these packets belong to the same TCP flow, the performance is reduced, since the packets are not in order [10]. This issue has been addressed in [11], where solutions are presented to evenly distribute flows on parallel links (which occur to be equal-weight paths).

6. CONCLUSIONS

For IP over WDM networks we made the classification of asymmetry in unidirectionality, path asymmetry, wavelength asymmetry and bandwidth asymmetry. We presented different realizations to provide asymmetry in these networks on a link and path basis.

We have shown where investment cost reductions are obtained when asymmetric links and paths are employed in the network. However, higher network management effort and partially also protocol changes are introduced. Therefore we propose further considerations to determine what kind of asymmetry can help to reduce the overall costs of a network. A quantitative analysis of the advantages of asymmetry is performed in future investigations.

Unidirectional and parallel links are likely to be gradually employed in IP networks. The routing weight assignment problem has to be reconsidered to allow for unidirectional links and multiple paths between router-pairs.

7. ACKNOWLEDGMENTS

This work was supported by the German Federal Ministry of Education and Research (BMBF) in the project TransiNet (`http://www.transinet.de/`).

The author would like to thank A. Autenrieth for his helpful discussions.

References

[1] K. Thompson, G.J. Miller, and R. Wilder. Wide-area Internet traffic patterns and characteristics. *IEEE Network*, 11(6), 1997.

[2] B.S. Arnaud. Architectural and Engineering Issues for Building an Optical Internet. In *All-Optical Networking: Architecture, Control, and Management Issues*. SPIE, 1998.

[3] R. Ramaswami and K.N. Sivarajan. *Optical Networks: A Practical Perspective*. Morgan Kaufmann Publishers, 1998.

[4] E. Duros, W. Dabbous, Izumiyama H., N. Fujii, and Y. Zhang. A Link Layer Tunneling Mechanism for Unidirectional Links. Internet-Draft `draft-ietf-udlr-lltunnel-04`, 2000.

[5] D.O. Awduche, Y. Rekhter, J. Drake, and R. Coltun. Multi-Protocol Lambda Switching: Combining MPLS Traffic Engineering Control With Optical Crossconnects. Internet-Draft `draft-awduche-mpls-te-optical-02`, 2000.

[6] D. Guo, L. Zhang, J. Fu, and R. Cheung. Extensions to RSVP-TE for Bi-directional Optical Path Setup. Internet-Draft `draft-sorrento-rsvp-bi-osp-00`, 2000.

[7] H. Zang, J.P. Jue, and B. Mukherjee. A review of routing and wavelength assignment approaches for wavelength-routed optical WDM networks. *Optical Networks Magazine*, Jan 2000.

[8] A. Bley, M. Grötschel, and R. Wessäly. Design of Broadband Virtual Private Networks: Model and Heuristics for the B-WiN. Technical report, Konrad-Zuse-Zentrum für Informationstechnik Berlin, 1998.

[9] W.B. Ameur, E. Gourdin, and B. Liau. Internet Routing an Topology Problems. In *Second International Workshop on the Design of Reliable Communication Networks (DRCN)*, 2000.

[10] C. Huitema. *Routing in the Internet*. Prentice-Hall, 2000.

[11] D. Thaler and C. Hopps. Multipath Issues in Unicast and Multicast Next-Hop Selection. Internet-Draft `draft-thaler-multipath-05`, 2000.

On Design and Architecture of an IP over WDM Optical Network Control Plane

Chunsheng Xin[1,2], Ti-Shiang Wang[1], Yinghua Ye[1], Myungsik Yoo[1]
Sudhir Dixit[1], and Chunming Qiao[2]
1. Nokia Research Center 5 Wayside Rd Burlington, MA01803 U.S.A
2. Department of Computer Science and Engineering, SUNY Buffalo, Buffalo, NY 14260, U.S.A

Key words: **IP over Optical Network, Control Plane, WDM**

Abstract: To demonstrate an IP-based routing and signalling approach in IP over WDM optical networks, we present the design and architecture of a WDM optical network control plane. With a modular design, the control plane effectively implements the key functionality such as routing, signalling, protection/restoration and quality of service support.

1. INTRODUCTION

The rapid growth of Internet and new digital services is driving the Internet service providers to provide huge bandwidth. With technical and economical feasibility, WDM optical networks are becoming an ideal infrastructure candidate in core networks and even access networks due to its potentially unlimited bandwidth. Although WDM optical networks are already in use to provide the point-to-point connection for a multiple-layer architecture, such as IP over ATM over SONET over WDM, to transport IP traffic, this approach has experienced high management cost and complexity for the service providers. Therefore, there is a consensus recently in the research community and industry to utilize a two-layer architecture, i.e. IP over WDM, to transport IP traffic directly over WDM optical networks. However, the rapid and effective bandwidth provisioning and protection/restoration remain the challenging issues for IP over WDM optical networks. Currently, there are many ongoing research efforts on the control requirements and mechanisms for IP over WDM optical networks, [1 – 9].

In this paper, we demonstrate how to use the existing IP protocols and mechanisms to introduce the intelligence into the optical crossconnect (OXC) in WDM optical networks by designing an IP-based optical control plane for the OXC. The control plane may be integrated into the same box of an OXC, or run on a separate IP/MPLS router, which is connected to the OXC. A dedicated and preconfigured wavelength, called *control channel,* between two neighboring OXCs, is terminated on both ends and used for the communication between the neighboring control planes (running on the neighboring OXCs). Each control message generated in a control plane and destined to a peering control plane goes through E-O conversion and is sent to the *control channel.* The control message goes through O-E conversion in the neighboring OXC and is received by the neighboring control plane. If this neighboring control plane is not the destination, the control message is relayed to the next hop control plane. This procedure continues until the control message arrives at the destination control plane.

This paper is outlined as follows. Section 2 introduces the general architecture of our proposed optical network control plane. A detailed description of a GUI component for the control plane is presented in section 2.1 and the function of resource management module is introduced in section 2.2. In section 2.3, we discuss the function of connection module. To illustrate the importance of survivability in optical network, section 2.4 presents the function of protection and restoration module in the control plane. Section 2.5 introduces the operation and function of the main module. Finally, in section 3, we conclude our current work and propose some future works.

2. GENERAL ARCHITECTURE

In this paper, we consider a mesh-type optical network without wavelength conversion capability. That is, the optical network that consists of edge and core optical nodes interconnected by fiber links is used to connect high-speed IP/MPLS client routers in IP networks, as shown in Fig. 1. Each optical node consists of an optical cross connect (OXC) and a control plane running upon the OXC, as shown in Fig. 2. The control plane is an IP-based controller that employs IP protocols to operate the underlying OXC. The border optical node can be connected by IP/MPLS routers (client routers). The peering control planes interact with each other to provide fast and effective provisioning (e.g. lightpath setup, teardown and abort), fault monitoring, protection and restoration in WDM optical networks. For example, in order to set up a lightpath, the source control plane works with peering control planes to compute an end-to-end routing path and configure the switching fabric of OXCs along the path by a signalling protocol.

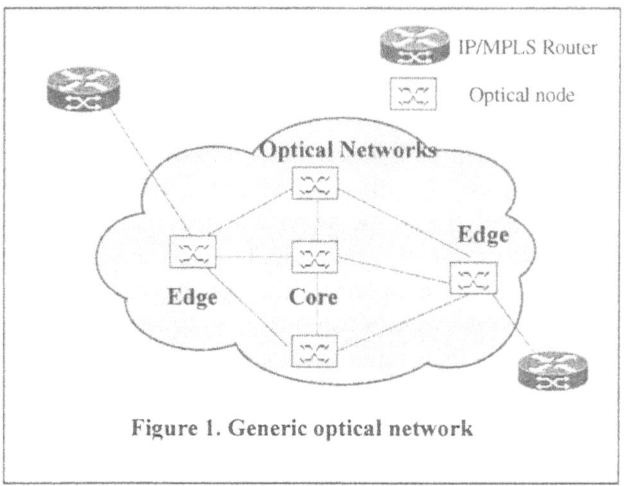

Figure 1. Generic optical network

Figure 1. Generic Optical Network

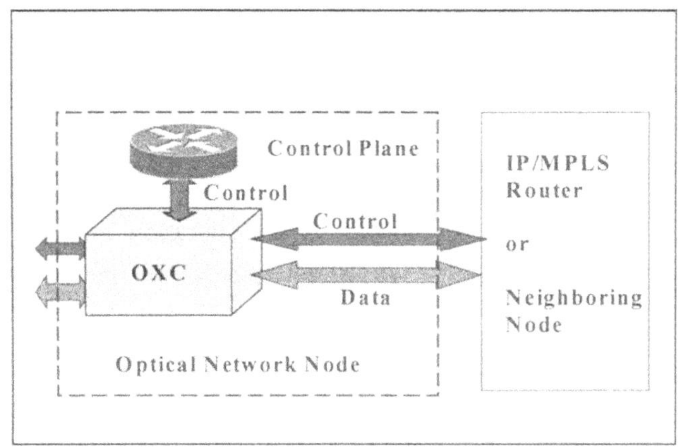

Figure 2. Architecture of an Optical Node

In terms of routing and signalling across the IP and optical boundary (User Network Interface, or UNI), three network models (peer model, overlay model and integrated model) have been proposed for IP and optical internetworking. In the peer model, a single routing protocol instance runs across both IP and optical domains, i.e. the OXC is a peer of a router. In the integrated model, separate routing protocol instances run in IP and optical domain respectively. However, partial routing information can be leaked between the routing protocol instances. For example, an inter-domain routing protocol (e.g. BGP) can be used to exchange the IP reachability information. In this scenario, the optical domain as a whole is a peer of the IP domain. In the overlay model, the routing and signalling in optical domain is also independent of the routing and signalling of IP domain, as in the integrated model. Furthermore, the separate routing protocol instance does not pass any routing information into each other. Therefore the border router (the router attached to optical domain) may have to register its address to optical domain and query the reachable external address (of another border router). After an initial virtual topology is formed over some border routers (their addresses are obtained through query), a routing protocol instance among the border routers can be run on this overlay topology.

In this paper, our focus is on designing an IP-based control plane for the optical domain. "IP-based" means that the control plane uses the IP protocols to control and operate the OXC in the optical network. Our objective is to demonstrate an approach to build an optical control plane using IP-based routing and signalling mechanism. As far as the internetworking with IP domain is concerned, the control plane does not differentiate the three UNI network models described above, but can be easily extended to support all of them. To demonstrate the generic functionality of the control plane without getting involved into the specifics of the UNI network models, we have designed a graphical user interface (GUI) component, called control plane GUI (CPGUI). First of all, CPGUI provides a user-friendly interface for the service provider to monitor, manage and operate the optical network. Second, CPGUI can emulate client routers and graphically demonstrate the key functionality of control plane, such as service provisioning and protection/restoration procedure. CPGUI can be deployed either as a centralized network management station or as a distributed management system on multiple nodes.

Fig. 3 shows the structure of a control plane and its interaction with CPGUI. Each control plane consists of four modules: main module (MM), connection module (CM), resource management module (RMM), and protection/restoration module (PRM). Socket connection is used to provide a communication infrastructure between the neighboring control planes, and between the control plane and CPGUI. The current implementation of socket connection is utilizing TCP stream socket.

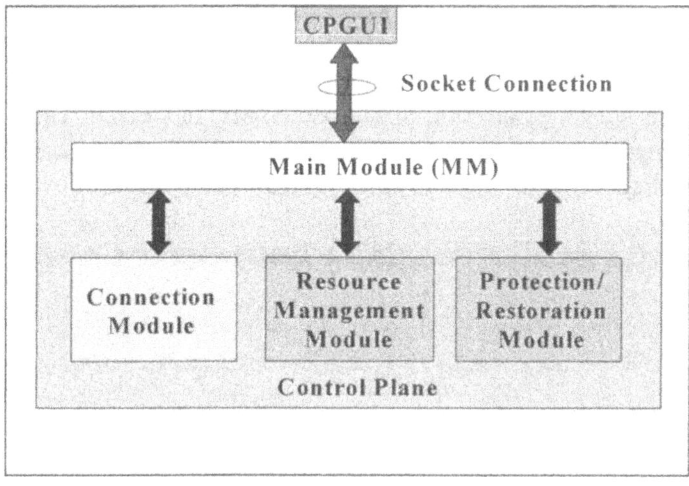

Figure 3. Structure of a control plane and CPGUI

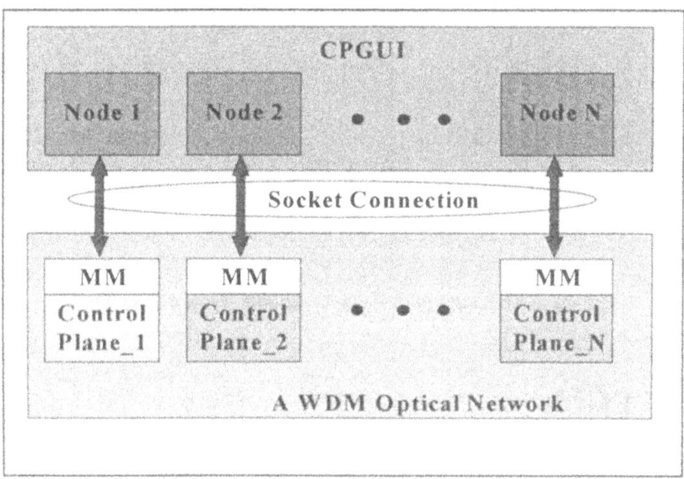

Figure 4. A network-wide view of control planes and CPGUI

CM is used for the lightpath signalling and maintenance (such as lightpath setup and teardown). TCP-based signalling is used by the control plane for its simplicity and flexibility. RMM is used for routing and wavelength assignment (RWA), topology and resource discovery, and service differentiation (QoS) support. PRM provides the functions of fault monitoring, and fast protection and restoration. The objective of the MM is to receive the incoming messages and works closely with other modules to process the requests. Fig. 4 shows a network-wide view of control planes and CPGUI. Each control plane is associated with an individual optical node. In addition to the socket connections among neighboring control planes (representing the physical connectivity of the optical nodes), each control plane has its socket connection to a "node object" in CPGUI. Thus, through real-time message exchange, CPGUI can provide dynamic network views and issue client requests (for example, lightpath setup and teardown).

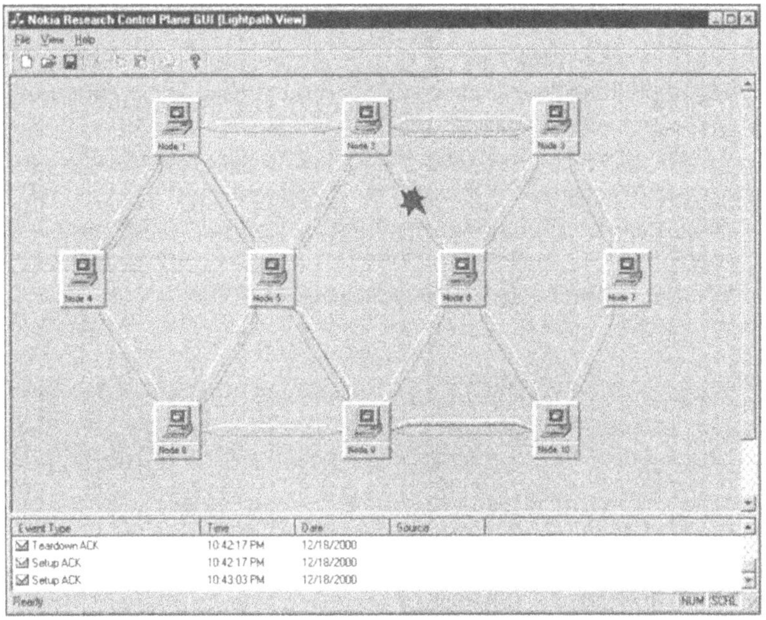

Figure 5. The Lightpath View and Event View of CPGUI

2.1 CPGUI

CPGUI is a tool to provide a graphical user interface to monitor, manage and operate the optical network through the information exchange with control planes. It can also be used to emulate client IP routers to request service provisioning, accept the network state information from the control plane and display the network state graphically.

In CPGUI, there are three network views: topology edit view, lightpath view and event view. In topology edit view, the network elements, such as node, link, and fiber can be created and edited to form the physical topology of the optical network. In lightpath view, CPGUI is able to perform the operations such as lightpath setup/teardown, lightpath physical path and status inquiry, lightpath protection path inquiry and its dynamic restoration procedure, and network element state inquiry, such as OXC switching state, fiber state and wavelength state. In event view, an event history list is displayed. An event is generated when CPGUI receives a message from any control plane. The message can possibly be either an acknowledgement (ACK) of a previous request invoked by CPGUI to that control plane, or some fault notifications. Based on our design, an event is displayed in a single row that includes a series of attributes, such as event type, event generation time and source. Fig. 5 shows the lightpath view (at the top) and event view (at the bottom). In Fig. 5, there is a link failure between node 2 and node 6. The traffic from node 1 to node 10 whose primary path travels through the failure link is shifted to its protection path (node 1, node 5, node 9, node 10). Fig. 6 shows the switch state of node 5 in Fig 5.The switch is configured to switch <port 2, lambda 1> to <port 3, lambda 1> and <port 4, lambda 2> to <port 5, lambda 1> for lightpath {node 1, node 10} and lightpath {node 8, node 3}, where node 8 and node 2 are connected to port 2 and port 3 respectively, and node 1 and node 9 are connected to port 4 and port 5 respectively.

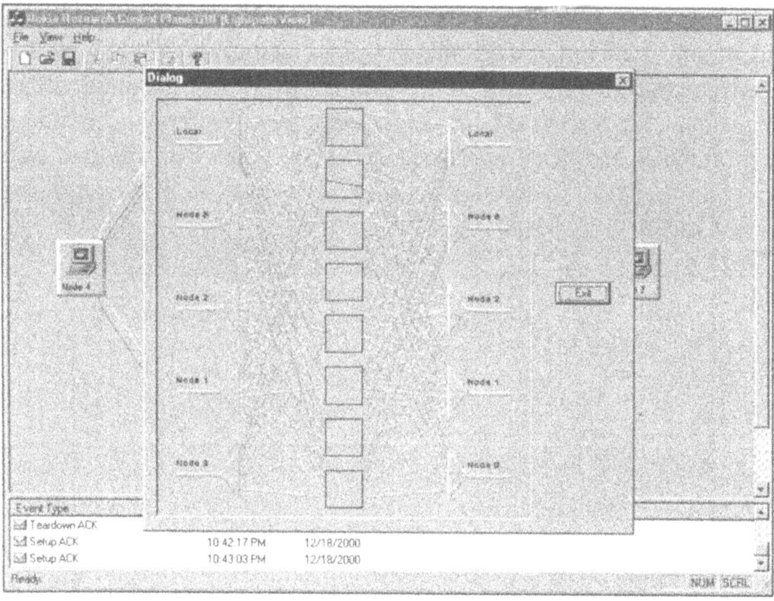

Figure 6. The Switch State of an OXC

2.2 Resource Management Module (RMM)

The main function of RMM is for RWA, topology and resource discovery, and QoS support. In the current implementation, the explicit source routing is used and the source control plane uses the shortest path first algorithm to compute an end-to-end routing path based on the network topology. The computed routing path is cached for the subsequent requests to save the per request routing computation overhead. The first-fit algorithm is used to assign the wavelength based on the resource availability along the computed routing path, which can be obtained from the resource database. The constraints based routing and other wavelength assignment will be added to enhance the resource management.

A topology connectivity matrix (TCM) is used to store the network topology. Fig. 7 shows a TCM of an optical network with N nodes, where each node is addressed by a "node ID" between 1 and N. Row vector i in the figure represents the link cost between node i and node 1 through N, where the last element is a sequence number extracted from the most recent topology update message (TUM) (to be discussed later). The link cost is generally relevant to the resource usage and the link cost "0" indicates the link is in fault or there is no link between two nodes. The active connectivity between a node and its neighbors is detected by the neighbor discovery mechanism (to be discussed later) and stored into a local connectivity vector (LCV), which is the same as a row vector in TCM with the last element excluded, "seq_num". The link cost in LCV changes with the consideration of the link usage.

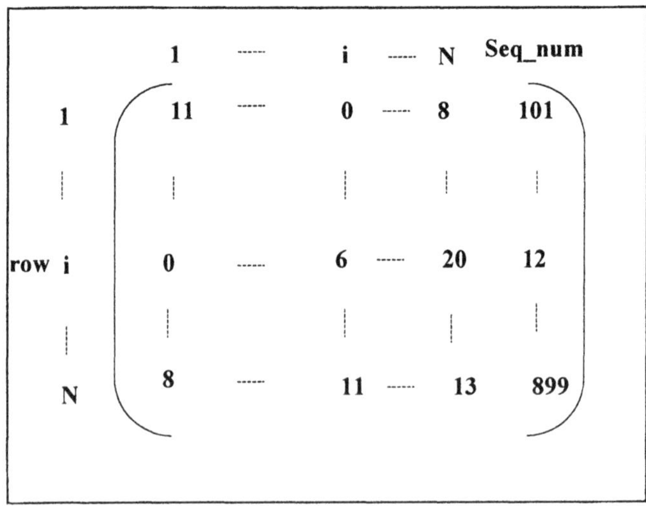

Figure 7. Topology Connectivity Matrix

Initially, when a control plane is started or rebooted, in order to build its own TCM, it will request its neighboring nodes for their TCMs. Then both this control plane and its neighboring control planes update their LCVs and broadcast their LCVs to the network. All other operational nodes in the network will accept those LCVs and update their TCMs. Within some period, the TCMs of all nodes will converge to a stable state that represents the current network topology.

To discover and maintain the active connectivity between the neighboring nodes, each control plane sends *keep-alive* messages periodically to *control channels* in all outgoing links. If a control plane cannot receive a *keep-alive* message from a neighboring control channel within some initially negotiated hold time, this link is considered as failed. The PRM will be invoked to process the fault. On the other hand, if a control plane receives a *keep-alive* from a failed link, the link is considered as restored. When the LCV is updated due to a link cost change, a link failure or restoration, it is broadcasted to the network using the same flooding mechanism as employed by OSPF.

To understand the flooding mechanism, without losing the generality, let us consider only one end of the failure link and assume this end is node A. Because of the link failure, the LCV of control plane A, named LCV_A, is updated. Then, a sequence number (SN_A) is acquired from a local sequence number generator. A topology update message, named TUM_A, which includes LCV_A, node A ID (its value is assumed to be ''k'') and SN_A, is generated in RMM and sent on the *control channels* of all outgoing working links. When control plane B (CP_B) receives TUM_A, RMM of CP_B determines if this is the first time to receive TUM_A by comparing the SN_A in TUM_A with the ''seq_num'' in the kth row vector ("k" is node A ID) in its TCM. If SN_A in TUM_A is larger than the "seq_num" in the row vector, it means this is a new message and RMM of CP_B overwrites k^{th} row vector of TCM with LCV_A and SN_A. Then RMM of CP_B sends TUM_A to CP_B's neighboring control planes. If the SN_A is smaller, it means the received TUM_A is an out-of-date message and CP_B deletes TUM_A.

If a control plane (denoted as CP_C) fails, every neighboring control plane of CP_C will not be able to receive the *keep-alive* message from CP_C. Then, every neighboring control plane broadcasts a TUM to the network. With all of these TUMs, any control plane can determine that CP_C failed and reset the row vector (representing the connectivity of node C to other nodes) in TCM.

RMM also manages resource availability exchange. At each control plane, the local resource availability is stored in a local resource table (LRT), as shown in Table 1. In an optical node, each port is attached with one single fiber and the maximum number of ports is assumed as "M". The ‘‘node ID’’ in the row of "port k" represents the neighboring node ID connected by the fiber attached to "port k". The "λ_i -status" in the row of "port k" indicates the state of the ith wavelength on the fiber attached to "port k", where $1 \leq i \leq W$. A wavelength can be in one of the following four states: “used and preemptable”, “used and non-preemptable”, “available”, “fault”. The “used and preemptable” state indicates this wavelength is being used by a low priority lightpath which can be pre-empted by a high priority lightpath.

Table 1. Local Resource Table

Port 1	node ID	λ_1 status
		...
		λ_W status
		...
...	...	...
		...
		λ_1 status
Port k	node ID	...
		λ_W status
		...
...	...	...
		...
		λ_1 status
Port M	node ID	...
		λ_W status

The global resource database is represented by a global resource table (GRT) which consists of a collection of LRTs of all nodes in the entire network. During the initialisation phase while a node starts or reboots, the establishment of initial GRT, the advertisement of LRTs and the GRTs convergence of the entire network are similar to the procedure in the topology discovery as described above. After the initialisation phase, a node is operational and ready to both accept LRTs advertised by other nodes and broadcast its own LRT when there is a change.

Another function of RMM is working with PRM and CM to support service differentiation. To demonstrate the QoS capability of the control plane, without loss of generality, we define three service types: high priority (mission critical), normal priority (protection sensitive), low priority (best-effort). According to the traffic characteristics and the end-to-end performance requirements, all client lightpath setup requests are classified into the three service types. The mission critical service needs a dedicated protection path and the establishment of both primary and protection lightpaths may preempt the existing lightpath of best-effort type. The protection sensitive service requires a shared protection path, but the establishment of both primary and protection lightpaths cannot preempt any existing lightpath. On the other hand, once the primary and protection lightpath of this service type have been established, both of them cannot be preempted. A protection sensitive service request may be rejected even at the source node if either a primary path or a protection path cannot be found. The best-effort service only requires restoration function and may be preempted by the mission critical traffic. Upon receiving a mission critical lightpath setup request, RMM computes the primary and protection lightpaths, which may preempt some wavelength resource from the existing lightpaths. When CM of an intermediate control plane receives the setup request for a mission critical lightpath and the assigned wavelength for the mission critical lightpath is being utilized by a best-effort lightpath, a lightpath abort message is sent out for the preempted best-effort lightpath and the wavelength is allocated to the mission critical lightpath.

Table 2. Lightpath Table Entry

Lightpath ID			status	Protection restoration	Input port ID	Output port ID	Lambda ID
Source Node ID	Destination node ID	Lightpath sequence number					

2.3 Connection Module (CM)

The main function of CM is lightpath signalling and maintenance. At each node, a lightpath table (LT), maintained by CM, is used to manage all lightpaths (originating, passing-through and terminating) over the OXC. Table 2 shows the structure of an entry in LT, named lightpath table entry (LTE). In each LTE, ''lightpath identifier'' (lightpath ID) includes the source node ID, destination node ID and lightpath sequence number. The node ID is used to address the optical node. Although IP addressing mechanism can be used, the current implementation uses a simple addressing mechanism such that each node is assigned a unique ID (an integer) and other network elements, e.g. fiber and wavelength, are addressed using a selector (an ID, but not globally unique). Lightpath sequence number is used to identify the lightpaths originating at each node. Thus, a lightpath ID uniquely identifies a lightpath in the network domain. The ''status'' attribute in LTE indicates the state of a lightpath (precreate, reserved, active, or delete). The "Protection/Restoration" attribute indicates the QoS type (or service type) as described in RMM section. The ''input/output port ID'' represents the ID of the incoming/outgoing port of the lightpath in the OXC.

The ''lambda ID'' indicates which wavelength is occupied in the outgoing port specified by "output port ID".

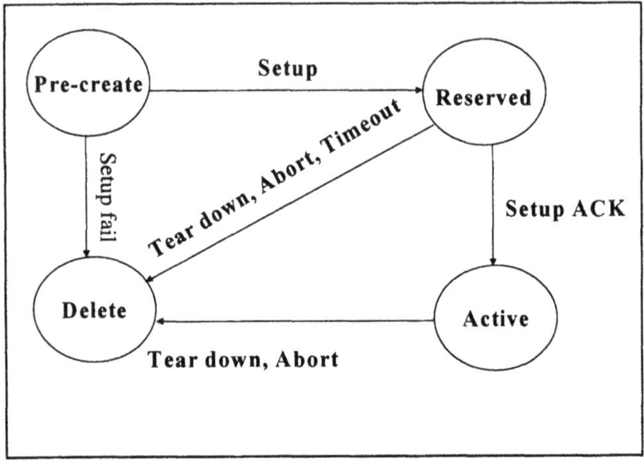

Figure 8. Lightpath State Transfer Graph

Fig. 8 shows the life cycle of a lightpath at each control plane. We assume that the lightpath is in the "precreate" state initially. When a lightpath setup is requested, CM starts to process the request and reserve the wavelength for this lightpath. If the reservation succeeds, CM allocates an LTE in LT and sets all attributes in the LTE, with the ''status'' attribute set to ''reserved''. When CM receives a setup ACK for this lightpath, the "status" attribute is set to ''active'' and the OXC switching fabric is configured for this lightpath based on the LTE information. Then the client traffic can be transported through this lightpath. If CM receives a teardown, abort, or timeout message for an existing lightpath (''reserved'' or "active" state), the LTE of this lightpath in LT will be cleared and the reserved wavelength will be released. Conceptually, the lightpath is deleted.

To illustrate the end-to-end signalling procedure, we take the example of lightpath setup and consider the signalling of the working path. As a lightpath setup is requested from CPGUI, RMM of the source control plane utilizes the TCM and GRT to compute an end-to-end routing path and assign an available wavelength for the lightpath. RMM then sends the request with the computed working path to PRM which handles the protection and restoration. For the working (primary) path, RMM of the source control plane reserves the assigned wavelength by checking whether the status of this wavelength in the associated entry of LRT (specified by the output port) is available. If yes, the status for the assigned wavelength is set to "used". Then the CM of the source control plane allocates a LTE in LT for this lightpath and sets all attributes in the LTE, with the status set to "reserved".

Afterwards, CM creates and sends a lightpath setup message, which includes the explicit routing path and the assigned wavelength computed by RMM, to the downstream control plane. Similarly, the RMM of the downstream control plane also reserves the assigned wavelength and updates LRT, and the CM allocates a LTE in LT, sets all attributes, with the status set to "reserved", and forwards the lightpath setup message to its own downstream control plane (specified by the explicit routing path). This procedure continues until the destination control plane receives the setup message and reserves the assigned wavelength. If this final reservation succeeds, a setup ACK is sent back to the source control plane along the routing path. The CMs of all intermediate control planes set the status attribute to "active" in the LTE of this lightpath and configure the OXC switching fabric based on the information of LTE. If the reservation fails at any control plane, e.g. the assigned wavelength is not available, a setup NAK is sent back to the source control plane. The intermediate control planes release the reserved wavelength and update the corresponding entry in the LT and LRT. The setup ACK or NAK will be sent to CPGUI by the CM of the source control plane.

2.4 Protection/Restoration Module (PRM)

PRM provides the function of protection/restoration path selection, fault detection and notification. Because of the similarity to compute protection path and restoration path, we take the protection path selection as an example to illustrate the merit of the PRM. Protection method includes fiber level protection (FLP) and channel level protection (CLP). In this paper, we consider to use channel level protection. CLP scheme will result in high network utilization. Because mesh network is rich in connectivity, the spare resources can be shared between multiple protection paths as long as they are not likely to simultaneously require it. We consider the end-to-end protection/restoration paths, and the disjointed-edge is selected as the restored path in PRM. This disjointed-edge path can provide fast recovery without diagnosing the fault location.

When PRM receives a protection/restoration request with the working path, it builds the Shared Risk Link Group (SRLG) of the working path. SRLG refers those resources which share the same physical ducts or fiber links as the working path. Then, PRM submits SRLG to RMM. RMM temporarily excludes SRLG from resource database (e.g. TCM and GRT) and computes the secondary lightpath as protection. Once the secondary lightpath is obtained, RMM will recover the SRLG in resource database, and send the secondary lightpath to PRM. According to the information from RMM, PRM will update its own protection/restoration routing table. In the protection case, the secondary path is pre-calculated. Once the source control plane detects (or is notified of) a failure, it will look up its own protection routing table, and send a lightpath setup request to CM to set up the protection/restoration lightpath. If the secondary lightpath cannot be computed by RMM (e.g. due to lack of resource), the lightpath setup request will be rejected and a setup NAK is sent to the CPGUI by RMM.

Fault detection is responsible for monitoring network status and initiating protection/restoration process. It includes hardware detection to detect the low layer impairments, such as loss of signal (LOS) and higher layer detection by using the link probing mechanism. All control planes along the working path are responsible for detection. The rapid fault notification also greatly affects the protection /restoration time. While detecting the failure, the control plane will send out a failure notification by transmitting a failure indication signal (FIS) to its upstream nodes. This notification is relayed hop-by-hop to the upstream control plane until the notification reaches the source control plane. Once the source control plane receives the FIS, it will check if the reliability attribute of lightpath is protection or restoration. If the protection path is requested, the source node will look up the protection routing table and invoke the CM. If the restoration path is requested, it will involve rerouting process. For instance, the protection/restoration path selection process is invoked in PRM to calculate the rerouting path to recover the working traffic.

2.5 Main Module (MM)

The main function of MM consists of initializing the control plane, including all modules, waiting for incoming messages or requests from its neighboring control planes, CPGUI, and invoking other modules to process the message or request. When a control plane is started initially or rebooted (due to some failure), at first, the control plane establishes the *control channels* among neighboring control planes (the neighboring information is manually configured) and between a control plane and CPGUI (if CPGUI is running), and create all tables and data structures, such as LRT, LCV, LT, GRT, TCM, etc. Then the control plane queries the neighboring control planes to obtain the correct data in those tables if it is necessary. After the initialization, MM begins accepting requests from CPGUI or the neighboring control planes and invoking other modules to process the requests.

3. CONCLUSION AND FUTURE WORKS

In this paper, we have demonstrated the concept of control plane for IP over WDM optical networks. The proposed control plane is capable of several key functions, such as routing, signalling and protection/restoration. With the help of CPGUI, the proposed control plane can graphically demonstrate the service provisioning and protection/restoration. In the future, we will explore the more effective routing algorithm, which can provide the desirable traffic engineering capability to improve the resource utilization in the optical network, and the more effective protection/restoration mechanism. We will also evaluate and compare the performance of different approaches for IP over WDM integration.

4. ACRONYM

ACK: Acknowledge
CM: Connection Module
CLP: Channel Level Protection
CPGUI: Control Plane Graphical User Interface
FIS: Failure Indication Signal
FLP: Fiber Level Protection
GRT: Global Resource Table
LCV: Topology Vector
LOS: Loss of Signal
LRT: Local Resource Table
LT: Lightpath Table
LTE: Lightpath Table Entry
MM: Main Module
MPLS: Multi-protocol Lambda Switching
NAK: Negative Acknowledge
OSPF: Open Shortest Path First
OXC: Optical Crossconnect
PRM: Protection/Restoration Module
QoS: Quality of Service
RMM: Resource Management Module
RWA: Routing and Wavelength Assignment
SRLG: Shared Risk Link Group
TCM: Topology Connectivity Matrix
TUM: Topology Update Message

5. REFERENCE

[1] N. Chandhok, etc. "IP over Optical Networks: A Summary of Issues", Internet draft, draft-osu-ipo-mpls-issues-01.txt, 2000

[2] B. Rajagopalan, etc. "IP over Optical Networks: A Framework", Internet draft, draft-many-ip-optical-framework-01.txt, 2000

[3] G. Wang, etc. "Extensions to OSPF/IS-IS for optical Routing", Intenet draft, draft-wang-ospf-isis-lambda-te-routing-00.txt, 2000

[4] D. Pendarakis, etc. "Routing Information Exchange in Optical Networks", Internet draft, draft-prs-optical-routing-00.ps, 2000

[5] S. Chaudhuri, etc. "Control of Lightpaths in an Optical Network", Internet draft, draft-chaudhuri-ip-olxc-control-00.txt, 2000

[6] D. Awduche, etc. "Multi-Protocol Lambda Switching: Combining MPLS Traffic Engineering Control With Optical Crossconnects", Internet draft, draft-awduche-mpls-te-optical-02.txt, 2000

[7] D. Basak, etc. "Multi-protocol Lambda Switching: Issues in Combining MPLS Traffic Engineering Control With Optical Crossconnects", Internet draft, draft-basak-mpls-oxc-issues-01.txt, 2000

[8] M. Krishnaswamy, etc. "MPLS control plane for Switched Optical Networks", Internet draft, draft-krishnaswamy-mpls-son-00.txt, 2000

[9] Z. Tang, etc. "Extensions to CR-LDP for Path Establishment in Optical Networks", Internet draft, draft-tang-crldp-optical-00.txt, 2000

[10] Y. Ye, M.Ali, T.S. Wang, and S. Dixit, " Design hybrid protection scheme for survivable wavelenth-routed optical transport networks", Proc. Of European Conference on Networks & Optical Communication, June 2000

[11] Y. Rekhter, etc. "A Border Gateway Protocol 4 (BGP-4)", RFC 1771

Analysis and dimensioning of interconnected single-layer "switchless" all-optical networks*

A. Bianco[1], N. P. Caponio[2], G. Galante[1], E. Leonardi[1] and F. Neri[1]

(1) *Politecnico di Torino, Torino, Italy*
{bianco,galante,leonardi,neri}@polito.it

(2) *Optical Technology Center (OTC), Torino, Italy*
currently with *Agilent Technologies, Torino, Italy*
paolo_caponio@agilent.com

Key words: All optical networks. Switchless networks. Multiple router interconnection.

Abstract: A "switchless" all-optical network requires neither electronic switches nor active optical cross-connects because the switching function is implemented at terminals via time and wavelength agile transceivers. Terminals are grouped into passive optical networks which are connected to a single passive wavelength router. One centralised controller grants the terminals access to the time/wavelength frame. This paper defines a simple non-uniform traffic pattern based on a locality principle and evaluates the hardware/software complexity required to cover the same user population with either one or four fully interconnected "switchless" networks.

1. INTRODUCTION

The European Union ACTS project SONATA explored the feasibility of a single-layer nation-wide all-optical network. The network design aims at exploiting the best features of photonic technologies, such as the very large bandwidth available on optical fibers, and the flexibility in modifying the

* This work has been partially funded by the European Union ACTS Programme within the Project SONATA.

end-to-end connectivity provided by optical channels, without trying to emulate in the optical domain the features available in electronic networks. With respect to electronic networks and to other, more conventional, optical network designs (such as wavelength routing networks based upon optical cross-connects), the proposed architecture avoids the need for large and fast switching nodes, drastically simplifying both the network structure and the layer architecture within the network. It is a very innovative design, in which new solutions specifically designed for the optical domain are devised to gain cost and performance benefits.

The SONATA architecture is said to be "switchless" because it is completely buffer-less and none of the nodes has to be equipped with either electronic switches or active optical cross-connects. This is possible because the switching function is removed from the nodes and distributed at the terminals where it is implemented by transceivers which can be quickly tuned on a large number of frequencies (also called wavelength channels).

Depending on the amount of traffic generated, either a street cabinet, or an IP router, or a LAN switch, or a workstation can be connected to the network through a terminal interface (or, more briefly, terminal) equipped with one fast-tuneable transceiver.

Terminals are partitioned in many large groups each of which is served by a passive optical network (PON). In each PON the traffic originated by terminals is multiplexed into wavelength channels (also called light-paths) via a WDM/TDM access protocol implemented by the time/wavelength agile terminal transceivers. The same holds for demultiplexing the traffic carried by light-paths into their contributing flows. All of the PONs are connected to one centralised passive wavelength router node (PWRN). The PWRN provides full connectivity among PONs linking each pair of PONs by means of one dedicated light-path.

A pool of wavelength converters arrays (WCA) directly connected to some extra PWRN ports allows both additional connectivity and network flexibility in adapting to any traffic pattern.

The access to network resources must be reserved end-to-end by terminals: a centralised network controller (NC) grants the terminals access to the time/wavelength frame, and manages the wavelength converters. The network is designed to provide both connection-oriented and connectionless services, in the sense that terminal requests can be for "persistent" or for "non-persistent" slot allocations.

The SONATA network feasibility, the components that should be used, the physical limitations that should be taken into account when dimensioning the network are discussed in [1]. The specification of the signalling protocol between terminals and the NC is given in [2], while a thorough description and performance analysis of the resource allocation algorithms to be used by the network controller is given in [3].

This paper defines a simple non-uniform traffic pattern and evaluates the hardware/software complexity required to cover the same user population with either one or four fully interconnected “switchless” networks. In particular, Section 2 describes the architecture of a single network and its extension to the case of L fully-interconnected networks. Section 3 defines a simple telephone traffic model and evaluates the call arrival rates. The two-level traffic distribution used here is based on a locality principle, i.e. topologically near terminals exchange among them a higher traffic rate than farther ones. In Section 4 both one and four fully interconnected networks are dimensioned for the traffic distribution defined in Section 3. The number of router ports and wavelength converters required is evaluated together with the number of calls received and processed by each network controller. Section 5 summarises on our results.

2. NETWORK ARCHITECTURE

This section describes the structure of the "switchless” networks utilised in the sequel of the paper. Section 2.1 reviews the architecture of a single router network, while Section 2.2 considers the case in which the network is split into L identical sub-networks. Each sub-network has its own passive wavelength router and its own network controller. The PWRNs are connected in a full mesh through some additional router ports attached to active interconnection boxes.

2.1 Single router network

Terminals are grouped into N identical PONs. Each PON sends its data to one input port of the passive wavelength router through a fiber and receives data from one output port of the PWRN via a different fiber (see left and right hand side of *Figure 1*).

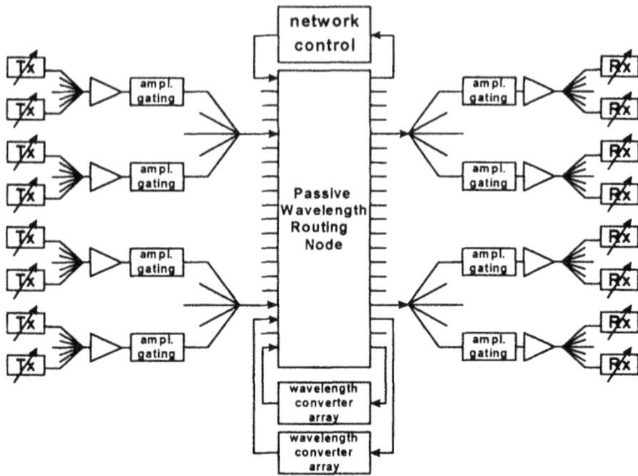

Figure 1. Single router network

The passive wavelength router implements a fully interconnected topology among its ports because each output port is reachable from any input port using a known (and fixed) wavelength channel. Both the input and the output side of the PWRN have 1 port connected to the network controller, N ports attached to the PONs and d ports (called "dummy" ports) through which it is possible to reach the arrays of wavelength converters.

Each terminal has one transmitter and one receiver quickly tuneable over $N+d+1$ different frequencies; therefore each fiber carries $N+d+1$ wavelength channels having a bandwidth of $B = 622$ Mb/s each.

The fast-tuneable transmitters, the wavelength converter arrays and the fast-tuneable receivers interconnected via the wavelength router behave as a three-stage switch. The NC controls the three-stage switch configuration (i.e. the logical topology of the network) by programming the wavelength converter arrays. In this way connectivity among PONs is given in two forms: via direct (or "wired") wavelength channels, and via channels re-circulating through wavelength converters (also called "programmable" channels).

Terminals can request both connection-oriented and connectionless services to one centralised network controller by means of a signalling protocol. The NC grants the terminals access to the time wavelength frame avoiding all conflicts among transmitters and receivers. More precisely the NC stores the requests received from the terminals during a time frame, processes them in the subsequent frame and then responds to the users in the following frame; only after this response there will be the transmission of data among end-users. Note that, as shown in *Figure 2*, these four phases can be pipelined and an extra delay is required between the reply frame and the transmission frame because of propagation delays in the network.

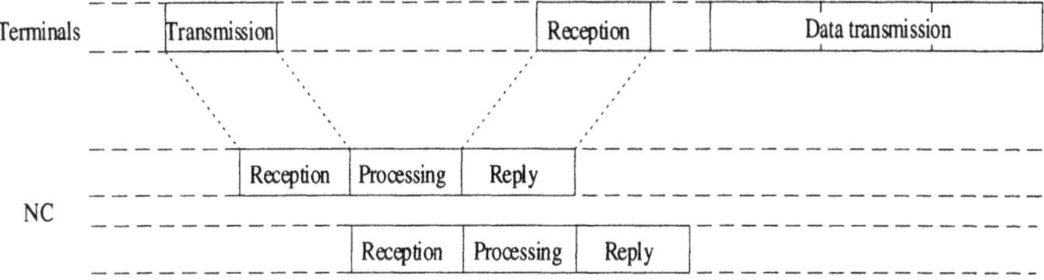

Figure 2. NC operations scheme

Four types of signaling packets are required:

- Connection-oriented/connection-less bandwidth requests (upstream)
- Acknowledge (reply) packets (downstream)
- Reconfiguration packets (downstream)
- Wavelength converter arrays programming packets (downstream)

Packets of the first kind are sent by users to the NC to request or release bandwidth. Packets of the second kind are sent by the NC to end-users to accept bandwidth requests and to identify allocated slots. Packets of the third kind may be sent by the NC to the end-users when a reconfiguration of the logical network topology takes place. Packets of the fourth kind, finally, are sent by the NC to the wavelength converter arrays to configure them each time the network logical topology is rearranged.

2.2 Multiple router network

Now we consider a multiple router network consisting of L identical sub-networks interconnected by a full mesh as shown in *Figure 3* for L=4. Each sub-network has its own network controller (NC i) and is served by its own wavelength router (PWRN i) which has N ports for direct PON connection, d "dummy" ports and r additional "remote" ports. Any router can send data to the other L-1 sub-networks by L-1 sets of r/(L-1) fibers, each passing through an interconnection box (SW i-j).

The network adopts a single-hop routing policy. This means that remote traffic, i.e., traffic whose source and destination terminals are not attached to the same router, is directly routed from the source router to the destination router, without traversing intermediate routers.

Each interconnection box (switch) must realise at least a full wavelength switching matrix to reach each PON of the destination sub-network from each PON of the source sub-network. These switching boxes can be implemented either all-optically, or using electro-optical conversions.

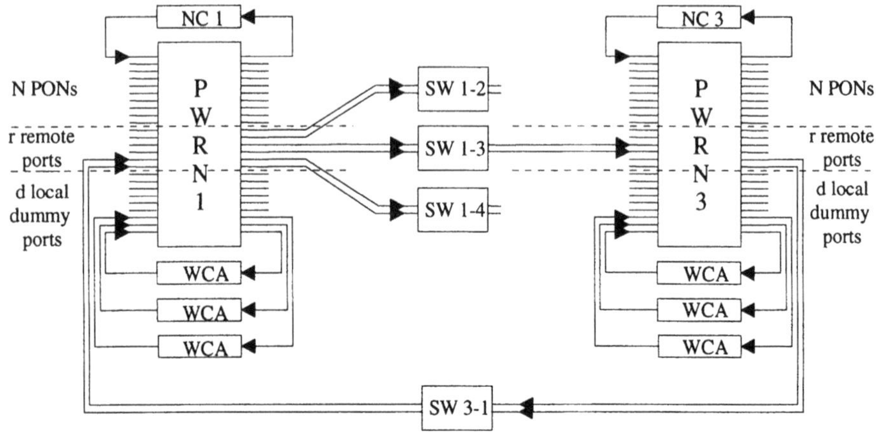

Figure 3. Multiple router network

If an interconnection box contains a complete time and wavelength switching matrix, the problem of allocating a new connection between sub-network i and sub-network j can be decoupled for the two routers, since a full T-λ switch can map any free slot belonging to any input wavelength to any free slot belonging to any output wavelength. Therefore NC i, to establish a new connection, has to allocate a suitable number of slots on a wavelength (i.e., on one of the $r/3$ remote ports) from the calling terminal to switch SW i-j, while NC j needs only to allocate the same number of slots on the wavelength from switch SW i-j to the called terminal, without even knowing the slot allocation map at NC i. In this case, a different process can handle the forwarding of requests on each of the r remote fibers.

The third possibility is to add to the time-wavelength matrix a space switching matrix, thence any time slot on any wavelength carried by an input fiber of switch SW i-j can be mapped to any slot belonging to any wavelength on any output fiber SW i-j, thereby granting the maximum interconnection flexibility to the system.

Local traffic, i.e. traffic exchanged among terminals belonging to the same sub-network, is handled by the network controller of the related sub-network using the signalling protocols described in Section 2.1.

Remote traffic management requires two levels of interaction: local signalling between the terminals and their own network controller, remote signalling between the two involved network controllers. The operations needed to establish a new connection between terminal t_i belonging to sub-network i and terminal t_j belonging to sub-network j are (see *Figure 4*) the following.

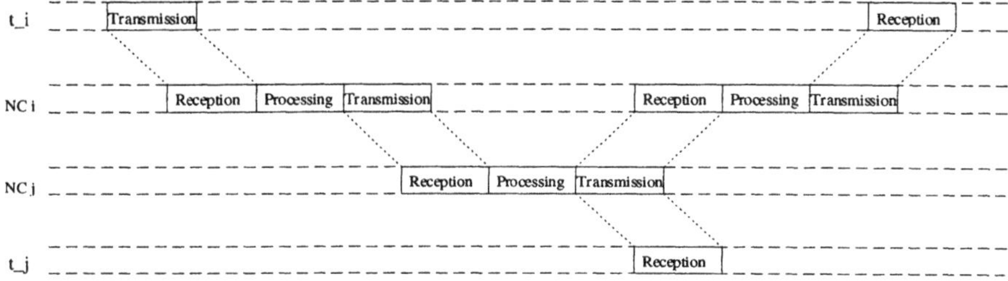

Figure 4. Delays implied in opening a inter-router connection

1. Terminal t_i sends a signalling request to NC *i*
2. The request is processed by NC *i* during the next time frame and, if it can be accepted, it is forwarded to NC *j* in the subsequent time frame
3. NC *j* processes the request in the time frame after its arrival and, if it can be scheduled, generates two signalling messages: one addressed to NC *i*, the other addressed to terminal t_j. These messages will be transmitted during the next time frame
4. The message is processed by NC *i* during the subsequent time frame and a signalling message is generated to be sent to terminal t_i during the next time frame.

3. TRAFFIC MODEL

We consider the traffic model that was taken as a reference in the SONATA project for the study of multiple-router networks. The approach taken in this paper can however be extended to other traffic models.

The population to be served by the networking infrastructure is made of T = 20,000,000 user terminals (which can be workstations, IP routers, street cabinets, etc.). These T terminals are divided into N_c = 40 geographical areas (which can be metropolitan areas, regions, or countries) called cells, containing T_c = 500,000 terminals each. Note that the terminals belonging to the same geographical cell may end up belonging to one PON or to more PONs, and also to one router or to more routers, depending on the network dimensioning.

The traffic is not uniform, in the sense that the amount of traffic exchanged between two terminals belonging to the same cell is different from the amount of traffic exchanged between two terminals belonging to different cells. We assume for simplicity that all cells are alike.

We assume that each terminal generates an average traffic b = 10 Mb/s, out of which 7.8 Mb/s are uniformly distributed over the users belonging to the same cell, and 2.2 Mb/s are equally divided among the remaining

19,500,000 terminals. Therefore each terminal sends an average of b_c = 7.8 Mb/s / 500,000 = 15.6 b/s toward each terminal of the same cell and b_o = 2.2 Mb/s / 19,500,000 = 0.113 b/s toward the remaining terminals. This traffic pattern is represented by the matrix in *Table 1*. Note that the aggregate network capacity is 200 Tb/s.

Table 1. The considered traffic matrix

Cell		1				2			...			40		
	Terminal	1	2	...	T_c	1	...	T_c	....			1	...	T_c
	1	0	b_c		b_c	b_o		b_o	b_o		b_o	b_o		b_o
	2	b_c	0		b_c	b_o		b_o	b_o		b_o	b_o		b_o
1	⋮	b_c	b_c	⋱	b_c	b_o	⋱	b_o	b_o	⋱	b_o	b_o	⋱	b_o
	T_c	b_c	b_c		0	b_o		b_o	b_o		b_o	b_o		b_o
⋮	⋮	⋮				⋮			⋮			⋮		

Under the assumption of telephone traffic with a rate of 64 kb/s and a call duration of 3 minutes, each call transmits 64 kb/s × 180 s = 11,520 kbits. The time interval between two successive calls between one pair of terminals belonging to the same cell is t_c = 11,520 kb / b_c = 11,520 kb/15.6 b/s = 738,461 s = 8 days 13 hours 7 minutes 41 s and the corresponding call arrival rate is $\lambda_c = 1/t_c = 1.35 \times 10^{-6}$ calls/s. Instead, the time interval between two successive calls between one pair of terminals belonging to different cells is: t_o = 11,520 kb / b_o = 11,520 kb / 0.113 b/s = 101,946,902 s = 3 years 84 days 22 hours 35 minutes 2 s corresponding to a call arrival rate $\lambda_o = 1/t_o = 9.8\ 10^{-9}$ calls/s.

4. NETWORK DIMENSIONING

Network dimensioning is performed by a decomposition of the two-level traffic matrix defined above into a uniform traffic matrix and a non-uniform traffic matrix. The uniform traffic portion is then routed through the "wired" channels of the passive wavelength router, while the non-uniform part is subject to wavelength conversion. The number of terminals in each PON (also called PON split) is chosen in order to fully utilise the bandwidth available in wavelength channels.

Since there are many ways of splitting the traffic matrix in two parts, we choose the one which minimises the number of wavelength converters. Therefore a matrix contains the maximum uniform component while the other captures the minimum non-uniform component as shown in *Table 2* and *Table 3*.

Table 2. Maximum uniform component of the traffic matrix

Cell		1				2			...			40		
	Terminal	1	2	...	T_c	1	...	T_c	....			1	...	T_c
	1	0	b_o		b_o	b_o		b_o	b_o		b_o	b_o		b_o
	2	b_o	0		b_o	b_o		b_o	b_o		b_o	b_o		b_o
1	⋮	b_o	b_o	⋱	b_o	b_o	⋱	b_o	b_o	⋱	b_o	b_o	⋱	b_o
	T_c	b_o	b_o		0	b_o		b_o	b_o		b_o	b_o		b_o
⋮	⋮	⋮				⋮			⋮			⋮		

Table 3. Minimum non-uniform component of the traffic matrix

Cell		1				2			...			40		
	Terminal	1	2	...	T_c	1	...	T_c	....			1	...	T_c
	1	0	b_c-b_o		b_c-b_o	0		0	0		0	0		0
	2	b_c-b_o	0		b_c-b_o	0		0	0		0	0		0
1	⋮	b_c-b_o	b_c-b_o	⋱	b_c-b_o	0	⋱	0	0	⋱	0	0	⋱	0
	T_c	b_c-b_o	b_c-b_o		0	0		0	0		0	0		0
⋮	⋮	⋮				⋮			⋮			⋮		

Note that this approach to network dimensioning does not take into consideration how flexible the network must be in terms of adaptability to traffic fluctuations. For example, if we consider a uniform traffic scenario, in which all terminals are alike and we do not make any difference between inter-cell and intra-cell traffic, the approach proposed here would lead to SONATA networks without wavelength converters (all the traffic would be routed through "wired" channels).

This approach is just one of several possible approaches to network dimensioning: other dimensioning criteria were proposed for SONATA networks. For example in [4] different criteria are proposed, permitting more flexibility in traffic allocation.

4.1 Single router network

The number s of terminals in one PON is computed imposing that the uniform portion of traffic $s(s\text{-}1) \times b_o \approx s^2 \times b_o$ generated either toward the same PON or toward each of the remaining PONs is equal to the bandwidth B of the "wired" wavelength channel that will carry it. In this way $s = \sqrt{B / b_o}$ = 74,250.8. Choosing s = 62,500 the number of PONs becomes $N = T / s$ = 320, so there are 320 PONs / 40 cells = 8 PONs / cell. This means that we are in the situation in which more than one PON is needed to attach the terminals belonging to the same cell, and each cell is completely served by a particular router.

The number d of wavelength channels passing through the wavelength converters is determined by the extra traffic generated by one PON both toward itself and toward the other 7 PONs belonging to the same cell, therefore $d = 8 \times (b_c\text{-}b_o) \times s^2 / B = 778.9 \approx 779$ channels. The total number of router ports (and thence wavelength channels) needed is $N + d = 1{,}099$.

Note that each router port attached to a PON carries up to $N + d = 1{,}099$ channels (we neglect here and in the sequel the channel used to communicate with the NC), while each router port attached to an array of wavelength converters carries $N = 320$ channels. Each array needs therefore $N = 320$ converters, and the total number of wavelength converters is $N_{wc} = N \times d = 249{,}280$.

The call rate from one PON toward itself is $s(s\text{-}1)\ \lambda_c = 5{,}289.6$ calls/s, while the call rate between two PONs belonging to the same cell is $s^2\lambda_c = 5{,}289.7$ calls/s; these two call rates will be assumed to be identical. The call rate between two PONs belonging to different cells is $s^2\lambda_o = 38.3$ calls/s. Therefore the call rate between a PON and the 20 millions of terminals in the network would be 8 PONs × 5,289 calls/s + 312 PONs × 38.3 calls/s = 54,261 calls/s and the call arrival rate of the whole network would be 320 PONs× 54,261 calls/s = 17,363,712 calls/s.

The overall memory requirement at the NC for the data structures needed to run the network control algorithms (both the time-scheduling and the logical topology design algorithm, see [3]) can be seen to be about 27 Gigabyte.

4.2 Multiple router network

We consider now the case in which the T terminals are divided into L=4 subsets of $T_s = T / L = 5{,}000{,}000$ terminals and each subset is connected to a different wavelength router. Thus, the first $T_s / T_c = 10$ cells are assigned to the first sub-network, cells from 11 to 20 are assigned to the second sub-network, cells from 21 to 30 are assigned to the third sub-network and, finally, cells from 31 to 40 are assigned to the last sub-network.

The number of users per PON s remains the same as for the single router case because it depends only on b_o and B. Moreover each sub-network has $N_s = N / L = 80$ PONs and still needs $d = 779$ auxiliary wavelength channels because each cell covers 8 PONs as in the single router network. Since each fiber connecting a “dummy” port to a wavelength converter array carries N_s channels, each sub-network needs $N_{wc} = N_s \times d = 62{,}320$ wavelength converters to route traffic internal to the sub-network.

The number of remote ports and, thence, the number of fibers to be used in the interconnection among sub-networks can be reduced at the price of adding some wavelength converters to the topology. Here we explore 3 alternatives:

a) no additional wavelength converters

b) some additional wavelength converters
c) some additional wavelength converters and shared use of the converters already connected to "dummy" ports.

In the first case, the number r of wavelength channels needed by a PON belonging to one sub-network to send its remote traffic to the remaining 3 sub-networks is given by $r = (L\text{-}1) \times T_s \times s \times b_o / B = 170.3 \approx 171$. The total number of router ports (and hence wavelength channels) needed is $N_s + d + r = 1{,}030$. The N_s PON ports carry $N_s + d + r = 1{,}030$ channels, while the number of channels carried by each of the r remote fibers (and by the d dummy ports) is $N_s = 80$. Therefore the number of wavelength converters needed in the interconnection boxes attached to each sub-network is $N_s \times r =$ 13,680 converters.

Another possibility would be to re-circulate part or all of the remote traffic through some additional wavelength converter arrays local to each router. By doing so, more wavelength channels will be accommodated on remote fibers, with the advantage of reducing the number of remote fibers, hence of interconnection costs. Since we need r ports to carry remote traffic, we can have r' remote ports attached to interconnection boxes and $r'' = r - r'$ extra dummy ports connected to wavelength converters. If we assume that these r'' extra dummy ports are dedicated to incoming and outgoing remote traffic, they carry $N_s + r'$ channels each; thus, the number of additional wavelength converters (outside the interconnection box) is $r'' \times (N_s + r')$. We have constraints on r' and r'': the number of remote channels is $r' \times (N_s + r'')$ and this must be at least equal to $r \times N_s = (r' + r'') \times N_s$. This means that r' can range between N_s and r. For example, we can assume that remote connections are re-circulated through $r'' = 90$ additional dummy ports; the number of remote ports is reduced to $r' = 81$, each carrying $N_s + r'' = 170$ channels. Therefore, the number of wavelength converters should be increased by $r'' \times N_s = 7{,}200$ for outgoing connections and by $r' \times r'' = 7{,}290$ for incoming connections. The number of wavelength converters needed in the interconnection boxes attached to each sub-network would be $r' \times (N_s + r'') = 13{,}770$. Note, however, that the number of router ports remains $N_s + d + r' + r'' = 1{,}030$ since $r = r' + r''$.

A more drastic reduction in the number of remote ports can be obtained by using all $d + r''$ dummy ports to route remote traffic. In this case r' can be reduced to the minimum $r' = r \times N_s / (N_s + d + r) \approx 13.3$ (assuming that the maximum number of channels per fiber is equal to the number of router ports $N_s + d + r$). Since r' must be an integer multiple of $L\text{-}1 = 3$, we can take $r' = 15$ and $r'' = 156$. The number of additional wavelength converters becomes $r'' \times (N_s + r') + d \times r' = 26{,}505$ while the number of wavelength converters needed in the interconnection boxes attached to each sub-network is $r' \times (N_s + d + r) = 15{,}450$.

Note that the numbers of wavelength converters given above in the hypothesis of all-optical implementation of the interconnection boxes can be arguably related to the number of necessary tuneable lasers required for an electro-optical implementation (in which received optical channels are converted to electronic streams, possibly stored, and later re-converted to optics with tuneable lasers).

Table 4 summarises the project parameters of one of the sub-networks for the three alternatives: the column labelled "# of extra λ converters" shows the number of wavelength converters needed to lower the number r' of interconnection fibers among routers. The column "Total # of λ converters" is the sum of the number given above, the number of wavelength converters connected to the "dummy" ports (always 62,320) and the number of wavelength converters put into the interconnection boxes. Note that the number of wavelength converters is dominated by the requirements of unbalanced intra-cell traffic.

Table 4. Architectural alternatives for interconnected routers

Conf	Remote traffic through ...	r'	r''	# of extra λ converters	# of λ converters in interconnection boxes	Total # of λ converters	# of λ on remote fibers
a	No dummy ports	171	0	0	13,680	76,000	80
b	Dedicated dummy ports	81	90	14,490	13,770	90,580	170
c	Shared dummy ports	15	156	26,505	15,450	104,275	1030

The call arrival rate between one PON and the other terminals of the same sub-network belonging to different cells is 72 PONs × 38.3 calls/s = 2,758 calls/s, therefore the call rate between a PON and the terminals belonging to the same sub-network is 8 PONs × 5,289 calls/s + 72 PONs × 38.3 calls/s = 45,070 calls/s. Finally the call rate between a sub-network and itself is 80 PONs × 45,070 calls/s = 3,605,568 calls/s. The call arrival rate between a PON and the 240 PONs belonging to the other sub-networks is 240 PONs × 38.3 calls/s = 9,159 calls/s, thence the call rate between two different sub-networks is 80 PONs × 9,159 calls/s = 732,792 calls/s. The total call arrival rate toward the network controller of a sub-network is 5,071,152 calls/s and consists of 3 contributions:

- 3,605,568 calls/s coming from the sub-network
- 732,792 calls/s going to the network controllers of the other sub-networks
- 732,792 calls/s coming from the network controllers of the other sub-networks.

Since the complexity of the NC grows much faster than linearly with the number of calls/s to be processed, the architecture with several interconnected sub-networks leads to significant savings in complexity at the NC.

Since each sub-network is exactly ¼ of the whole network, the upstream and downstream signalling bandwidth requirements of one sub-network are exactly ¼ of the ones of the single router network.

The overall memory requirement at the NC of each sub-network for the data structures needed to run the time-scheduling and logical topology design algorithms can be seen to be about 7 Gigabyte, while each interconnection box needs at least 100 Megabyte of memory to store the time/wavelength switching matrix.

5. CONCLUSIONS

The European Union ACTS project SONATA devises an advanced transport architecture for a single-layer all-optical WDM network covering a nation-wide area. Its outstanding features are a very simple passive star topology built around a passive wavelength routing node and terminals equipped with fast-tuneable transceivers. The architecture is said to be "switchless" because no explicit switching devices are used; the only elementary switching function resides in the tuneability of the wavelength converter arrays around the central node (i.e. the PWRN) and of the terminals' transceivers.

In this paper we describe the structure of a single-router network and outline the steps required to extend it to a multiple router architecture. Then we dimension two network architectures consisting of one and four interconnected routers for the same simple non-uniform telephone traffic pattern.

On the one hand, the single router network has 320 PONs and 779 wavelength converter arrays connected to "dummy" ports, so it requires a 1,099 ports router and 250,000 wavelength converters. While the network controller hardware must be equipped with 27 Gigabyte of memory and has to be fast enough to process 17,000,000 terminal requests/second. Although this is a very large complexity, the aggregate capacity is very large (order of 200 Tb/s), and the network can be considered as a nation-wide telecommunications infrastructure.

On the other hand, in the interconnected architecture, each sub-network has 80 PONs, requires 171 "remote" ports and still needs 779 wavelength converter arrays, thus each sub-network is built around a 1,030 ports router. The number r' of "remote" ports (and therefore of fibers) used for interconnections among routers can be varied between 171 and 15 simply increasing the number of wavelength converters used from 300,000 to 400,000. Each network controller hardware must be equipped with 7 Gigabyte of memory and has to be fast enough to process 5,000,000 terminal requests/second.

In this way we have shown that it is possible to reduce the tremendous hardware complexity (which scales more than linearly with the number of terminal requests/second) of the single router network controller by distributing it over more satellite network controllers at the price of increased delays for non-local traffic and higher costs for passive wavelength routers, wavelength converter arrays and interconnection fibers.

REFERENCES

[1] N.P.Caponio, A.M.Hill, F.Neri, R.Sabella, *Single-Layer Optical Platform Based on WDM/TDM Multiple Access for Large-Scale "Switchless" Networks*, European Transactions on Telecommunications, Special Issue on WDM Networks, January 2000.

[2] A.Bianco, E.Leonardi, M.Mellia, M.Motisi, F.Neri, *Specifications of signalling functions in SONATA*, contribution to deliverable D9 (see [4]) of the ACTS Project SONATA AC-351, Politecnico di Torino, 1999.

[3] A.Bianco, E.Leonardi, M.Mellia, F.Neri, *Network Controller Design for SONATA, a Large Scale All-Optical Passive Network*, IEEE Journal on Selected Areas in Communications, Special Issue on Protocols and Architectures for the Next Generation Optical Networks, October 2000

[4] ACTS Project SONATA AC-351, *Final Report on System Studies*, Deliverable D9, 2000.

IP DIFFERENTIATED SERVICES OVER A WDM PASSIVE OPTICAL STAR

Josué Kuri
Ecole Nationale Supérieure des Télécommunications
46, rue Barrault. 75634 Paris CEDEX 13, France
Tel: (+33) 1.45.81.75.70 Fax: (+33) 1.45.81.31.19
kuri@enst.fr

Maurice Gagnaire
Ecole Nationale Supérieure des Télécommunications
46, rue Barrault. 75634 Paris CEDEX 13, France
gagnaire@enst.fr

Abstract

Important recent advances in the field of opto-electronic devices make WDM optical networks a promising solution for the next generation Internet. Such technologies can be considered in the long-haul, metro and access sections of carrier's infrastructures. The main issue in this paper is the design of a Metropolitan Area Network based on a WDM passive optical star well suited for IP differentiated services. Our proposal focuses on both the MAC protocol and buffer management at the end nodes. Various simulation results outline the capacity of our proposal to satisfy the IP Diffserv [1] QoS requirements, as they are specified by the IETF. For that purpose, both Expedited Forwarding [3] and Assured Forwarding [4] are investigated.

Keywords: Diffserv, MAN, MAC protocol, WDM, Passive Optical Star.

Introduction

The passive optical star (POS) is a popular topology for single-hop broadcast-and-select LANs and MANs because of its superiority over other topologies (e.g., bus) in distributing the transmitted power equally to all nodes without inducing much excess loss [6]. Several Media Access Control (MAC) protocols have been proposed to coordinate data transmission over parallel WDM channels on this network. The main issue in this paper is the design of an efficient MAC protocol for a POS with support of differentiated services. The protocol incorporates service disciplines and buffer management mechanisms to support specific Diffserv's functional elements called Per-Hop-Behaviors (explained latter in this section).

The Differentiated Services architecture (Diffserv or DS) aims to provide scalable service differentiation in the Internet. A service [1] is defined as set of significant characteristics of packet transmission in one direction across a set of one or more paths within a network. These characteristics may be specified in quantitative or statistical terms of throughput, delay, jitter and/or loss, or may otherwise be specified in terms of some relative priority of access to network resources.

The Diffserv architecture defines a topological and functional distinction between nodes. *Boundary nodes*, which connect the DS-capable network to another network (either DS or not DS-capable), perform packet classification and traffic conditioning functions (e.g., metering, marking, shaping, policing). Traffic entering the network is classified and possibly conditioned by these nodes; it is then assigned to different relatively coarse collections of packets called *behavior aggregates* (hereafter, the term *aggregate* is used to refer to *behavior aggregates*). Within the DS network, aggregates are serviced by *interior nodes* in such a way that a specific, externally observable, forwarding behavior is guaranteed to each aggregate at each hop. This forwarding behavior is called *Per-Hop-Behavior* (PHB) (actually, both boundary and interior nodes must be able to apply the appropriate PHB to aggregates) and may be specified in terms of the relative access priority to the node's resources (e.g., buffer space, bandwidth) with respect to other PHBs, or in terms of its relative observable traffic characteristics (e.g., allocated link bandwidth, delay, loss) [1]. PHBs are usually specified as groups (PHB group) that share a common constraint applying to each PHB within the group. A single PHB defined in isolation is a special case of a PHB group.

Services meaningful to network users may defined in the Diffserv framework by applying classification and conditioning rules at the network boundary to create aggregates, and coupling these rules with specific PHBs. The Diffserv working group of the IETF currently works on the definition of the *Per-Domain-Behavior* concept, which is the expected treatment that an identifiable or target group of packets will receive from "edge-to-edge" of a DS domain. A particular PHB (or, if applicable, a list of PHBs) and traffic conditioning requirements are associated with each PDB. Measurable, quantifiable attributes such as jitter and delay are associated with each PDB and these can be used to describe what will happen to packets of that PDB as they cross the DS network [2].

Currently, a formal specification is available for two PHB (groups): the Expedited Forwarding (EF) PHB and the Assured Forwarding (AF) PHB group.

The EF PHB [3] is defined as a forwarding treatment for a particular aggregate, say the EF aggregate. The aggregate's *departure rate* in an EF-capable node must be equal to or greater than a configured rate for the aggregate. The EF aggregate should receive this rate independent of the intensity of any other traffic attempting to pass across the node. Furthermore, it should average at least the configured rate when measured over any time interval equal to or longer than the transmission time of a maximal-sized packet at the configured rate, whereby the maximal packet size is given by the MTU (Maximum Transmission Unit) of the output link.

The AF PHB group [4] (hereafter, referred to as the AF PHB) is a mean to offer different levels of forwarding assurance for IP packets. Four AF classes are defined, where each class is allocated a certain amount of forwarding resources (buffer space and link bandwidth) in the interior nodes. Within each AF class, packets are marked with one of three possible drop precedence levels. In case of congestion, the drop precedence of a packet determines its relative importance within the class. Thus, the forwarding assurance of a packet belonging to an AF class depends on 1) the amount of forwarding resources allocated to the class, 2) the current load of the class and 3) in case of congestion within the class, the drop precedence of the packet. An AF-compliant node should implement four AF classes. Packets in one AF class must be forwarded independently from packets in another AF class. The node must allocate a configurable, minimum amount of forwarding resources (buffer space and bandwidth) to each implemented AF class. Each class should be serviced in a manner to achieve the configured service rate over both small and large time scales.

It should be noted that the EF and AF PHBs are not a mandatory part of the Diffserv architecture, i.e., a node is not required to implement these PHBs in order to be considered Diffserv-compliant.

The rest of the paper is organized as follows: a description of the network architecture is presented in Section 1. The MAC protocol is described in Section 2. Section 3 shows numerical results about QoS parameters obtained through simulations. Finally, Section 4 presents our conclusions and future work.

1. NETWORK CONFIGURATION

We consider a WDM network with N nodes connected through a central passive star coupler (Figure 1). The system has a total of $N+1$ channels, N channels for data transmission and one for control information. The $\lambda_i(1 \leq i \leq N)$ data channels and the λ_c control channel are *slot synchronized* (Figure 2).

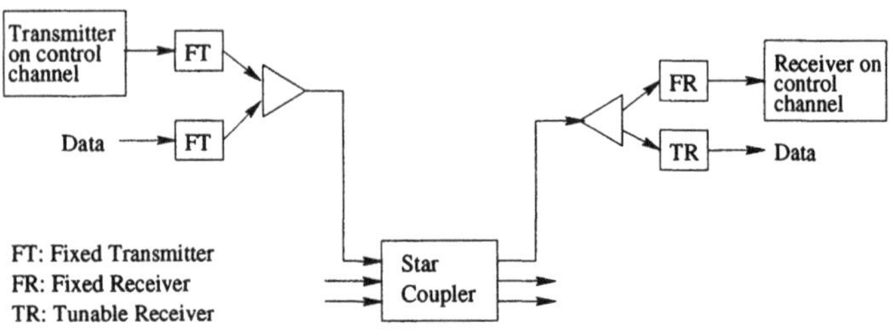

Figure 1 Passive optical star network architecture.

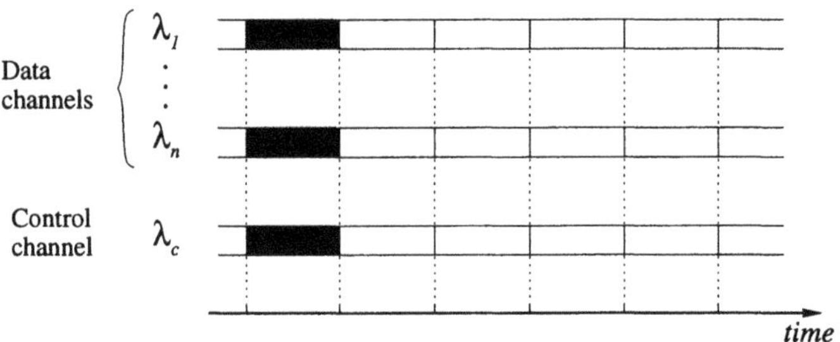

Figure 2 Synchronized slotted structure of the passive optical star.

Each node N_i is equipped with a transmitter fixed at a unique wavelength λ_i for data transmission. A tunable receiver with negligible tuning latency (i.e., the tuning time is significantly smaller than the slot time) and capable of covering all λ_i channels is used to receive data packets. In addition, each node is equipped with a transceiver fixed at the control channel λ_c. We denote the maximum propagation delay (time) permitted by the system between any pair of nodes by d_p. Furthermore, we assume that distance ranging is performed by all the nodes.

One of the weak points of the POS topology is the risk of failure of passive coupler at the center of the star. An original topology based on multiple interconnected POSs proposed in [5] prevents this drawback by providing protection and restoration techniques for POSs.

2. MAC PROTOCOL

Several collision-free MAC protocols have been proposed for POS networks (e.g., [7, 8, 9]). These protocols are interesting in metropolitan sized networks, where non-negligible propagation delays make collision handling expensive. Reservation-based collision-free protocols (e.g., [10, 11]) assign wavelengths to the logical channels between transmitters and receivers on a slot-by-slot basis. Key issues such as the load of the signaling channel, the efficiency of the scheduling algorithm and the impact of propagation delay have been widely investigated for these protocols. In this paper, additional aspects including buffer management mechanisms and service disciplines are considered in order to support service differentiation for IP traffic.

2.1. DATA UNIT FORMAT

Slots in the data and control channels are all L bits long. A *data unit* is the smallest unit of transmission and fits into a slot. A sequence of data units is called a *packet*. The maximum length of a packet is M data units. Upper layer packets longer than M are split into several MAC-level packets. As depicted in Figure 3, the control channel contains a signaling sub-channel and a reservation sub-channel, which are time-multiplexed on λ_c. Thus, each data unit in the control channel is divided into two fields, SIG and RES, that correspond to the signaling and reservation subchannels, respectively (the utilization of these fields is explained in the following subsections). The SIG field is S bits long and is divided into N mini-slots of length $s = S/N$, one for each node. The RES field is $R = L - S$ bits long and contains a variable number of *reservations*. A reservation consists of three sub-fields: the destination address, the packet length (in data units) and the Diffserv's PHB ID (PID). The binary representation (base 2) of the values for these fields fits in $\log_2(N)$, $\log_2(M)$ and $\log_2(P)$ bits, respectively. With the total length of a reservation being $r = \log_2(N \cdot M \cdot P)$, up to R/r reservations fit into a RES field.

2.2. SIGNALING

Reservation-based protocols use a signaling mechanism to convey the transmission requests among the nodes. Scalability is a major concern in such protocols because the bandwidth requirement in the signaling channel is in general a polynomial function of the number of nodes. The SRD (Signaling-Reservation Decoupling) protocol [12] alleviates this problem by distributing information more efficiently in the signaling channel. The

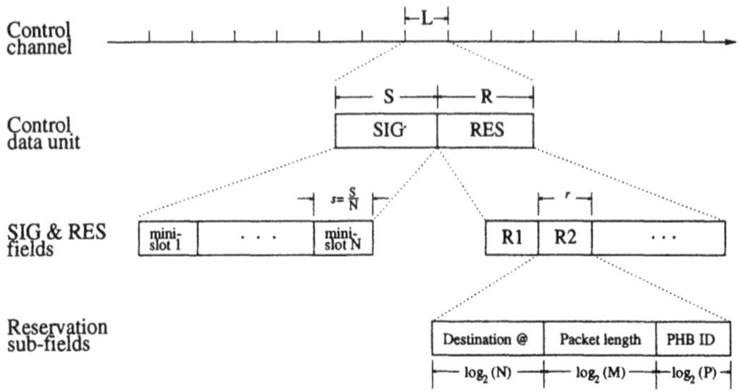

Figure 3 Format of a control data unit

basic idea is to decouple the information indicating that a new packet has to be sent, from the detailed information referring to this packet, namely, the destination address and the length of the packet (in data units). This mechanism allows the signaling channel to be dynamically allocated in such a way that active nodes (those with packets to send) can access the portion of the signaling channel currently unused by inactive nodes. In what follows we present an extended version of the SRD protocol, referred to as SRD_{DS} (Diffserv-SRD). The protocol consists of the following steps repeated by each node at each time slot (see Figure 4):

1 In time slot t, each node N_i in the network posts the number of packets, if any, that it received for transmission during $t-1$. The binary representation (base 2) of this number is put into the mini-slot assigned to N_i in the SIG_t field. At the end of t, SIG_t contains the filled mini-slots of all nodes; it propagates then through λ_c.

2 At slot $t+d_p$ (the current slot at the time the SIG_t field has completely propagated), the SIG_t field is received and every node is thus aware of how many packets the other nodes whish to transfer.

3 In slot $t+d_p+1$, the nodes fill the RES_{t+d_p+1} field with reservations for the packets they announced in the SIG_t field. A Round-Robin service discipline determines their order: each node is allowed to post one reservation in turn. If all nodes have posted their first reservation and if there is still more space available in RES_{t+d_p+1}, another round is started, until there is no other reservation to post or no more space left in RES_{t+d_p+1}. The signals that did not lead

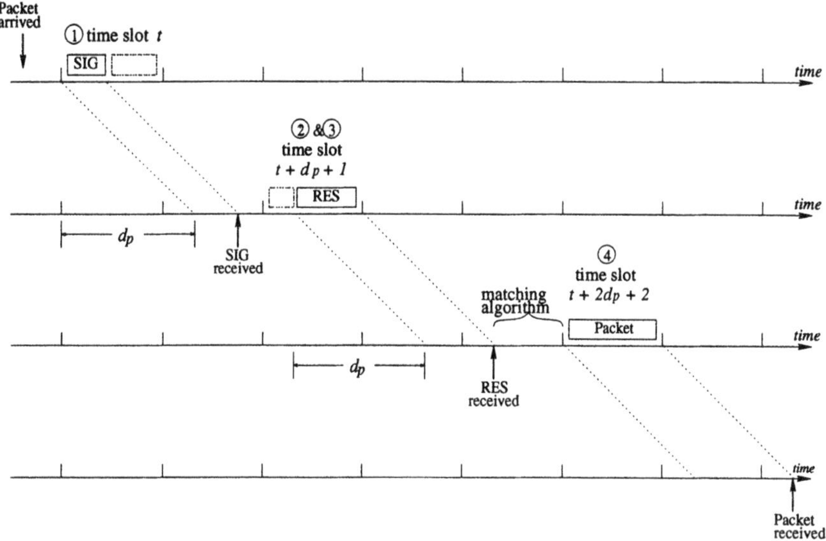

Figure 4 Time diagram of a SRD_{DS} packet transmission (in this case, the packet fits in a single data unit).

to a reservation are stored at each node in a FCFS queue and considered in the next slot.

4 Once the RES_{t+d_p+1} field propagates to all nodes, at slot $t+2d_p+2$, the nodes become aware of the destinations and lengths of data packets that arrived at their respective transmitting nodes before time slot t. This information may be conveniently represented as a *backlog matrix*, $B(t) = (b_{ij}(t))_{i,j=1}^{n}$. $b_{ij}(t)$ represents the set of packets present up to time slot t in node i that are destined to node j and to be sent on wavelength λ_i. Furthermore, each $b_{ij}(t)$ entry is a FCFS queue.

2.3. ARBITRATION

At step 4 of the previous algorithm, all the nodes use a common *matching algorithm* on matrix $B(t)$ to determine the data packets that each of them must transmit, if any, in the following slot. The matching algorithm picks a set of packets among those represented in $B(t)$ such that no two packets in the set have the same source node or destination node. Thus the simultaneous transmission of the set of packets results in no contentions. We use an extended version of the Random Scheduling Algorithm (RSA) [13] for multiple classes of traffic to compute the matching. The original RSA algorithm was modified so that when the first data unit of a packet longer than a single data unit is scheduled for transmission (i.e., it is matched), all the remaining data units of the packets will be scheduled with no interruption. To support multiple classes of traffic, the algorithm uses five different backlog matrices $B_{c \in 0..4}(t)$, where $B_0(t)$ represents the EF traffic and the other the traffic of each AF class, from 1 to 4. The algorithm is presented in Fig. 5.

```
/* At each time slot */
 1. Set T = {T_i : 1,...,N}; R = {R_j : 1,...,N}; C = {C_k : 1,...,4}
    /* transmitters, receivers, AF classes */
 2. B_c(t) <- B_0(t) (EF backlog matrix)
 3. Randomly select (and check) an unchecked T_i
 4. Randomly select (and check) an unchecked R_j, i ≠ j
 5. if b_{i,j}(t) ≠ ∅ and ¬(invt(T_i) || invt(R_j))
    /* invt(a): a is involved in a transmission. */
    5.1 Schedule the next packet in the b_{i,j}(t) queue.
    5.2 invt(T_i) <- true; invt(R_j) <- true.
        /* T_i and R_j remain in this state until all the
           data units of the packet are transmitted. */
    5.3 Remove T_i from T and R_j from R.
 6. Goto 4 until R = ∅ or all receivers were checked
 7. Goto 3 until T = ∅ or all transmitters were checked
 8. if C ≠ ∅ and T ≠ ∅ and R ≠ ∅
    8.1 C_k <- f(C); B_c(t) <- B_{C_k}(t)
    8.2 Remove C_k from C.
    8.3 Uncheck all elements in T and R.
    8.4 if {∃ (m,n) | b_{m,n}(t) ≠ ∅} Goto 3 else Goto 8
 9. end
```

Figure 5 The packet-aware multiple class RSA matching algorithm

Step 8.1 of the algortihm is executed 4 times at most. On each execution, the function $f(C)$ selects a particular class C_k. The *proportion* of times (with respect to the total number of times the algorithm is executed) that C_k is selected in the first execution of the step is equal to the proportion of resources allocated to the class; the same applying for the second, third and fourth executions of the step.

2.4. BUFFER MANAGEMENT AND SERVICE DISCIPLINE

In what follows, we describe the buffer management mechanisms and the service discipline considered in the MAC protocol to support the EF and AF PHBs requirements as defined in RFC 2598 [3] and RFC 2597 [4], respectively.

It should be noted that a PHB is defined as the externally observable forwarding behavior applied at a Diffserv-compliant *node* to a Diffserv behavior aggregate [1]. The POS for which our MAC protocol is designed for, is a broadcast-and-select, *single-hop* network. So, we consider the whole POS network as topologically and functionally equivalent to a *interior node* of a multi-hop network, where the PHBs are implemented. Furthermore, the "externally observable" forwarding behavior is considered in the POS at the egress of the network, that is, at the sink nodes (Figure 6). Because of this fundamental consideration, we assume that the traffic entering the network is already classified into aggregates and conditioned.

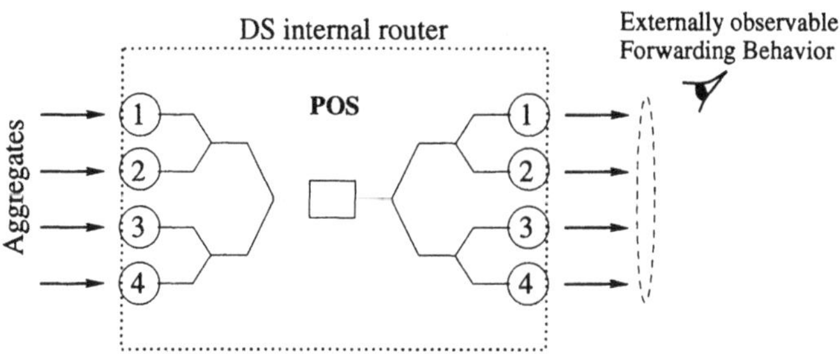

Figure 6 The POS as a DS interior node.

In order to meet the specific PHB requirements, we begin by identifying the resources to be allocated in the POS to the EF and AF aggregates. Recall from the presentation of the signaling protocol that, for a packet to get transfered, it must be first *queued* at the source node; then signaled in the *SIG field*. Once the SIG field is broadcast, the packet must access a dynamically allocated reservation in the *RES field* to convey its detailed information. When the reservation reaches all the nodes, it must be scheduled by the *matching algorithm* before the packet is finally transmitted (i.e., all the data units that conform the packet). From this recapitulation we identify the required resources: the transmission queue of the source node, a share of the signaling channel and the scheduling of the packet by the matching algorithm.

The transmission queues in a node N_i are arranged as depicted in Figure 7. There is one queue dedicated to EF traffic and four queues for AF traffic, one per AF class. Each queue, either EF or AF, is logically divided into three sections by two pointers (see the AF class 4 queue at the bottom of the figure). The leftmost section contains the packets waiting to be signaled i.e., the packets waiting to be indicated in the next SIG field (the node's mini-slot in the next SIG field will contain the number 2). The section in the center contains the packets waiting for a reservation in the RES sub-channel. Finally, the rightmost section contains the packets waiting to be scheduled (i.e., waiting for a "match" of the common matching algorithm) before being transmitted. Thus, the right side pointer moves to the right when a packet is signaled (SIG) and the left side pointer moves also to the right when the reservation (RES) of a packet has been sent.

The sources feeding the node N_i may be grouped into $N-1$ sets according to their destination node N_j, where $i \neq j$ and $1 \leq j \leq N$. Each $S_{i \to j}$ set may contain zero or more EF sources and zero or more sources of each AF class.

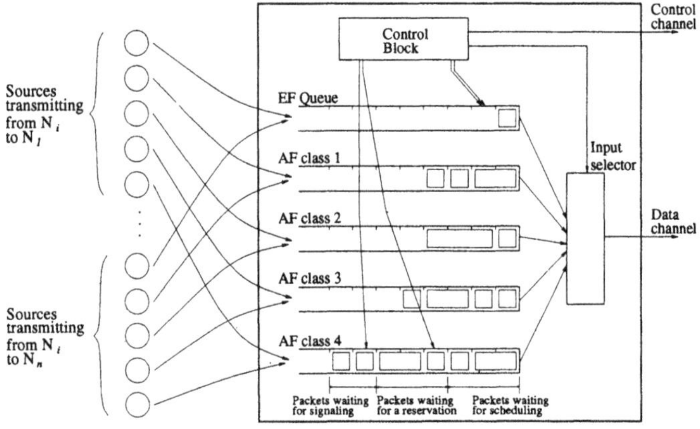

Figure 7 SRD_{DS} node configuration with EF/AF support.

The buffer management mechanism must be implemented in the queues in order to avoid potential congestions. Upon a packet arrival, the mechanism determines whether the packet is accepted or not. The EF queue implements tail dropping (dropping in this queue should *never* occur because the arriving traffic is supposed to be rate-limited and shaped before entering the network. Furthermore, it is served with static priority), whereas AF queues use a Random Early Drop (RED) algorithm [14]. The RED algorithm aims to achieve high throughput and low delay, mainly under responsive traffic (e.g., TCP), by controlling the average queue size. The average size is kept low, while fluctuations in the actual size are allowed to accommodate bursty traffic and transient congestions. The RED algorithm calculates the average queue size q_a at each packet arrival using a low-pass filter with an exponential weighted moving average. The q_a value is compared to two queue size thresholds min_{th} and max_{th}. When $q_a < min_{th}$, the packet is accepted. When $q_a \geq max_{th}$, the packet is dropped. If $min_{th} \leq q_a < max_{th}$, the packet is dropped with probability $p_a = f(q_a, P_{max})$, where P_{max} is the maximum dropping probability. We extend the basic RED algorithm to accommodate three levels of dropping precedence within the AF queue, namely, low precedence (green), i.e., packets have low probability of being dropped; medium precedence (yellow) and high precedence (red). A tuple $< min_{th}, max_{th}, P_{max} >$ is associated with each dropping precedence level (Figure 8) in such a way that the total buffer space allocated to the AF class is a function of the values of the three tuples.

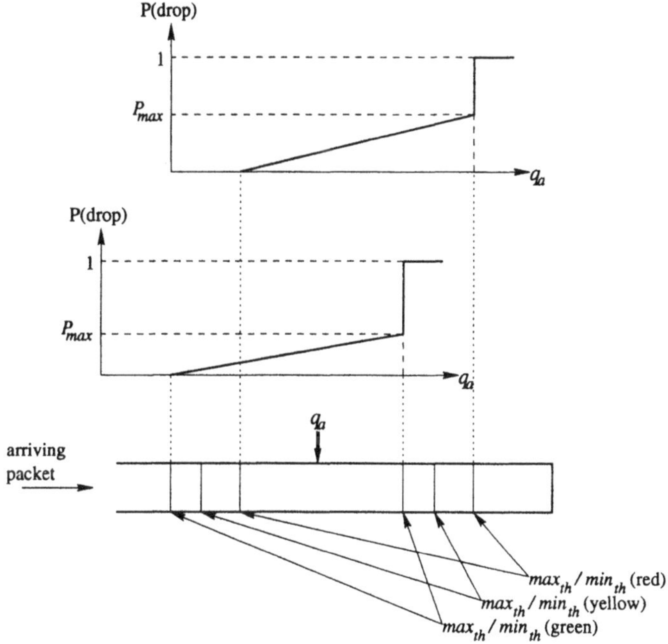

Figure 8 RED queue with three dropping precedence levels.

Once an arriving packet is accepted in the queue, it must be signaled through the control channel. At this point, the node counts the total number of packets that arrived at *all* its queues in the previous slot and posts this value in its assigned mini-slot of the SIG field. When the SIG field is received, the RES field of the next slot is filled up with reservations for the signaled packets. If all nodes have packets to transfer, each node is allocated one out of N reservations in the RES field. Each node must in turn distribute its *allocated signaling* among the packets of its EF and AF queues. This distribution has to be done so that the bandwidth requirement of each PHB, namely, the EF and the four AF classes, is conveniently signaled. Thus, the service rule is first to post reservations for packets in the EF queue up to exhaustion. The remaining reservations are then distributed among the packets in the AF queues using a Weighted Round Robin service discipline, where the weight of each queue corresponds to the share of bandwidth for the respective AF class. Here, note that the average size of packets in an AF class must be considered when calculating the weight of its queue; otherwise, if two queues have the same weight but different average packet size, the one with the largest average will get actually more bandwidth.

The final step of the MAC protocol, before a packet is actually transmitted, is the scheduling of the intended transmission by the matching algorithm. The Random Scheduling Algorithm must also satisfy the bandwidth requirements of both PHBs. The algorithm first attempts to schedule transmissions of EF packets. The AF transmissions are scheduled, according to the weight of their respective class, only when there are no more possible matches in the EF backlog matrix $B_0(t)$, and there are still sources and destinations available (see Figure 5).

2.5. TRAFFIC CONDITIONING AND MARKING

The traffic entering the network must be conditioned in order to meet the expected requirements. Regarding the EF traffic, it is clear that, since it is served with static priority, it must be hard-limited to a configured peak rate and shaped when entering a node, so that bursts are not injected in the network and the AF aggregates do not starve. Starvation implies high access delay and high packet loss of AF traffic due to congestion.

For the AF PHB, a certain amount of forwarding resources (buffer space and bandwidth) is allocated to each class. More precisely, the *signaling allocation* of the node and the transmissions scheduled by the RSA algorithm for AF traffic are proportionally distributed among the four AF classes. Based on the proportion of allocated resources, a *nominal rate* for each AF aggregate is specified. A rate meter [15] is used to calculate the instantaneous rate of packets arriving at a given AF queue. If the measured rate is below or slightly above the *nominal rate*, the arriving packet is marked with the lowest dropping precedence (green).

If the rate is above the nominal rate but below a first threshold, it is marked with medium precedence (yellow); finally, if the rate is above this threshold, the packed is marked with the highest dropping precedence (red). Once the packet has been marked, the RED algorithm is applied to determine if the packet is either accepted or not in the AF queue, based on the average queue size and the queue size thresholds defined for the color of the packet.

3. PERFORMANCE EVALUATION

In this section, we experimentally evaluate the ability of the MAC protocol to meet the requirements of the EF and AF PHBs. In addition, we present results illustrating the impact of the interaction of EF and AF PHBs on traffic metrics such as packet loss.

3.1. SIMULATION AND TRAFFIC MODELS

Table 1 shows the parameters of the network considered in the simulations.

Table 1 Network parameters used in simulations

Parameter	*Value*
Nodes	8
Channel rate	2.5 Gbit/s
Slot size	384 bits
Propagation delay	10 slots

The sources feeding the EF aggregate generate fixed-length packets of 48 bytes at constant bit rate. The sources generating traffic for the AF classes are characterized by an empirical model of LAN traffic introduced in [16]. Packet interarrival time is assumed to be exponential, with a mean rate λ and the packet length is distributed as in tables 2 and 3. For an EF source, its offered load is the ratio of its constant bit rate to the rate of the data channel over which it send its packets. For AF traffic, the offered load of a source is the ratio of its average rate to the rate of the data channel over which it sends its packets.

The packet's length distribution of table 2 represents typical WWW traffic, whereas the distribution in table 3 represents a traffic pattern typical of bulk transfers (e.g., FTP, e-mail, etc.).

Table 2 WWW traffic's packet length distribution (bytes)

Packet length	64	128	256	512	1024	1518
Probability	0.6	0.06	0.04	0.02	0.25	0.03

Table 3 FTP traffic's packet length distribution (bytes)

Packet length	64	128	256	512	1024	1518
Probability	0.06	0.06	0.04	0.02	0.25	0.57

3.2. EF CONFORMANCE

We first verify the bounds of the rate that can be guaranteed to EF sources. We set $N = 4$ nodes in the network as active, each one with sources transmitting to the other three active nodes; thus, there are $N(N-1)$ possible unidirectional $\mathrm{EF}_{i \to j}$ transmissions in the network. Sources generate traffic at 2500 Mbit/s (i.e., the channel rate), which means that all the EF queues are always backlogged. Figure 9 depicts the rate attained by the EF transmissions over time (measured in sink nodes).

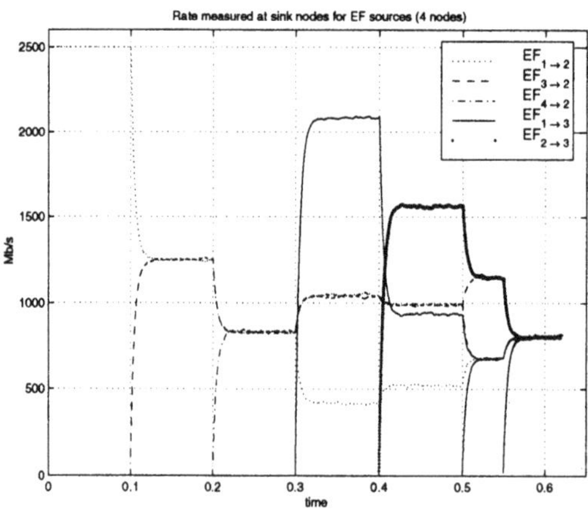

Figure 9 Rate of EF sources measured at sink nodes (N=4).

The source feeding the EF transmission from node 1 to node 2, $EF_{1\to 2}$, is turned on at $t = 0$ s. Since it is the only active transmission, it gets all the resources in the network and attains the *maximal rate* of 2500 Mbit/s. On $t = 0.1$ s, the source feeding the $EF_{3\to 2}$ transmission is turned on. At this point, the receiver of node 2 must be tunned back and forth to the λ_1 and λ_3 channels (i.e., the channels transmitters 1 and 3 are fixed to, respectively) so, each transmission attains a half of the maximal rate. On $t = 0.2$ s, the $EF_{4\to 2}$ transmission becomes active and the receiver of node 2 is "shared" among the three transmissions; each transmissions gets a third of the maximal rate ($\approx$ 833.33 Mbit/s). On $t = 0.3$ s, the $EF_{1\to 3}$ transmission is active. Here, note that the rate of both $EF_{3\to 2}$ and $EF_{4\to 2}$ increases, whereas $EF_{1\to 2}$ achieves a lower rate. This is because now the transmitter of node 1 must be "shared" between $EF_{1\to 2}$ and the recently introduced $EF_{1\to 3}$ transmission. Consequently, $EF_{1\to 2}$ gets a lower rate ($\approx$ 416.66 Mbit/s), whereas $EF_{3\to 2}$ and $EF_{4\to 2}$ take advantage of the released bandwidth at receiver 2. On the other hand, the introduced transmission $EF_{1\to 3}$ attains a rate of $\approx$ 2083.3 Mbit/s which, added to the rate of $\approx$ 416.66 Mbit/s of the $EF_{1\to 2}$, yields the maximal transmission rate (2500 Mbit/s) of the node 1's transmitter. On $t = 0.4$ s and $t = 0.5$ s, other two transmissions become active. Finally, on $t = 0.55$ s, the six remaining transmissions are activated. At this point, the *minimum guaranteed rate* of $C \cdot 1/(N-1)$ is attained for each possible connection in the network.

According to RFC 2598, EF traffic should receive the configured rate independent of the intensity of any other traffic attempting to transit the node. We are interested on the impact of AF traffic on EF connections. Figure 10 shows the instantaneous rate of EF sources observed at sink nodes. The sources generate data-unit-long packets at a) 200 Mbit/s, b) 300 Mbit/s, and c) 400 Mbit/s with WWW background load. The global load (EF and AF) is fixed to 72 % of the network capacity (the load offered by all the sources over all data channels). In a), the measured rate shows little jitter, whereas the rate in c) is highly variable. In the MAC protocol, the rate of a transmission directly depends on the number of matches per time unit that the RSA algorithm allocates to it. Let us recall that a match of a transmitter-receiver is infeasible if either the transmitter or the receiver is involved in another transmission at the time that the match is attempted. This occurs when:

1 The transmitter/receiver was assigned in a *previous* time slot to another transmission which has not finished yet.

2 The transmitter/receiver was already assigned to another transmission, either EF or AF, in the *current* time slot.

Note that in the first case, an EF transmission can be blocked only by an AF transmission since all the EF packets are one data unit long (i.e., they are transmitted in one slot time). In the second case, an EF transmission can be blocked only by another EF transmission because of the static priority mechanism. The probability of both events proportionally increases with the rate of the EF transmissions.

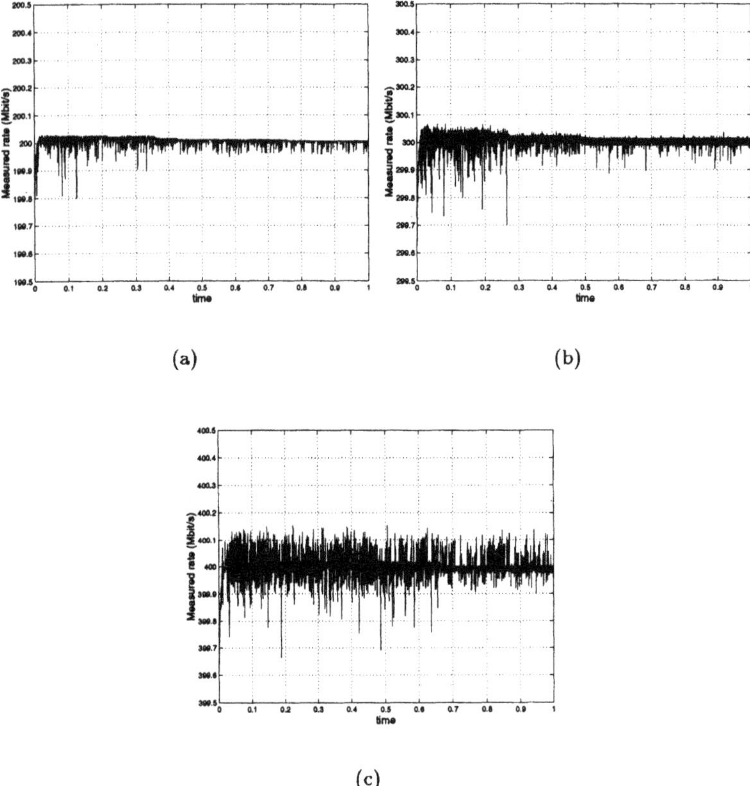

Figure 10 Rate of EF sources with background traffic measured at sink nodes. a) EF=200 Mbit/s, b) EF=300 Mbit/s, c) EF=400 Mbit/s. $\rho = 0.72$

3.3. AF CONFORMANCE

An AF-compliant node must allocate a configurable, minimum amount of forwarding resources (buffer space and bandwidth) to each implemented AF class.

We carried out experiments to evaluate the relative priority of access to network resources of the four AF classes. We set four AF source for each possible unidirectional transmission in the network (with $N = 8$ nodes, there are 56 transmissions), thus a total of 224 sources feed the network. The offered load is equally divided among the sources. In the nodes, the four AF classes are allocated the same amount of buffer space. To evaluate the proportional allocation of resources (i.e., signaling channel and RSA matches), we set the following proportion to each class:

Table 4 Proportional allocation of forwarding resources among AF classes.

AF class	*weight*
1	0.4
2	0.3
3	0.2
4	0.1

Clearly, in a scenario where all the AF classes receive the same traffic load, a class with a relatively low proportion of resources will be more susceptible to congestions than classes with more allocated resources, yielding higher values of packet loss probability. The loss probability of a class is associated with its forwarding assurance.

Figure 11 shows the packet loss probabilities of the four AF classes. The left side of the figure shows the probabilities when the AF traffic follows the packet's length distribution of Table 2 (WWW traffic). The right side depicts the probabilities when bulk transfer's packet length's distribution is used. We observe in both cases that a differentiation among classes in terms of forwarding assurance can be attained by the protocol. We observe further that bulk transfers achieve lower drop probabilities than WWW traffic.

3.4. EF/AF INTERACTION

Useful behavioral distinctions are mainly observed when multiple behavior aggregates compete for buffer and bandwidth resources in the network. The proposed mechanism for the EF PHB (i.e., priority queue) allows unlimited preemption of other traffic. We evaluate the damage that EF traffic could inflict on AF traffic in order to determine the limits that must be imposed on it (e.g., maximum rate). Concretely, preemption of AF traffic by EF traffic induces, in the worst case, significant congestion in the AF queues of the node. This prob-

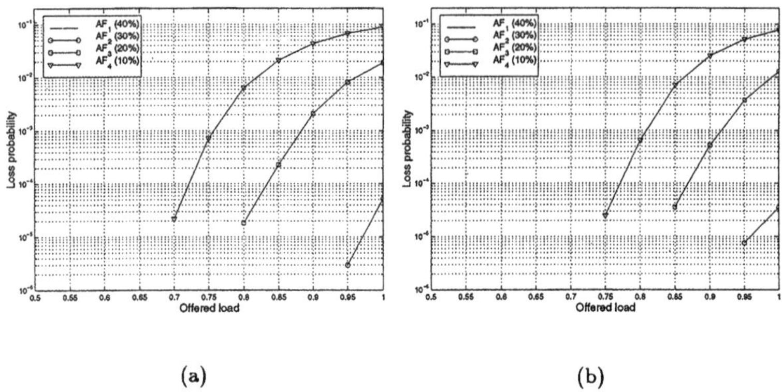

Figure 11 Packet loss probability of the AF classes with a) WWW traffic and b) FTP traffic.

lem implies an increment of the packet loss probability. For the results in Figure 12, class AF_1 is allocated 40% of forwarding resources (i.e., signaling channel and RSA matches), class AF_2 30%, class AF_3 20% and class AF_4 the remaining 10%. The RED parameters are the same for the three dropping precedence levels of the four AF queues $< min_{th} : 22, max_{th} : 32, P_{max} : 0.5 >$. Traffic for each AF class is set to 12.5% of the offered load, thus, the four AF classes receive altogether 50% of the offered load. EF traffic is injected in 5% steps of the total offered load and the packet loss probability is observed for the AF classes at each step.

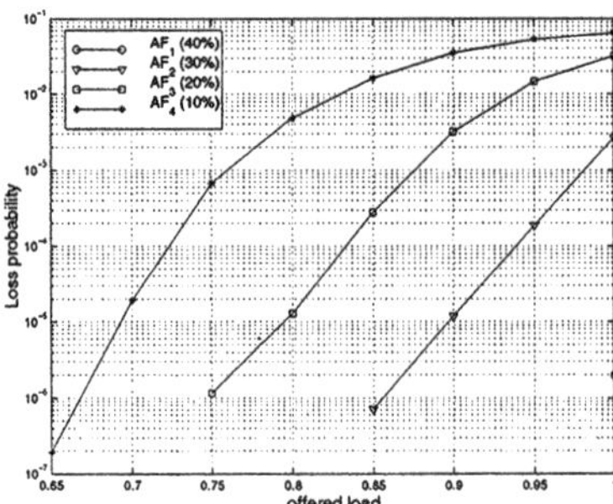

Figure 12 Packet loss probability of the AF classes with background EF traffic.

By comparing the loss probability values of Figures 11 and 12, we observe that the EF aggregate induces significant congestions in the AF queues and, consequently, packet loss.

4. CONCLUSIONS AND FUTURE WORK

The new Internet generation requires high capacity networks able to provide a certain level of QoS guarantees both in the core and metropolitan access networks. In this paper we have presented a MAC protocol for a WDM POS with support for the EF and AF PHBs as defined in the IETF Diffserv architecture. Signaling, arbitration, buffer management and service discipline functions interact to support service differentiation. These various issues have been investigated through simulations.

Regarding the EF PHB, the maximum rate that can be guaranteed to a particular EF transmission in the network depends on the number and rate of other transmissions concurrently using the transmitter and receiver involved in the transmission. Moreover, jitter induced in the EF transmissions is mainly due to the intensity of the EF traffic itself, rather than to the intensity of other traffic.

For the AF PHB group, the proportional allocation of forwarding resources (i.e., the signaling channel and scheduling of transmissions) among classes has been evaluated. Under the same load conditions (i.e., each AF class receiving the same amount of traffic) and the same buffer space allocation, packets belonging to classes with more allocated resources experience lower loss probability than packets belonging to the classes with less resources. In this context, two traffic patterns have considered: WWW and bulk data transfers.

The main contribution of this paper is the introduction of service differentiation support in the MAC protocol. A critical aspect that must be further studied is the RSA scheduling algorithm. Its random nature induces distortion in the observed forwarding behaviors. Deterministic scheduling algorithms should be also considered.

Finally, the optimization of the RED's parameters for buffer management in this network must be investigated under more realistic traffic conditions (e.g., responsive/non-responsive traffic). Usually, RED is utilized to simplify the congestion control job required in transport protocols like TCP. Each time a packet is dropped, the concerned TCP session must adjust its rate. TCP flows in an aggregate may be punished for their good behavior in the presence of non-congestion-sensitive flows like UDP within the same aggregate.

In our coming studies, the service disciplines and buffer management techniques used in this paper will be further exploited for the design of Diffserv edge routers in the context of a cloud of optical switch-routers.

References

[1] S. Blake, D. Black, M. Carlson, E. Davies, Z. Wang and W. Weiss: An Architecture for Differentiated Services, IETF RFC 2475, December 1998.

[2] K. Nichols and B. Carpenter: Definition of Differentiated Services Per Domain Behaviors and Rules for their Specification, work in progress, (`draft-ietf-diffserv-pdb-def-00.txt`), June 2000.

[3] V. Jacobson, K. Nichols and K. Poduri: An Expedited Forwarding PHB, IETF RFC 2598, June 1999.

[4] J. Heinanen, F. Baker, W. Weiss and J. Wroclawski: Assured Forwarding PHB Group, IETF RFC 2597, June 1999.

[5] M. Gagnaire and T. Stroesslin: A WDM Network with Flexible Topology for IP and ATM Traffic Transport, 4th IFIP ONDM'2000 Conference, Athens, Greece, February 2000.

[6] R. Ramaswami and K. Sivarajan: Optical Networks: a practical perspective, Morgan Kaufmann Publishers, 1998.

[7] D. Levine and I. Akyildiz: PROTON: A Media Access Control Protocol for Optical Networks with Star Topology, IEEE/ACM Transactions on Networking, vol. 3, no. 2, April 1995, pp. 158-169.

[8] M. Guizani: High-Speed Protocol for an All-Optical Packet Switched Metropolitan Area Network, International Journal of Network Management, vol. 7, no. 1, January/February 1997, pp. 9-17.

[9] P. Humblet, R. Ramaswami and K. Sivarajan: An Efficient Communication Protocol for High-Speed Packet-Switched Multichannel Networks, IEEE Journal on Selected Areas in Communications, vol. 11, no. 4, May 1993, pp. 568-578.

[10] D. Levine and I. Akyildiz: A Reservation and Collision-Free Media Access Protocol for Optical Star Local Area Networks, Proceedings of IEEE GLOBECOM 1994, San Francisco, CA, November/December 1994, pp. 567-571.

[11] E. Dinan and M Gagnaire: PCSA: A MAC Protocol for Single-hop Wavelength Division Multiplexed Photonic Networks, 1999 Symposium on Performance Evaluation of Computer and Telecommunication Systems, Chicago IL, July 1999, pp. 84-90.

[12] T. Stroesslin and M. Gagnaire: A Flexible MAC protocol for all-optical WDM Metropolitan Area Networks, Proceedings of IEEE IPCCC 2000, Phoenix, AZ, February 2000, pp. 113-117.

[13] R. Chipalkatti, Z. Zhang and A. S. Acampora: Protocols for Optical Star-Coupler Using WDM: Performance and Complexity Study, IEEE Journal on Selected Areas in Communications, vol. 11, no. 4, May 1993, pp. 579-589.

[14] S. Floyd and V. Jacobson: Random Early Detection Gateways for Congestion Avoidance, IEEE/ACM Transactions on Networking, vol. 1, no. 4, August 1993, pp. 397-413.

[15] David D. Clark and Wenjia Fang: Explicit Allocation of Best-Effort Packet Delivery Service, IEEE/ACM Transactions on Networking, vol. 6, no. 4, August 1998, pp. 362-373.

[16] Nada Golmie, Sandrine Masson, Gerard Pieris and David Su: Performance Evaluation of MAC Protocol Components for HFC Networks, SPIE Proceeding, vol. 2917, 1996, pp. 120-130.

Part Six

Wavelength routing and on-demand circuit provisioning

Wavelength assignment in optical networks according to traffic requirements and transmission impairments

M. Moreschini, F. Matera, M. Settembre
Fondazione Ugo Bordoni
via B. Castiglione 59, 00142 Roma, Italy
tel.+39-06-54802215, fax +39-06-54804402, e-mail mat@fub.it

Key words: networks, optical systems, FWM, XPM, routing, WDM, wavelength

Abstract: We describe a novel algorithm that designs wavelength-routed optical telecommunication networks taking into account both the traffic demands and the physical limits due to the WDM signal transmission. Some results for a metropolitan ring are reported.

1. INTRODUCTION

The explosion of the Internet phenomenon shows that in the next years telecommunication networks will operate with Terabits/s capacities. Optical technology can satisfy such a requirement thanks to its advances and in particular to the development of the Wavelength Division Multiplexing (WDM) technique. At the moment it is accepted that the novel high capacity networks should be based on wavelength routing devices [1]. This way all the traffic from node i to node j can be carried by directly using one or more dedicated wavelengths. However due to some physical limitations and cost reasons, the number of the wavelengths could be insufficient to obtain all the direct optical paths that are required by the traffic requirement, so electronic mu-demultiplexing is needed to allow low rate nodes to share the same wavelength. As a consequence, for a network design, traffic grooming and wavelength assignment algorithms are required. At the moment several algorithms have been already proposed, but at our knowledge, none of these takes into account the physical impairments of the networks, especially the ones due to the signal propagation. Also for metropolitan networks the fibre impairments could result much degrading for the WDM transmission and in particular the Kerr nonlinearity can manifest as a strong limitation both in the G.652 fibres in terms of Cross Phase Modulation (XPM) and in G.653 in terms of Four Wave Mixing (FWM) [3-4].

The aim of this work is to show an algorithm that permits to assign the wavelengths in an optical networks taking into account both the traffic requirements and the propagation limitations. Such an algorithm is used to design a metropolitan optical network connecting seven nodes by means of a fibre ring considering different values of the total traffic.

2. ALGORITHM

Our algorithm, based on the one reported in ref. [1], permits to achieve a matrix [W] describing the wavelength connections among the nodes taking into account traffic demand and physical impairments. We suppose that each node has some electronic Add Drop Multiplexers (ADMs) [2] and a Wavelength ADM (WADM). Each ADM is used to aggregate low rate signals onto a single high rate signal that is carried on a wavelength representing a path. In each node the wavelengths can be mu-demultiplexed by means of the WADM, and if the information carried by a dropped wavelength is not directed to the node it is retransmitted on other wavelength.

First of all we define the following input network parameters: the traffic matrix [T], the channel bit rate R, the number of nodes, N, and the physical parameters of the links connecting the nodes.

The algorithm mainly consists of two parts: the transmission and traffic routine.

Transmission routine. Analytical calculations permits in many propagation cases to evaluate the maximum capacity versus distance. In the presence of FWM and XPM the maximum number of wavelength, w_{MAX}, versus distance, can be analytically evaluated by using the methods reported in ref. [3] that permit to evaluate the Q factor. We remember that to have a bit error rate lower than 10-9, under some hypothesis [4], the Q factor has to be higher than 6. We calculate the number w_{MAX} versus distance by assuming the condition that each channel has to assume a Q>6 in a power interval between Pmin and 10 Pmin, where Pmin is the minimum power that permits to overcome to the noise effects.

The function w_{MAX} (n_c), where nc is the number of crossed node, is used as constraint of the traffic routine.

Traffic routine. This routine is composed by some steps that consists in finding the logical topology that permits to route the traffic with 1, 2, 3.. hops and each time we verify if the wavelength distribution in each fibre link can be permitted by the link limitations that have been previously evaluated.

First of all we calculate the matrix [W(0)] whose elements w(0)ij are given by the integer number corresponding to [tij/R]. Starting from [W(0)], we can calculate for each wavelength, s, the number of wavelengths, nmet-s(n_c), that are met during its path. By analysing previous simulations we have observed that the degradation of each wavelength path can be taken into account by defining the function, nW-s(n_c), obtained by arranging in a decreasing order the values nmet-s; a correct design of the network is obtained when for each s and nc the value of nW-s is always lower than w_{MAX}.

If such a condition is not verified the wavelength assignment will be changed looking for a new configuration taking into account the requirement of w_{MAX} (n_c). By means of this procedure, if the physical conditions are compatible with the traffic requirement, a matrix [W(0)] will be obtained.

After this step we can find the wavelength assignment for the remaining traffic [T(1)]=[T]-R[W(0)]. In particular according to ref. [1] the assignment can be obtained by solving the following simple linear program

$$F(h1) = \max \sum_{i=1}^{N} \sum_{j=1}^{N} w_{ij}{}^{(1)} T^{(1)}{}_{ij} \qquad i \neq j \tag{1}$$

where wij(1)=1 if a wavelength is available between i and j (0 otherwise) and by considering some constraints.

We have assumed two constraints, the first is that each node always has at least two input-output wavelengths and that the total number of wavelength has to be lower than a value M_{h1}. As in the previous step, the wavelength assignment has to satisfy the physical conditions.

At this point if all the traffic has been not routed we will continue with further steps considering other hops and assuming other values M_{h2}, M_{h3}.., up to finish all the traffic in agreement with the physical requirements. In particular, in each step, first we try to encapsulate the traffic in the wavelengths that have available circuits (155 Mb/s, STM-1) and subsequently for the remaining traffic we consider another simple linear program as reported in ref. [1]. For a fixed number of hops, to have an efficient wavelength assignment, the values Mh have to be minimised.

At the end of this routine, if the transmission conditions are compatible with the traffic demand, we will have the final matrix [W]= [W(0)]+ [W(1)]+ [W(2)]+..

3. RESULTS

We have used such an algorithm to design an optical WDM ring network that connects seven high capacity nodes. The optical part of each node consists of a WADM and an EDFA amplifier that compensates for the fibre loss (0.25 dB/km) and for the other loss of the node devices (10 dB). The total length of the ring is 42 km, the average distance among the nodes is 7 km. Such a physical structure is compatible with the geographical distribution of the main telecommunication centres of Rome. We assume that the ring, composed by two fibres (one is for providing protection), can encompass both G.652 and G.653 fibres. For G.562 fibres the chromatic dispersion is 16 ps/nm/km, while for G.653 fibre the chromatic dispersion is 1 ps/nm/km in the centre of the WDM signal bandwidth. We have assumed a traffic matrix $[T]=\Psi[\Gamma]$, where Ψ is a coefficient ranging between 1 and 10, and $[\Gamma]$ is a 7x7 matrix whose row elements are: 0, 20, 5, 8, 8, 8, 6; 20, 0, 15, 30, 22, 10,3; 5, 10, 0, 6, 6, 6, 7; 8, 11, 5, 0, 8, 8, 7; 9, 10, 5, 8, 0, 8, 10; 10, 10, 4, 8, 8, 0, 9; 7, 2, 6, 9, 9, 9, 0. According to these data, the total traffic of the network varies between 38.3 and 383 Gb/s. We assume a channel bit rate equal to 10 Gb/s with a frequency spacing among the channel of 50 GHz and optical amplifiers with a flat gain in a bandwidth of 30 nm.

In these propagation condition, in G.652 fibres, XPM and FWM have a negligible contribution and also the pulse broadening due to the chromatic dispersion is not relevant. In G.653 links FWM for can induce a relevant degradation and this is shown by the function w_{MAX} that assumes the values of 72, 27, 12 and 4 for nc equal to 3, 4, 5 and 6 respectively. It means that for G.652 links the wavelength assignment is only given by the traffic algorithm, while for G.653 links the transmission routine has a relevant role. In particular for G.653 links, the function w_{MAX} shows as the propagation of many wavelengths having long paths should be avoided and it suggests to decrease the length of the paths at expense of an increasing of the wavelengths. In our cases all the traffic can be routed by considering at maximum 5 hops (h5).

In fig. 1 we report the required number of wavelengths versus the total traffic considering 1 (h1), 2 (h2) and 5 hops. In the G.652 fibres, in the h1 case, the number of wavelengths is high for any traffic condition, since each node has to be connected to all the other nodes. Conversely, when the traffic is high the difference between h1, and h5 is not so relevant since many wavelengths are required to carry the heavy traffic. For G.653 links due to the values of w_{MAX}, the configuration with h1 cannot be used, and, especially for high traffic, a larger number of wavelengths is required with respect to the G.652 case, to avoid the presence of many long wavelength paths. However too wavelengths can induces a high FWM effect limiting the wavelength assignment as shown in fig. 1 for h2 for a traffic higher than 250 GHz.

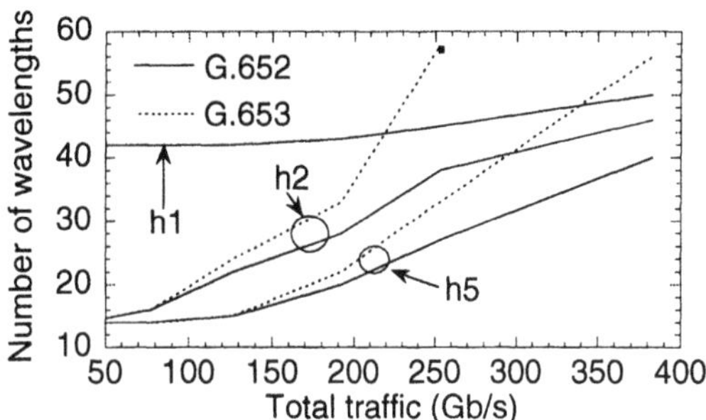

Fig. 1: Number of wavelengths vs total traffic for the optical ring network with R=10 Gb/s.

The problems of the network encompassing G.653 fibres can be overcome by doubling the fibres and introducing two counterpropagating directions in the rings; this way the wavelength is transmitted on the ring with the direction that permits to have the shortest path between the nodes.

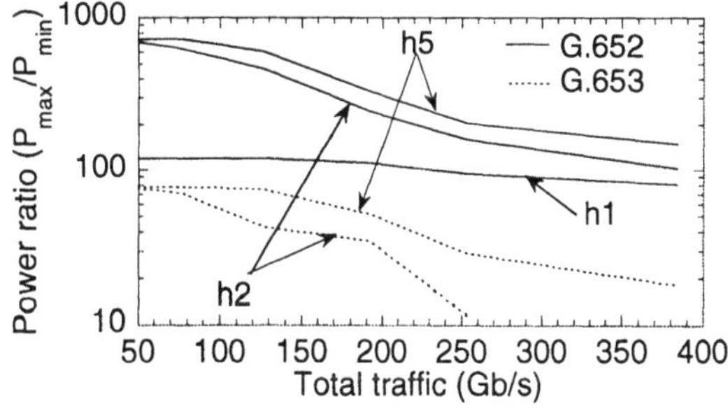

Fig. 2: transmission tolerance in terms input power per channel with the wavelength assignment of fig. 1.

The transmission performance of the network have been tested by means of a simulation code based on the split step method. For each network configuration we have evaluated the Q factor versus input power for each wavelength path; in conditions of good performance Q is higher than 6 a power interval between Pmin and Pmax , where Pmax is the maximum input power permitted by the nonlinear effects. Among all the paths the worst ratio ρ=Pmax/Pmin is found and it is reported in fig. 2. The figure 2 shows how the performance decreases by increasing the traffic and that, for the considered network, a better transmission quality can be achieved by using the G.652 fibres.

The results shown in fig. 1 for G.652 fibres can be used for a network with the same physical parameters, a lower channel bit rate and a reduced traffic; for an instance the same wavelength assignment can be obtained for a channel bit rate of 2.5 Gb/s and a total traffic ranging between 9.6 and 96 Gb/s.

4. CONCLUSIONS

We have described an algorithm that permits to assign the wavelengths in an optical network taking into account both the traffic and transmission requirements. As an example we have reported the results obtained for a metropolitan network in which the FWM effect can show severe impairments.

ACKNOWLEDGEMENTS

Work carried out under the agreement between Fondazione Ugo Bordoni and Italian Communications Administration (ISCTI) and in the framework of the European COST266 and of the IST ATLAS project.

REFERENCES

[1] S. Banerjee, J. Yoo, C. Chen, "Design of wavelength-routed optical networks for packet switched traffic" J. of Lightwave Technology, 15, 1636-1646 (1997).
[2] A. L. Chiu, E. H. Modiano, "Traffic grooming algorithms for reducing electronic multiplexing costs in WDM rings networks" J. of Lightwave Technology, 18 , 2-12 (2000)
[3] S. Ten, K. M. Ennser, J. M. Grochocinski, S. P. Burtsev, V. L. daSilva, "Comparison of four-wave mixing and cross phase modulation penalties in dense WDM systems" Proc. of OFC'99, San Diego (CA), ThC4-1/43 (1999).
[4] F. Matera, M. Settembre, Journal of Lightwave Technology, "Comparison of the Performance of Optically Amplified Transmission Systems" 14, 1-12 (1996).

Dynamic Wavelength Provisioning in DWDM-Based Optical Network

Abdallah Shami, Chadi Assi, and Mohammed Ali

Dept. of Electrical Engineering; City College of the City University of New York,
{shami, assi, eeali}@ees1s0.engr.ccny.cuny.edu

Key words: IP/WDM, RWA, Lightpath Provisioning.

Abstract:

The problem of dynamic provisioning of optical channels in a DWDM optical network is considered. Methodologies and associated algorithms for lightpath computation are outlined. We present and compare the performance of two different schemes for provisioning of the optical channels. The first scheme is a simple semi-dynamic constraint-based routing and wavelength assignment (RWA) algorithm that compute solutions to three sub-problems: (a) The routing problem; (b) The constraint-based shortest route selection problem; (c) The wavelength assignment problem. The second scheme is a more dynamic algorithm where both components, routing and wavelength assignment, are considered simultaneously. A new cost definition of optical links is introduced and two new path selections schemes are presented for the full-dynamic RWA. The performance of theses selections algorithms is compared by simulation.

1. INTRODUCTION

If predictions of exponential growth in Internet-related traffic over the next decade hold true, a network 100 times the size of today's voice network will be needed-with voice relegated to a 1 percent minority share of a 99 percent data-dominated network. What kind of technology can best meet these demands to deliver such a "network of networks"; take the most widely talked about data networking protocol-IP-and combine it with the ultimate in bandwidth provisioning-DWDM-to realize the ultimate "blend" of hot technologies. The vision of a hybrid IP-centric optical Internet have captured the imagination of researcher and network planners alike since the rapid and successful commercialization of DWDM in the early-mid 1990s. Such a network will take advantage of the massive raw bandwidth afforded by dense wavelength division multiplexing (DWDM) in its core.

Communication system suppliers are advertising DWDM transmission systems with capacities greater than 1 T/s over a single fiber by means of multiplexing more than a hundred channels at 10 Gb/s each. Much of this capacity growth has been in point-to-point long-distance backbone transmission. Over the last few years, DWDM has proven to be a cost-effective means of increasing the bandwidth of installed fiber plant. While the technology originally only served to increase the size of the fiber spans, it is quickly becoming the foundation for networks that will offer customers a new class of high-bandwidth and broadband capabilities.

No doubt about it, DWDM-based optical networking will change every thing. The myriad innovations in dense wavelength-division multiplexing (DWDM) technology that increase the number of wavelengths, generate more bandwidth, send multi-gigabit signals faster and longer distances, dense WDM add/drop multiplexers (ADMs), wideband optical amplifiers, and those enabling optical layer cross-connects (OLXCs) - with the ability to add, drop, and in effect construct wavelength-switched and wavelength-routed networks are moving us towards this point. These innovations are now beginning to shift the focus more toward optical networking and network-level issues. As such, it presents an attractive opportunity to evolve DWDM technology toward an optical networking infrastructure with transport, multiplexing, switching, routing, survivability, bandwidth provisioning, and performance monitoring, all supported at the optical layer.

Now that the basic building blocks are available for building such a "network of networks", the key innovations will come from adding intelligence that enables the inter- working of all the network elements (Routers, ATM switches, DWDM transmission systems and (OLXCs). The IETF has already addressed the interworking of routers and optical switches through the Multi-protocol lambda switching (MPλS) initiative [1]. The main goal of this initiative is to provide a framework for real-time provisioning of optical channels, through combining recent advances in Multi-protocol label switching (MPLS) traffic-engineering control plane with emerging optical switching technology in a hybrid IP-centric optical network.

Dynamic wavelength provisioning, that is the focus of this paper, will enable service providers to light up new connections faster than their competitors. Activation times will drop from months to minutes. While the idea of bandwidth-on-demand is certainly not new, existing networks do not support real-time provisioning. Current provisioning of bandwidth is painstakingly static and cumbersome process. Provisioning a cross-country OC-48 service is extremely manual, involves complex network planning and roll-out activities, and generally takes several months to accomplish.

This paper considers the problem of dynamic provisioning of optical channels in a hybrid IP-centric DWDM-based networking model. Methodologies and associated algorithms for dynamic lightpath computation are outlined. We present and compare the performance of two different schemes for dynamic provisioning of the optical channels. The first scheme simplifies the complex problem of dynamic provisioning of optical channels by using a simple semi-dynamic constraint-based routing and wavelength assignment (RWA) algorithm that compute solutions to three sub-problems: (a) The routing problem; (b) The constraint-based shortest route selection problem; (c) The wavelength assignment problem. Unlike the conventional RWA problem used in most algorithms, which is often de-coupled into the routing sub-problem and wavelength assignment sub-problem, the second scheme integrate both the routing and wavelength assignment sub-problems into a single dynamic constraint-based routing problem. Both algorithms are based on a fully distributed implementation.

The remainder of this paper is organized into four sections. Section 2 presents an overview for the routing and wavelength assignment. Section 3 outlines the network modeled. In Section 4, we present and compare the performance of two different schemes for dynamic provisioning of the optical channels. Section 5 presents the simulation results. Finally, Section 6 offers summary and conclusion.

2. ROUTING AND WAVELENGTH ASSIGNMENT PROBLEM:

Given a set of connections, the problem of setting up lightpaths by routing and assigning a wavelength to each connection is called the Routing and Wavelength-Assignment (RWA) problem [2]. Typically, connection requests may be of three types: static, incremental, and dynamic [2]. With static traffic the entire set of connections is known in advance, and the problem is then to set up lightpaths for these connections in a global fashion while minimizing network resources utilization such as the number of wavelengths or the number of fibers in the network. The RWA problem for static traffic is known as the Static Lightpath Establishment (SLE) problem. In the incremental traffic case, connection requests arrive sequentially, a lightpath is established for each connection, and the lightpath remains in the network indefinitely. For the case of dynamic traffic, a lightpath is set up for each connection request as it arrives, and the lightpath is released after some finite amount of time. The objective in the incremental and dynamic traffic case is to set up lightpaths and to assign wavelengths in a manner that minimizes the amount of connections blocking, or that maximizes the number of connections that are established in the network at any time. This problem is referred as Dynamic Lightpath Establishment (DLE) problem. It can easily be shown that the optimal RWA problem is NP-complete by using the results of [3] on static lightpath establishment and by restricting the general problem to tree topologies. An integer programming formulation of the optimal RWA problem in presence of deterministic traffic can be found [4], while in [5] a similar formulation combined with randomized rounding has been presented.

In setting up a lightpath in the case of dynamic traffic, a route must be selected and a wavelength must be assigned to the lightpath. The combined routing and wavelength assignment is difficult to solve and it can be simplified by de-coupling the problem into three separate sub-problems:

3. THE NETWORK MODELED

In the network model considered here, IP routers are attached to an optical core network. The optical network consists of multiple optical layer cross-connects (OLXCs) interconnected via WDM links in a general mesh topology. The IP are clients of the optical network and are connected to their peers over dynamically switched optical paths (lightpaths) spanning potentially multiple OLXCs. The optical network essentially provides point-to-point connectivity between routers in the form of fixed-bandwidth circuits (lightpaths). The topology considered here is assumed to change dynamically. Consequently, an integrated constraint-based RWA protocol is required to for dynamic provisioning of lightpaths.

A lightpath is a fixed bandwidth connection between two network elements such as IP routers established via the OLXCs. Two IP routers are logically connected to each other by a single-hop channel. This logical channel is the so-called lightpath. A continuous lightpath is a path that uses the same wavelength on all links along the whole route form source-to-destination.

Each OLXC can switch high-speed optical signals (e.g. OC-48, OC-192) at a given wavelength from one of its input ports to one of its output ports, without undergoing Opto-electronic conversion. The switching fabric may be purely optical or electrical or a combination. In this work, we assume that the switching fabric is purely optical and that none of the OLXCs has wavelength conversion capability (all-optical wavelength-conversion is an immature/expensive technology). Each adjacent node pair is connected through a bi-directional physical link consisting of two pairs of unidirectional fiber.

The next section focuses on various approaches to routing connection requests in an optical network under dynamic traffic.

4. DYNAMIC PROVISIONING OF OPTICAL CHANNELS

4.1 Semi-Dynamic constraint-based RWA Scheme

To simplify the complex RWA problem, we develop a simple dynamic constraint-based routing and wavelength assignment algorithm that compute solutions to three sub-problems:

- The routing problem.
- The constraint-based shortest route selection problem.
- The wavelength assignment problem.

A: The routing problem:

The semi-dynamic approach is based on the static routing, where each source-destination pair is assigned a set of pre-computed paths, and this set may be searched in a fixed or adaptive order to find an available path at the time of connection request. Static routing based RWA is a constraint path selection in which a path is selected from the predetermined set of alternate candidate paths. For example, these candidate paths may include the shortest-path route, the second-shortest-path route, the third-shortest-path route, etc.

The static routing based RWA provides simplicity of control for setting up and tearing down lightpaths, and it may also be used to provide some degree of fault tolerance upon link failure.

We studied the performance of two routing selection algorithms *the k link-disjoint paths* and the *k best routes*.

A-1: The k link-disjoint paths:

In this first approach, the alternate routes should be link-disjoint. Two alternate routes between a source S and a destination D cannot share any links. This routing approach provides a straightforward approach to handling protection. The set of shortest paths (k) between all node pairs is computed in advance and stored in routing tables at each node. Figure 1.a illustrates a primary route (solid line) from node A to node D, and an alternate route (dashed line) from node A to node D for the source destination pair A and D, in this case we have only two alternate link-disjoint-path routes.

A-2: The k best routes:

This approach determines the k best routes for a source-destination pair. Two alternate routes should differ in at least one link. The following "shortest-route" method is useful for obtaining k best routes when k is small. Given the shortest route from node i to node j, the length for each link is set, in turn, to infinity. The shortest-route problem is then solved for each such case. The best of these routes is the desired second best route in the original network. The method can be extended to the k*th* best route but it very quickly becomes computationally unfeasible. For example, to obtain the third best route, assume that the best route consists of *m* links and the second best route consists of *n* links. Then, if the first and second best routes consist entirely of different links, the maximum number of shortest path problems to be solved will be of order *mn*, the best will be the third best route. In our case, we assume that the alternate routes are computed and stored off-line; hence, the computational time is not important. Figure 1.b illustrates the 3-best routes from node A to node D.

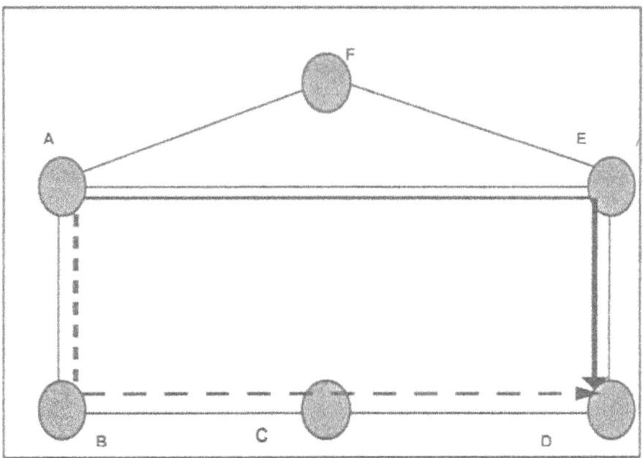

Figure 1a.Shortest and second shortest path from node A to node D (link-disjoint paths)

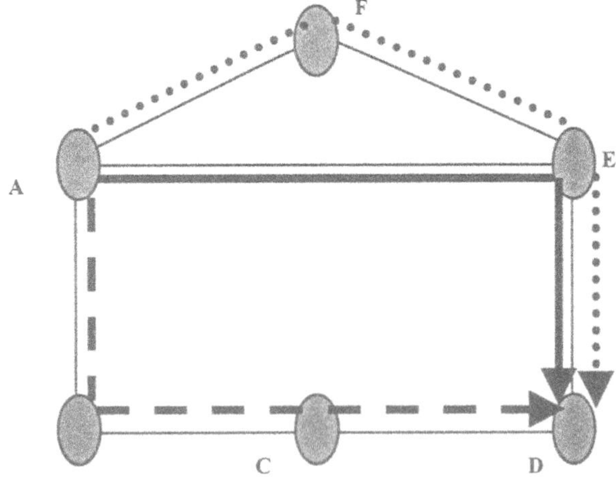

***Figure 1b**: the kth best routes from node A to node D (routes at least differ in one link)*

B: The constraint-based shortest route selection problem

To optimize the utilization of the network resources, the route with the most average number of available wavelengths (least-congested path) per the entire route is then selected out of the k-alternate routes. This is achieved by adding the number of available wavelengths at each link along the entire route and then dividing this number by the total number of links comprising the entire route. This algorithm uses the information stored and updated, via the link state advertisement protocol, at the source router's Traffic-Engineering Database (TED).

C: The wavelength Assignment problem:

For a given request, once a constrained shortest path is selected out of the k alternate routes, a wavelength assignment algorithm is invoked (on-line) to assign the appropriate wavelength across the entire route. Several heuristics wavelength assignment algorithms can be used such as Random wavelength assignment (R), First Fit (FF), Most Used (MU), least-loaded (LL), and Max-Sum algorithms [6]. The implementation of these algorithms requires the propagation of information throughout the network about the state of every wavelength on every link in the network. In our simulation, we used the MU since it is found to the best performance [6]

4.2 Dynamic Constraint-based RWA Scheme

In dynamic routing, the route from a source node to a destination node is chosen dynamically, depending on the network states determined by the set of connections that are currently in progress. We considered the RWA problem in multi-fiber networks under dynamic traffic (calls arrive to and depart from the network randomly). Mapped into the layered-graph, the difference of RWA single-fiber and multi-fiber networks is significant.

In single-fiber networks, the algorithm attempts to find a path with the lowest cost for a connection request. The cost associated with each edge may be based on physical length of the corresponding link if the propagation delay is a concern, or the cost can be set to 1 for all edges if the number of nodes in a lightpath is to be minimized. If one edge is used, its cost will be infinite. In multiple-fiber networks the definition of cost function for each edge is much more complex. A good algorithm should utilize the network resources effectively (fibers and wavelengths). In layered-graph there are different cost functions corresponding to different resource assignment strategy.

We model the physical WDM network consisting of w wavelengths on each link as a graph G, which in turn is represented by w wavelengths-graphs (or $WG's$), one for each wavelength. More specifically, every link in G can be thought of as being decomposed into w links, one in each graph WG_i $(1 \le i \le w)$.

The algorithms are proposed for dynamic routing and wavelength assignments in Multiple-fiber wavelength division multiplexing all-optical network. By virtue of layered-graph, routing and wavelength assignment can be considered simultaneously.

Figure 2 represents a simple network, where two neighboring OXCs are connected by N fibers. Figure 3 represents the layered graph model for the network shown in figure 2 with two wavelengths per fiber. We assign a weight to each link at every directed graph WG_i, a weight basically is the inverse of the number of available fibers between two nodes; hence, initially all links are initialized with equal weights.

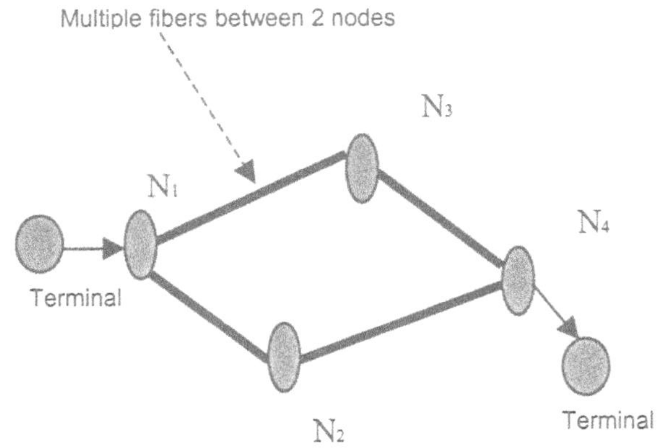

Fig. 2. The sample network model

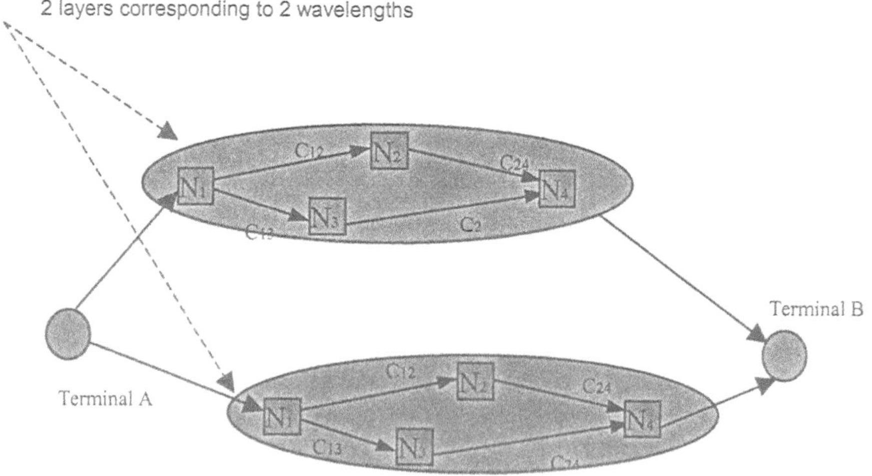

Fig. 3. The layered-graph model corresponding to Fig. 2.

Strategies and algorithms for dynamic routing and wavelength assignments:

On a plane of connectivity for a given wavelength, we first calculate the shortest path between any pair of OXC's. Given *w* wavelengths in the network, we calculate *w* (if possible) paths connecting the two end-points of the connection. With each path we associate a parameter total cost and the path with the "best" cost is then assigned to the connection. As a result, we get a combination of <*Pathi, Wavelengthi*> , i = 1,..,w, where we need to select an element with the best cost.

The definition of best cost is critical for the connection assignment. In this section we describe three heuristic algorithms to select a path among the w layers. These heuristics can all be implemented as on line algorithms.

A: minimum total cost path:

The cost of a path between a source s and a destination d on wavelength λ_j is $C_{sd}^{\lambda_j}$ with $C_{sd}^{\lambda_j} = \sum_{i=1}^{n} C(L_i^{\lambda_j})$ and $<L_1, L_2......, L_n>$, is the combination of links for the path between s and d. $C(L_i^{\lambda_j})$ is the cost of link

$$C(L_i^{\lambda_j}) = \begin{cases} \dfrac{1}{F - O(L_i^{\lambda_j})} & \text{if } O(L_i^{\lambda_j}) \prec |F| \\ \infty & \text{if } O(L_i^{\lambda_j}) = |F| \end{cases}$$

With F is the number of fibers per link and $O(L_i^{\lambda_j})$ is the number of current occupied channels on link $L_i^{\lambda_j}$. The path with the minimum total cost (cheapest) $C_{sd}^{\lambda_i}$ is selected and assigned to the connection.

B- Minimum average cost:

The cost along the path is given by:

$$\overline{C_{sd}^{\lambda_j}} = \frac{\sum_{i=1}^{n} C(L_i^{\lambda_j})}{1/n}$$

With n is the total number of links along the path between the source s and the destination d. the path selection is based on searching for a path with the least average cost and minimum number of hops among the shortest paths found on all directed graphs.

Figure *4.a* shows an example for illustration; clearly the connection is routed here on λ_j, although the path on λ_i has smaller total cost however the average total cost on λ_j is smaller. We assign here the path with minimum average cost hence avoiding assigning longer paths, thus minimizing the waste of resources.

C- Future-based cost:

Another approach for the selection of a path could be based on finding the future cost if the assignment were to happen on this path and the path with minimum future cost is a good candidate to be assigned.

We define $C_{sd}^{\lambda_j, future} = \sum_{i=1}^{n} C(L_i^{\lambda_j, future})$

Where $C(L_i^{\lambda_j, future})$ is obtained by decrementing the number of channels on each link by one. The future cost algorithm forwards connections on links where more resources are available. It tries to balance the load all over the network. Figure *4.b* shows an example for illustration, clearly the connection is routed here on λ_i. After assigning a path, we change the weight of the links along this path on the appropriate graph by basically decrementing the number of available channels on a link by one.

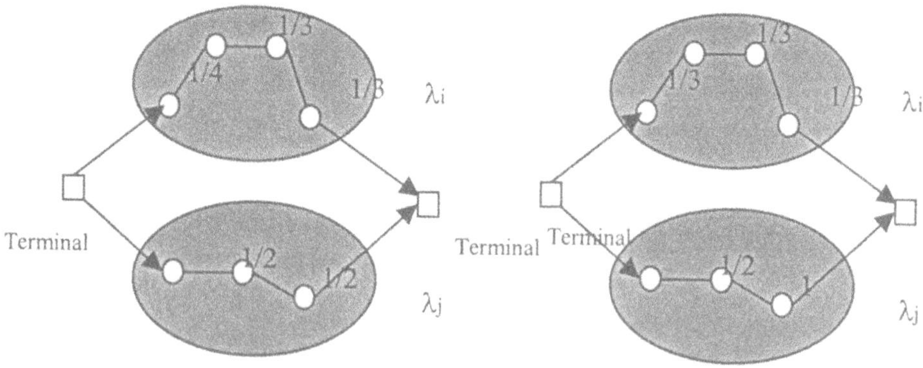

Fig. 4 Illustration example

5. SIMULATION RESULTS AND COMPARISON:

We evaluate the performance of the proposed algorithms on the NSFNet network shown in figure 5; the NSFNet consists of 14 nodes and 21 links. We assume that each link in the network has N fibers, and each fiber has the same number of wavelength (W). A connection between 2 nodes is assumed to be bi-directional, 1 link for each direction.

We use a dynamic traffic model in which call requests arrive at each node according to a Poisson process with a network arrival rate λ. An arrival session is equally likely to be destined to any node in the network. The session holding time is exponentially distributed with mean $1/\mu$. The blocking probability is the metric used to evaluate the network performance.

First we examine the performance of the dynamic routing and wavelength assignment and compare the performance against the static routing (discussed in section 3) with k = 4 (number of alternate paths). Figure 6 shows the performance improvement that adaptive routing introduces. The simulation is run on the NSF network where we have 1 fiber/link between 2 adjacent nodes and 8 wavelengths per fiber.

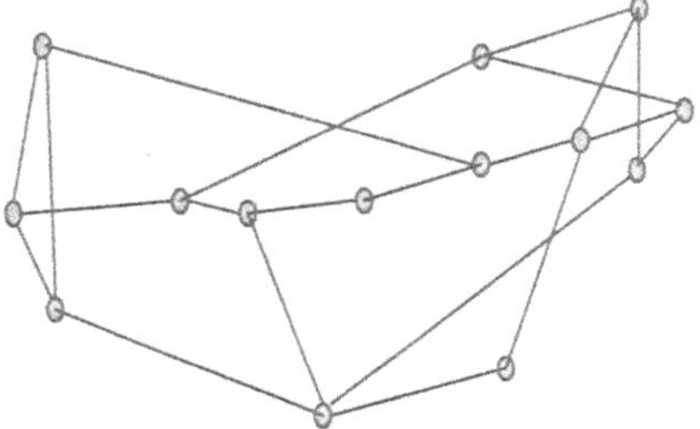

Fig. 5. NSFNet network

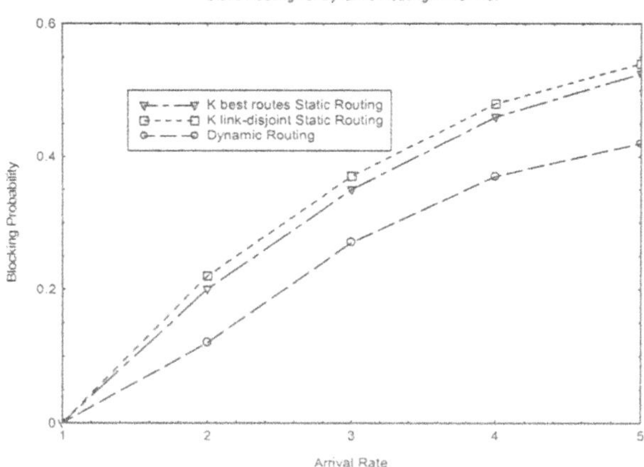

Fig. 6. Dynamic vs. Static Routing

Clearly, the k-best route scheme shows better performance over the k-link disjoint scheme. This is due to the fact that the first one provides a richer set of routes than the second one that always gives as many routes as the number of interfaces (which is most of the time less than k).

However, the simulation shows a better performance for the dynamic case. In the first case (static) a route between any s-d pair remains unchanged as it is computed and saved in the routing table for the ingress node. Now although wavelength assignment might exhaust the resources along this path, still new calls going from s to d have to follow the same path causing the links along the path to be more vulnerable and eventually it will be very difficult for a call between s-d to find a single wavelength available along the path. This problem is avoided in the full-dynamic case since the routing decision is based on the current network-state.

Two path selection schemes have been suggested in this paper. The first one selects the shortest path on each wavelength and associates an average cost with every path. The path among all these paths with minimum average cost is chosen. This also tries to choose paths with smaller number of hops. The second one introduces a future cost metric and selects the path with minimum future cost. This, on the other hand, routes a connection on a path where we have most resources.

Figure 7 and 8 shows the simulation results for the NSF network with 2 fibers per link (figure 7) and 4 fibers per link (figure 8) and 4 wavelengths per fiber. Figure 7 shows the results for the case where 2 fibers per link. Clearly, the path selection process based on the future cost performs the best, since always connections are routed around vulnerable part of the network. Figure 8 shows the same measurements but here we have 4 fibers per link. In this case the performance of the future cost selection process is still the best along with the average cost selection scheme. Since too many resources are in the network, avoiding longer paths or routing around the vulnerable part of the network will give the same performance.

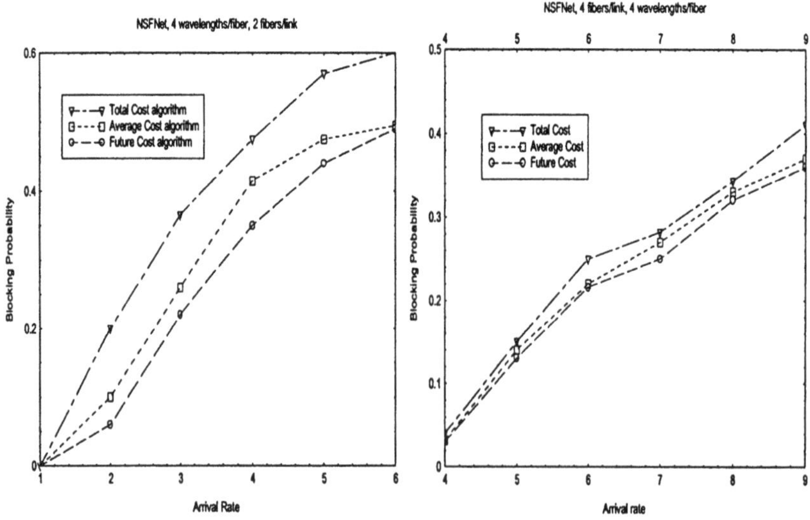

Fig. 7. *Dynamic RWA schemes in 2 fibers/link network*

Fig. 8. *Dynamic RWA schemes in 4 fibers/link network*

6. CONCLUSION

In this paper we study the problem of dynamic lightpath computation in all-optical networks with the objective of maximizing network resource utilization. We first compare the semi-dynamic scheme and the full-dynamic RWA. The results indicate that full dynamic achieves better performance. Also, simulation results show that in semi-dynamic approach the k-best route scheme achieves better performance over the k-link disjoint scheme.

Then, We propose two on-line path selection schemes, for the full-dynamic RWA, that haven't been proposed. Simulation results shows that Future Cost Based Selection is more suitable than the others since connections are always routed on links with more available resources.

In our future work we intend to extend our study to hit in deep the traffic-engineering problem and the real time provisioning and extend the cost parameter to include the overhead introduced by signaling.

References:

[1] N. Ghani, "Lambda-Labeling: A framework for IP-over-WDM using MPLS", Optical Networks Magazine, Vol. 1, No 2, April 2000, pp45-58.

[2] H. Zang, J. Jue, and B. Mukherje, "A Review of Routing and Wavelength Assignment Approaches for Wavelength-Routed Optical WDM Networks", Optical Networks magazine, V1, Number 1, January 2000

[3]. J. Chlamtac, A. Ganz, and G. Karmi, "Lightpath Communications: An approach to high bandwidth optical WAN's", IEEE Trans. Commun., Vol. 40, pp.1171-1182, July 1992

[4] A. Mokhtar and M. Azizoglu, "Adaptive wavelength routing in all optical networks", IEEE/ACM trans. Net, Vol. 6, pp197-206, April 1998.

[5] D. Banerjee and B. Mukherjee, "A practical approach to routing and wavelength assignment in large WDM routed networks," IEEE J. Select. Areas.Comm., Vol. 14, pp.903-908. June 1996.

[6] M. Kovacevic and A. Acampora, "On wavelength translation in all optical networks," Proceedings of IEEE INFOCOM'95, Boston, Ma, April 1995.

[7]A. Birman and A. Kershenbaum, "Routing and Wavelength Assignment Methods in Single-Hop All Optical Networks with blocking,"Proc., IEEE INFOCOM '95, Boston, MA, vol.2, pp.431-438,April 1995.

BIOGRAPHIES

Abdallah A. Shami received the B.E. degree in electrical and electronics engineering from the Lebanese University, Beirut, in 1997, the M.S. degree in electrical engineering from the Saint Joseph University in Lebanon in 1998. He is currently pursuing the Ph.D. degree in electrical engineering at CUNY, NY. In summer 2000 he was an intern at Sorrento Networks, San Diego, where he developed software for the RWA and wavelength conversion in optical networks. His current research interests are in the area of optical networking, IP over WDM and specifically traffic engineering issues and real time provisioning.

Chadi M. Assi received the B.E. degree in electrical and electronics engineering from the Lebanese University, Beirut, in 1997, the M.S. degree in electrical engineering from the City University of NY in 1999. He is currently pursuing the Ph.D. degree in electrical engineering at CUNY, NY. In summer 2000 he was an intern at Sorrento Networks, San Diego, where he developed software for the RWA and wavelength conversion in optical networks. His current research interests are in the area of optical networking, IP over WDM and specifically traffic engineering issues and real time provisioning.

Mohamed A. Ali received his Ph.D. degree in electrical engineering from the City College of the City University of New York in 1989. From 1980 to 1982, he was a Project Engineer at the Gulf Electrical and Mechanical Corporation, Cairo, Egypt. He joined the faculty of electrical engineering at the City College of New York in 1989, where he is currently a Professor. His research interest cover the fields of optical device technology, high speed optical communication systems and architectures, optical switches and high-performance IP routers, and WDM transmission and networking. His major interests are in the areas of computer simulation and modeling of high-speed communication systems. He has published over 60 papers in professional journals and international conferences. Dr. Ali received the NSF Faculty Career Development Award.

LIGHTRING: A DISTRIBUTED AND CONTENTION-FREE BANDWIDTH ON-DEMAND ARCHITECTURE

James Cai
Nortel Networks/Qtera Corporation
3201 N. Greenville Ave, Richardson, TX 75082, U.S.A.
James.Cai@qtera.com

Andrea Fumagalli
The University of Texas at Dallas
P.O. Box 830688, MS EC33, Richardson, TX 75083, U.S.A.
andreaf@utdallas.edu

Abstract: As opposed to current SONET/SDH solutions that provide static tributary signals, modern access and metro networks require bandwidth-dynamic architectures that can efficiently handle the transmission of users' data bursts. This paper presents the LightRing, a bandwidth-efficient multi-wavelength ring architecture that allows user to set up and tear down optical circuits on-demand. A unique distributed multi-token reservation mechanism is used in the LightRing to set up optical circuits in a contention-free manner. i.e., once initiated the reservation is always completed successfully. A tell-and-go approach is therefore possible whereby data transmission is initiated while the circuit is being set up. The LightRing's distributed control yields fast set-up times that are below the ring round trip propagation time, fair blocking probability and high bandwidth utilization even in presence of relatively short optical bursts of data.

Key words: LightRing, WDMring, bandwidth on-demand, distributed control, multi-token reservation mechanism, Routing and Wavelength Assignment (R-WA), optical burst switching

1. INTRODUCTION

The increasing demand for Internet connectivity and network applications drives the current explosion of network traffic volume worldwide.

It is expected that this exponential growth of traffic volume will continue in the foreseeable future. Optical fiber communication technology based on Wavelength Division Multiplexing (WDM) has been successfully employed as the major means to cope with the traffic volume growth. While WDM technology has already revolutionized the backbone network by enabling unprecedented increases in the leveraged capacity of a single fiber, a parallel revolution is now taking place in the access and metro networks.

One of the most critical challenges in designing today's access and metro networks is the fact that bandwidth demands have been consistently exceeding the most aggressive network planning prediction. In addition, the individual user's traffic burstiness makes static bandwidth reservation (e.g., SONET/SDH like) neither bandwidth efficient nor adequate to delay-sensitive traffic. This situation has generated an increasing interest towards all-optical networks that are capable of allocating network resources, i.e., bandwidth, in a dynamic way. Such networks must be able to reserve the necessary bandwidth on-demand, just prior to the transmission of the user's data burst. Once the transmission of the burst is completed, the reserved bandwidth is promptly released and made available to other burst transmissions.

In order to be of practical use, the bandwidth on-demand concept requires few but fundamental features. Three of the features are:

- fast set-up time of the optical circuit (or lightpath)
- fair blocking probability irrespective of the lightpath span (i.e., the number of fiber lines the lightpath is routed through)
- high bandwidth efficiency, i.e., the fraction of reserved bandwidth actually used to transmit data.

To understand how challenging it is to achieve these three features in the same architecture at once, one must observe that users' requests for lightpath are unpredictable and may occur simultaneously at distinct and geographically separated source nodes. As a result, concurrent lightpath requests may compete with one another for the same common resources, i.e., the available wavelengths in the network, and originate a number of unsuccessful attempts as some requests may be blocked by others that booked the resources first. In this scenario, it is thus possible to experience long set-up times, unfair blocking probabilities — lightpaths with longer span are more likely to be blocked as they require successful wavelength reservation on each and every fiber line they are routed through — and bandwidth waste due to the unsuccessful attempts to establish lightpaths.

Approaches so far proposed to solve the Routing and Wavelength Assignment (RWA) problem in dynamic WDM ring — access and metro rings are of particular interest due to the already existing fiber layouts — can be classified as either centralized [1, 4] or distributed [3, 5].

With a centralized approach, the source node sends the request for a lightpath to a special node called *controller*. The controller keeps track of the available network wavelengths and serves the nodes' requests on a First-Come First-Serve (FCFS) basis, by assigning one of the available wavelengths to the incoming request. The resource contention is thus resolved at the controller. In a unidirectional ring, latency of the signaling required between the source and the controller to set up and eventually tear down the lightpath is proportional to the ring latency, i.e., ring round trip propagation time. This latency may considerably delay the set-up time and reduce the bandwidth efficiency in metro applications.

With a distributed approach, every source must solve the RWA problem for its own newly requested lightpaths. One way to achieve this objective is to allow every node to keep track of the network-wide wavelength availability. The RWA problem is then solved based on shared global information [3]. In another approach, each node makes use of a local routing table for each wavelength which specifies the next hop and the cost associated with the shortest path to each destination on this wavelength [5]. Since multiple nodes may concurrently try to assign the same wavelength to distinct lightpath requests, resource contention may arise. Both distributed approaches need a positive acknowledgment to complete a reservation successfully. In case of reservation failure due to resource contention, multiple set-up attempts are required for the same lightpath. Therefore, at least one round trip time between source and destination, i.e., the ring latency in the case of unidirectional ring, is necessary to complete the lightpath set-up prior to beginning data transmission.

This paper presents the *LightRing* architecture whose objective is to overcome the drawbacks of the centralized and distributed reservation mechanisms so far proposed to handle on-demand bandwidth reservation in multi-wavelength ring. The LightRing architecture solves the RWA problem with a distributed approach and at the same time provides a contention-free reservation mechanism. It resorts to a unique distributed multi-token control according to which, access to each wavelength is controlled by a *wavelength specific signaling-token* that is circulated among the nodes connected to the ring in a round robin fashion. A lightpath is set up and torn down on a given wavelength only when the corresponding token is acquired by the source. While circulating along the ring, token broadcasts the lightpath status information to every node of the ring.

Contrary to all conventional wavelength assignment algorithms whereby an available (somehow optimal) wavelength is sought for each given lightpath request, the LightRing distributed control seeks the lightpath request — stored in the so called Best-Fit-Window (BFW) of the source's transmission queue — that optimally fits the resource availability on the wavelength identified by the arriving token. The bandwidth efficiency achieved by the proposed wavelength reservation mechanism is proportional to the number of requests that the reservation mechanism can choose from, thus it is proportional to the size of the BFW. Contrary to most of the existing reservation mechanisms whose complexity is proportional to the number of wavelengths, the complexity of the LightRing reservation mechanism is proportional only to the size of the BFW. The LightRing architecture thus scales well when the number of wavelengths increases.

The proposed LightRing reservation mechanism achieves better network performance than the exiting centralized and distributed reservation mechanisms by applying a tell-and-go reservation mechanism, that once initiated — i.e., acquiring of a token whose corresponding wavelength provides available resources from source to destination — is contention-free and always successful. Wavelength assignment is made locally by each source without requiring an hand-shake protocol with intermediate nodes, destination node or controller. With LightRing, the impact of ring latency on the network performance is mitigated because the set-up time of the lightpath may be significantly less than the ring latency. As a matter of fact, the larger the number of wavelengths, the shorter the expected set-up time becomes, since more tokens circulate in the ring and source is given more frequent opportunities to set up a lightpath.

In summary, as demonstrated in the following sections, due to its various unique features, the LightRing yields fast lightpath set-up times, fair blocking probability and efficient bandwidth utilization under a variety of data burst sizes.

2. SYSTEM DESCRIPTION

The network in question is a single fiber ring network that connects N nodes. The network makes use of W data channels and one control channel, for a total of $W + 1$ wavelengths. The optical signal on the control channel does not go through the node and it is separately handled by a control receiver and a control transmitter. For each data channel

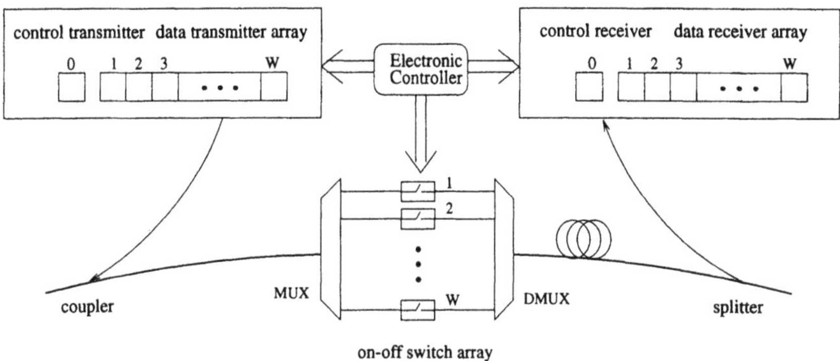

Figure 1 Node architecture.

a node has one fixed transmitter, one fixed receiver and one optical on-off switch connected as shown in Figure 1. This architecture allows the node to transmit and receive data independently (and simultaneously) on any data channel. The on-off switches are used to control the flow of optical signals through the node and prevent signal re-circulation in the ring. Once transmitted by the source node, the optical signal is removed from the ring by the destination node.

A transmission buffer is provided at each source to queue the outstanding lightpath requests. The nodes activities are regulated by an electronic control that determines the state for each on-off switch and the wavelength used to set-up any given lightpath request. The electronic elaboration of the control signal is done in parallel while the optical signal in the data channel propagates through the fiber delay line that connects the splitter to the demultiplexer.

The LightRing architecture does not require wavelength converters.

2.1. WAVELENGTH RESERVATION MECHANISM

To resolve resource contention arising between concurrent attempts to set up a lightpath originating at distinct sources, a wavelength reservation mechanism is required. In the LightRing architecture, this mechanism is based on a distributed multi-token control. Each data channel is assigned a token that is circulated among the nodes using the control channel, and regulates the access to the corresponding data channel. A total of W tokens are circulated and uniformly distributed along the ring perimeter. A lightpath may be set up and torn down only when a token is acquired by the lightpath source. Since only one node at a time is allowed to make a reservation on a given wavelength, LightRing is contention-free. While circulating along the ring, token broadcasts the source and destination of the lightpath being established to all nodes of the ring so that no other node will attempt to set up a lightpath on the same wavelength that overlaps in space with the one being established. As the token propagates towards the destination, each intermediate node sets one of the on-off switches to create the lightpath. The token is shortly followed by the data transmitted on the lightpath under construction, thus achieving the "tell-and-go" feature of the proposed reservation mechanism.

Upon reception of a token, source node first updates its local control table that keeps track of the available resource on the wavelength associated with the token, i.e., the fibers (or the ring segments) in which the wavelength is not assigned to any of the existing lightpaths in the network. Then it seeks a lightpath request that can be set up using the available resources on that wavelength. Since multiple requests may be awaiting transmission in the source transmission queue, an algorithm must be designed to determine which lightpath request (if any) shall be served first.

A Best-Fit-Window (BFW) algorithm is proposed to choose the lightpath request based on the following two observations.

Intuitively, a long-span lightpath request is more likely to be blocked than a short-span lightpath request, especially under medium and high workload. This is due to the well-know fragmentation problem in which short-span lightpaths already established on the wavelength in question block newly generated lightpath requests that span across multiple fibers. To counterbalance this effect the BFW algorithm selects the lightpath request with the longest span in the transmission queue that fits into the currently available fibers. If multiple requests exist with the same span, the oldest one is selected. An additional advantage of the BFW algorithm is to circumvent the head-of-the-line blocking that occurs when the oldest request in the queue, i.e., the first one in line, is blocked.

To limit the variance of the request waiting time in the queue, the BFW algorithm gives priority to those requests that due to blocking cannot be served within a chosen time interval. This objective is achieved by combining the best fit window scheme and the FCFS scheme as follows. Newly generated requests are first served using the BFW scheme. If the request cannot be served for a time interval longer than a threshold, it is then served using the FCFS scheme.

The BFW algorithm is formally described in the next two sections.

2.1.1 Definition of Parameters. The following parameters are defined to provide a formal description of the BFW algorithm.

BFW : maximum number of lightpath requests in the transmission queue considered by the BFW algorithm;

r_i : ith lightpath request stored in the transmission queue, $i = 0$ being the oldest one, i.e., the head-of-the-line request;

s_i : span of request r_i;

R_{TxQ} : $\{r_i \mid r_i \text{ in transmission queue }\}$;

R_{BFW} : $\{r_i | r_i \in R_{TxQ}, 0 \leq i < BFW\}$;

t : current time;

$t_i^{(a)}$: arrival time of r_i in R_{TxQ} (time at which r_i is stored in the transmission queue);

$t_i^{(s)}$: beginning of the service time of r_i (time at which r_i is removed from the transmission queue);

$t_j^{(t)}$: arrival time for token j;

$t_i^{(q)}$: time spent by r_i in R_{TxQ};

$t_i^{(w)}$: time spent by r_i in R_{BFW};

$e^{(w)}(t)$: estimated expected time spent in R_{BFW} by requests (this quantity may vary with time, t);

$d^{(w)}(t)$: soft deadline for r_i leaving R_{BFW} (this value is compared with the time spent by the request in R_{BFW});

R_{LATE} : $\{r_i | r_i \in R_{BFW}, t_i^{(w)} > d^{(w)}(t)\}$ (set of requests whose time in R_{BFW} exceeds the soft deadline);

b_j : number of fibers between source and the first downstream node (including source) that is currently using wavelength j (if wavelength j is not used at all, $b_j = N$);

$l_j(i)$: $b_j - s_i$ (a negative number indicates that $r(i)$ overlaps with lightpath(s) already set up);

$Fit(R, \lambda_j)$: $\{r_i | r_i \in R, l_j(i) \geq 0\}$;

$BestFit(R, \lambda_j)$: $\{r_i | r_i \in R, l_j(k) \geq l_j(i), \forall r_k \in R\}$;

$FCFS(R)$: $\{r_i | r_i \in R, t_i^{(a)} < t_k^{(a)}, \forall r_k \in R\}$;

2.1.2 Best-Fit-Window Algorithm. It assumed that no information is available on the life time of each lightpath request. The lightpath will be torn down once the transmission of data has been completed, but the network is not aware of how long the transmission will last.

The BFW algorithm is defined as follows:

- only r_i in R_{BFW} can be served upon a token's arrival

- arriving requests are dropped when no space is available in the transmission buffer R_{TxQ}.

At source, the following BFW algorithm is used to determine the lightpath request to be served next:

1. Upon arrival of token j, $t = t_j^{(t)}$ —
   set up lightpath for request r_i
   If $(Fit(R_{LATE}, \lambda_j) \neq \Phi)$ {
   $r_i = FCFS(Fit(R_{LATE}, \lambda_j))$
   release token j with current lightpath info
   set up r_i and begin data transmission
   } else if $(Fit(R_{BFW}, \lambda_j) \neq \Phi)$ {
   $r_i = BestFit(R_{BFW}, \lambda_j)$
   release token j with current lightpath info
   set up r_i and begin data transmission
   } else
   pass token j onto next node;

2. Upon setting up r_i, $t = t_i^{(s)}$ —
   i) remove r_i from R_{TxQ} and R_{BFW}; ii) set $e^{(w)}(t) = \beta e^{(w)} t_{i-1}^{(s)} + (1 - \beta) t_i^{(w)}$, where β is a value between 0 and 1, that is used to weight the previously measured waiting times. *beta* is chosen to be close to 1;
   iii) set $d^{(w)}(t) = \alpha e^{(w)}(t)$, where $\alpha = \frac{d^{(w)}(t)}{e^{(w)}(t)}$ is a constant chosen by the designer to determine the soft deadline that triggers the FCFS scheme. α is chosen to be greater than 1 but close to 1.

3. ANALYTICAL MODELS

Two analytical models are presented that help better understand the fundamental difference between the proposed reservation mechanism and the existing ones. The models are derived to assess the achievable throughput as a function of the expected size of the data burst and provide a preliminary performance comparison. For the LightRing reservation mechanism it is assumed that the size of the best fit window is $BFW = 1$. Being this the worst possible scenario for the LightRing architecture, the result will be considered as a lower bound on the achievable throughput for the proposed architecture.

The models are derived from the analysis of blocking probability developed by Barry and Humblet in [2] for a WDM network in which lightpaths are set up one at a time. In Barry's model the signaling latency is ignored. Barry's model allows to compute an approximate value for the blocking probability of any given lightpath request under a given network throughput. It must be noted that Barry's model does not cover the case in which lightpaths with relatively short life time are dynamically established and torn down.

The key variables in Barry's model are P_l — the probability that a lightpath ends and drops at any given node, and P_n — the probability that a lightpath starts at any given node on any given wavelength. The two variables have the following relationship

$$P_n = \frac{\rho P_l}{1 - \rho(1 - P_l)} \tag{1}$$

where ρ is the wavelength utilization factor. In absence of wavelength conversion the blocking probability is

$$P_b = [1 - (1 - P_n)^H]^W \tag{2}$$

where H is the number of hops of the lightpath, or span, and W is the number wavelengths in every fiber.

The above model is used to derive the blocking probability for the WDM ring architecture of interest. In the case of unidirectional WDM ring, $P_l = 2/N$. An iterative technique is used to derive the blocking probability for a randomly chosen lightpath request. Let $y = P_b' - P_b$ be the gap between the computed blocking probability at two consecutive iterations — P_b' is the result of the next iteration using P_b. Under saturation load, the following steps are used to derive the blocking probability:

the plane y versus P_b is used –
1. set network load $d = 1$, $P_{b0} = 0$ and $P_{b1} = 1$;
2. $\rho 0 = d(1 - P_{b0})$; $\rho 1 = d(1 - Pb1)$;
3. compute P_{b0}' and P_{b1}' from Eq. 1 and Eq. 2;
4. $y0 = P_{b0}' - P_{b0}$ and $y1 = P_{b1}' - P_{b1}$;
5. connect point $(P_{b0}, y0)$ and point $(P_{b1}, y1)$ with a straight line, and let P_b be the intersection of this line with axis $y = 0$;
6. compute $y = P_b' - P_b$, where P_b' is derived from Eq. 1 and 2 using $\rho = d(1 - P_b)$; if y has the same sign as y0, $P_{b0} = P_b$, otherwise $P_{b1} = P_b$;
7. go back to step 2 until $|y|$ is less than the desired resolution.

The achievable throughput is then derived taking into consideration the blocking probability, P_b, and the lightpath utilization, $E[\eta_a]$, that is the fraction of time during the lightpath life time that is actually used to transmit data

$$thr = E[\eta_a](1 - P_b) \tag{3}$$

The value of $E[\eta_a]$ is a function of the chosen reservation mechanism and the probability density function of the burst transmission time, $g_a(t)$. In the next two sections $E[\eta_a]$ is derived for both the LightRing reservation mechanism and the other existing mechanisms under the assumption that the burst transmission time is exponentially distributed with average a. In the derivation, the switching latency of the on-off switches is assumed to be negligible when compared to the ring latency, D.

3.1. LIGHTRING RESERVATION MECHANISM

In the LightRing architecture, data transmission begins at the same time instant when the lightpath is established. Once data transmission is completed, the lightpath is torn down as soon as the token that corresponds to the wavelength assigned to the lightpath reaches the source again, thus

$$E[\eta(a)] = \sum_{n=0}^{\infty} \int_{nD}^{(n+D)} \frac{t}{(n+1)D} g_a(t)dt. \tag{4}$$

If the burst transmission time is exponentially distributed with average a, then

$$\begin{aligned} E[\eta(a)] &= \sum_{n=0}^{\infty} \int_{nD}^{(n+D)} \frac{t}{(n+1)D} \frac{1}{a} e^{-t/a} dt \\ &= \sum_{n=0}^{\infty} \frac{a}{D(n+1)} (e^{-nD/a} - e^{-(n+1)D/a}) \\ &+ \sum_{n=0}^{\infty} \frac{1}{n+1} (ne^{-nD/a} - (n+1)e^{-(n+1)D/a}) \\ &= 1 + (e^{D/a} + \frac{a(1-e^{D/a})}{D}) \ln(1 - e^{-D/a}) \end{aligned} \tag{5}$$

3.2. OTHER RESERVATION MECHANISMS

As already discussed in the Introduction, the other exiting reservation mechanisms require at least one round trip time within the ring to complete the reservation. In the case of centralized reservation mechanism, it is exactly one round trip time. In the case of distributed reservation mechanism, one round trip time is the lower bound on the time required to successfully set up a lightpath. Only the centralized reservation mechanism is therefore considered here as it represents the most efficient solution among the exiting ones excluding the LightRing.

Observing that the lightpath is considered to be set up as soon as the controller assigns a wavelength to it, and it is torn down by the controller as soon as a disconnect signal is received from the source, the life time of the lightpath equals the burst transmission time plus the ring latency.

Thus, in general

$$E[\eta(a)] = \int_0^\infty \frac{t}{D+t} g_a(t)dt. \tag{6}$$

and for the exponentially distributed transmission time

$$E[\eta(a)] = \int_0^\infty \frac{t}{D+t} \frac{1}{a} e^{-t/a} dt. \tag{7}$$

4. PERFORMANCE RESULTS

The performance results presented in this section are obtained either from simulation or from the analytical models derived in Section 3. The simulation time is chosen to achieve a confidence interval of 5% or less at 98% confidence level. Unless otherwise stated, the network under consideration is a WDM ring with 32 wavelengths and 16 nodes evenly distributed over 80 km of fiber. Each wavelength supports fixed transmission rate of 10 Gbps. For demonstration purpose, the data bursts are assumed to be generated according to a Poisson arrival process and have size that is exponentially distributed. Traffic is uniformly distributed, implying that the source and the destination nodes of a newly generated data burst are randomly chosen.

Results obtained on the blocking probability confirm the fair behavior of the LightRing architecture with respect to all possible lightpath spans. This is due to the fact that newly generated data bursts are dropped only when they cannot be stored in the already full transmission queue. Consequently, data bursts are dropped irrespective of their source-destination pair.

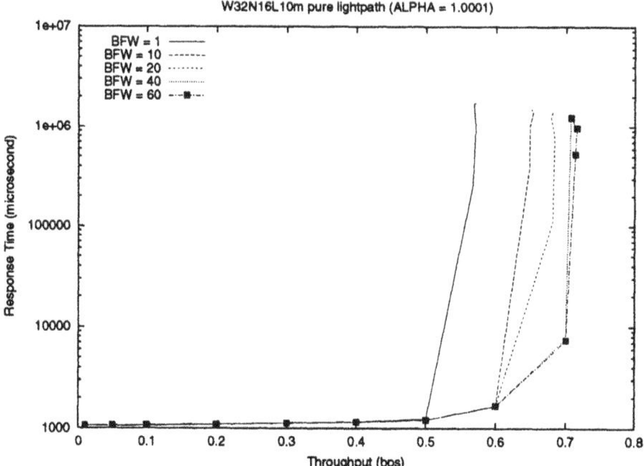

Figure 2 LightRing: expected response time versus throughput

Simulation results in Figure 2 plot the expected response time versus throughput obtained for various sizes of the BFW. Response time is defined as the sum of the waiting time in the transmission queue and the transmission time of the burst. Throughput is normalized to be equal to one when all wavelengths in all fibers are always used to transmit actual data. The expected burst length is chosen to be 10 Mbit, thus requiring an expected transmission time of 1 ms. The figure shows that response time and control complexity can be traded for throughput by varying the BFW size. The achievable throughput increases around 30% as the size of the BFW grows from 1 to 40. However, further increase of the BFW size does not yield any significant throughput gain, but it increases the complexity of the BFW algorithm. At low and medium workload the expected response time is less than the response time of the centralized reservation mechanism that is bounded from below by the sum of the ring latency and the expected burst transmission time, i.e, 1.4 ms. This fact indicates that the expected time required to establish a lightpath in the LightRing architecture can be well bellow the ring latency.

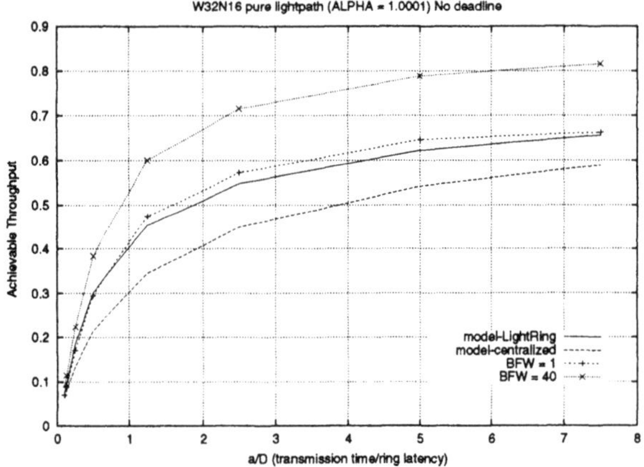

Figure 3 Achievable throughput, thr, versus expected burst transmission time normalized to the ring latency, a/D

Figure 3 depicts the achievable throughput versus the expected burst transmission time normalized to the ring latency, a/D. (For example, $a/D = 1$ indicates an expected burst transmission time of 400 μs in a 80 km ring.) Two curves are derived from simulation, respectively, for the case of LightRing with $BFW = 1$ and $BFW = 40$. The other two curves, respectively, for the LightRing architecture with $BFW = 1$ and the centralized reservation mechanism, are obtained using the models described in Section 3. The analytical results of the LightRing closely match the simulation results for $BFW = 1$, thus cross-validating the analytical model approximations and the simulation results. As expected, the LightRing architecture outperforms the centralized reservation mechanism for a variety of burst sizes. Theoretically, the two analytical curves converge when the burst length approaches infinity. In the LightRing architecture with $BFW = 40$, bandwidth efficiency is further improved for most of the burst lengths shown in the figure.

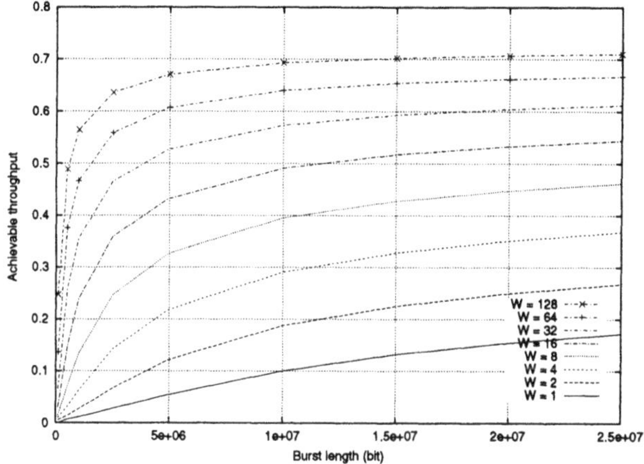

Figure 4 Achievable throughput, *thr*, versus expected burst length for various numbers of wavelengths but fixed total fiber bandwidth of 80 Gbps

Figure 4 shows the achievable throughput of the LightRing architecture as a function of the expected burst length for a variety of numbers of wavelengths. Results are derived from the analytical model. The total bandwidth is fixed at 80 Gbps, thus any increase in the number of wavelengths, proportionally reduces the transmission rate on each individual channel. Figure 4 indicates higher bandwidth efficiency when the number of channels increases. This is due to two factors, namely 1) with a larger number of wavelengths (and tokens) the source acquires tokens more frequently and 2) with a lower channel transmission rate, short data bursts can still be efficiently transmitted in relatively large rings.

5. SUMMARY

The LightRing architecture was presented in which a distributed and contention-free multi-token reservation mechanism is used to set up optical circuits on-demand. By performing a tell-and-go reservation of wavelengths, the LightRing architecture yields fast set-up time and efficient bandwidth utilization even in presence of relatively short bursts of data, e.g., bursts whose expected transmission time is 1 ms in a 80 km ring.

Among other interesting features, the LightRing architecture is compatible with optical packet switching [6] and its performance improves with the growing number of wavelengths, consistently with the current technological trend. Complexity of the reservation mechanism is not a function of the number of wavelengths, and can be varied to trade response time for bandwidth efficiency. Finally, the LightRing architecture is compatible with emerging protocols devised to provide bandwidth reservation in the optical layer, e.g., MPλS, and yields fair blocking probability irrespective of the circuit span in presence of uniform traffic.

References

[1] M. Kovaevic and A. Acampora, "Benefits of Wavelength Translation in All-Optical Clear Channel Networks," *IEEE JSAC*, Vol. 11, No. 5, June 1996.

[2] R. A. Barry and P. A. Humblet, "Models of Blocking Probability in All-Optical Networks with and without Wavelength Changers," *IEEE JSAC*, Vol. 14, No. 5, June 1996.

[3] R. Ramaswami, A. Segall, "Distributed Network Control for Optical Networks," IEEE/ACM Trans. on Networking, Vol. 5, No. 6, Dec. 1997.

[4] H. Zang, J. Jue and B. Mukherjee, "A Review of Routing and Wavelength Assignment Approaches for Wavelength Routed Optical WDM Networks," *SPIE Optical Networks Magazine*, Vol. 1, No. 1, Jan. 2000.

[5] H. Zang, et. *al.*, "Connection Management for Wavelength-Routed WDM Networks," in *Proc. IEEE Globecom'99*, Rio de Janeiro, Dec. 1999.

[6] A. Fumagalli, J. Cai and I. Chlamtac, "A Token Based Protocol for Integrated Packet and Circuit Switching in WDM Rings," in *Proceedings of IEEE Globecom'98*, Sydney, Nov. 1998.

Distributed discovery of wavelength paths in multi-service WDM networks

Gerald Franzl and Admela Jukan
Vienna University of Technology, Institute of Communication Networks, Favoritenstrasse 9/388, A-1040 Vienna, Austria, http://www.ikn.tuwien.ac.at/IKN

Key words: wavelength routing, QoS-routing, multipath, multiple constraints, optical network service.

Abstract: This work proposes a distributed method for wavelength channel provisioning in multi-service WDM networks, where subject to constraints on wavelength routing such as transmission quality or resource availability, the best of the multiple feasible paths can be found. The numerical examples and simulation study of multi-service wavelength-routed networks under dynamic traffic conditions have shown the applicability of our methods for the optical network services yet to emerge.

1. INTRODUCTION

The on-demand provision of wavelength circuits in a service-specific fashion is becoming increasingly important due to the ever-growing deployment of IP networks and the requirements for quality-of-service(QoS)-delivery within transport layers. The wavelength-routed transport networks are nowadays characterised by an excessive architectural and operational diversity, in particular with regard to technology of optical components, internal architecture of optical nodes and control protocols. In a networking scenario, where internetworking with a variety of candidate client networks (e.g. SDH, IP, ATM) is an assumption, while dealing with heterogeneous architectures and optical technologies, efficient and technology-independent algorithms and protocols for dynamic and service-differentiated wavelength circuit provisioning will play a key role [1].

Due to the several intrinsic properties related to the wavelength-routed connections, the maintenance of the network state information is becoming an important issue. The additive nature of signal degradations, which limits the cascadeability of optical components, as well as the type-of-user-dependent accommodation of wavelength connections (e.g. SDH-suited optical cross-connects), are some of the reasons that make the global (centralised) routing essentially impractical, particularly for open network structures handling various services in a heterogeneous environment. In addition, there is a growing necessity to leverage and co-ordinate the wavelength routing control with IP/MPLS distributed routing control. For that purpose, the distributed *path-selection* computation among intermediate nodes between source and destination arises as a solution for a scalable and accurate wavelength channel provisioning.

In this paper, we propose a generic approach to distributed connection set-up in QoS-routed WDM networks applicable to services yet to emerge in regular network operation as well as for wavelength-routed restoration. The proposed method relies on service-specific characterisation of optical network elements, such that the network state information, related to transmission quality, reliability, manageability and traffic conditions, is autonomously maintained. We use the term network element (NE) in a broader sense: it stands for any physical resource or *autonomously manageable* group of resources along optical paths (e.g. fibres, amplifiers, filters, switches, transceiver, regenerators, links, nodes, etc.) that might affect the wavelength channel performance associated with a certain optical network service. With our method, a choice between a number of selected feasible paths is made based on service-dependant transmission quality requirements and network constraints. This method can also yield the solution to the constraint-based routing problem, i.e. provision of service-specific guarantees under quality constraints, as it is necessary for the MPLS control plane, a feature that is still missing from the existing architectures.

2. NETWORK MODEL

We model the WDM network $\boldsymbol{G}(N,L,\Lambda,\boldsymbol{S})$ with the number of nodes N, number of WDM links L, a pool of possible wavelengths $\Lambda=\{\lambda_1,\lambda_2,\ldots,\lambda_F\}$, and a set of supported optical network services $\boldsymbol{S}=\{S_1,S_2,\ldots S_P\}$. We keep to the definition of optical network service S_i to be the handling of optical signals carrying data originating from optical network clients, such as SDH or IP, where *optical* QoS is guaranteed by appropriate allocation of particular wavelengths on concatenated physical resources, i.e. from transmitters via fibres and nodes to receivers, by which service-specific (optical) requirements on transmission quality, survivability and management functions are taken into account.

We consider the total number of network elements H, where

$$H := \sum^{N,L} \sum^{T} h_k \text{ , k=1,2...T,}$$

h_k referring to a network element (NE) of type k within a node N or link L from $\mathbf{G}$. Since T different network elements per node or per link are considered at maximum, H is upper-bounded by $(LT+NT)$. For simplicity, a network element associated with a node or link from $\mathbf{G}$ is denoted as $h_k^{N,L}$, k=1,2,...T, for example $h_1^{N_1}$ being transmitter (Tx) at N_1, $h_1^{N_2}$ being Tx at N_2, $h_2^{N_2}$ being receiver (Rx) at N_2, $h_5^{L_2}$ being fibre amplifier at L_2, etc. Upon a request for a wavelength-routed connection, for a certain service type all NEs (up to H totally) relevant to that service might be considered.

For the purposes of our distributed path discovery and selection, each NE is characterised by the so-called *Service-specific Wavelength Set*, Λ_{SWS}, defined as the following [1]:

> *For each network element $h_k^{N,L}$, k=1,2,..T, relevant to the performance of a service $S_r \in \mathbf{S}$, r=1,2,..P, we define a Service-specific Wavelength Set (SWS), $\Lambda_{SWS}[S_r, h_k^{N,L}] \subseteq \Lambda$, such that by allocating a wavelength $\lambda_m \in \Lambda_{SWS}[S_r, h_k^{N,L}]$ on that network element, the requested optical network service is provided with sufficient transmission quality and all necessary management and surveillance functions.*

In other words, for path searching and wavelength allocation, only the wavelengths from Λ_{SWS} on a certain NE are considered. For example, if for a certain optical network service, e.g. S_1, the Λ_{SWS} for the transmitting laser is defined as $\Lambda_{SWS}[S_1,\text{Tx}]=\{\lambda_6, \lambda_7\}$, for S_1 accommodation with quality guarantees, only these two wavelengths can be allocated, even when the tuneable transmitter range includes more wavelengths, e.g. $\{\lambda_1, \lambda_2, ...\lambda_8\}$, which might be suitable for other service types.

The actual task of dynamic wavelength channel provisioning is to find a path from source to destination node using the minimum amount of network resources under the constraints on various metrics for wavelength routes such as SNR, dispersion, jitter, crosstalk, link load, availability or restorability. Hence, an accurate network state information $\mathbf{a}[h_k^{N,L}, S_i, \lambda_i]$ has to contain those parameters related to concatenated NEs along the routed path, which will make it possible to consider the service-specific requirements, defined as a vector of **quality bounds**, such as

$D(S_i)$ = [*max. allowed transmission quality degradation (d);*
max. acceptable (monetary) cost (c);
min. sufficient availability / reliability (r);
min. sufficient manageability/signalling (m)].

The various parameters of the network state information ***a***, may be additive, multiplicative or restrictive in nature [2]. Thus transmission degradation and monetary costs can be summed along the path, reliability for concatenated components is the product of each single reliability, while restrictive attributes such as residual link capacity may excluded channels from the path discovery and selection.

3. AN EXAMPLE WDM NETWORK

For illustration, let us consider an example WDM network in a bidirectional ring configuration with five nodes and two wavelengths per link, capable of accommodating two service classes, S_1 and S_2 (Fig. 1). Assume that S_1 is required between N_1 and N_3, for which the max. transmission quality degradation (e.g. degradation of SNR) should not exceed 30 dB, i.e. $d_{max}(S_1)=30$. For this service type, we consider the following manageable, S_1-specific NEs: Tx(N_1), Rx (N_3), and network nodes and links, with their associated service-specific wavelength sets and quality properties. The network state information is expressed as a parameter-vector [*d,r*], corresponding to transmission quality degradation and reliability. Here, rather than considering two additive quality metrics (e.g. transmission degradation and cost), as the second attribute, the reliability is considered, calculated as the product of reliabilities of the concatenated NEs along the path [3]. The restrictive attributes are automatically taken into account by the definition of Λ_{SWS}: the wavelengths outside Λ_{SWS} are simply excluded from the path discovery (e.g. for service S_1 the wavelength λ_1 cannot be used along the links L_3, L_4, L_5). The quality attributes related to switches, filters, fibres, amplifiers, etc. are accumulated to a single "link" set, described by a common attribute, e.g. [$d(L_1)$, $r(L_1)$].

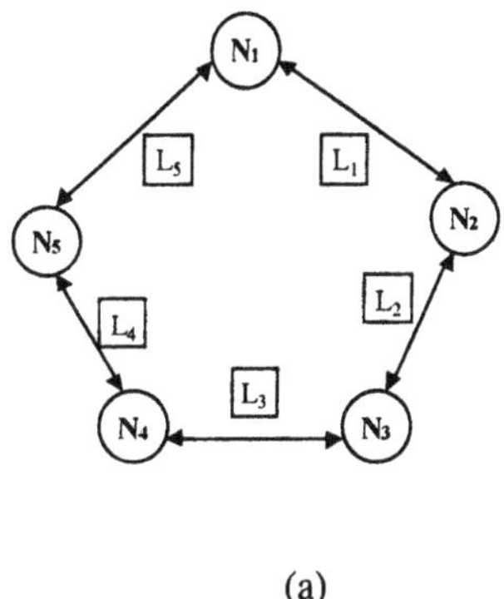

(a)

S_1-specific NE		d[dB]	$r\times100$ [%]
Tx @N_1	λ_1	4	0.95
	λ_2	6	0.99
Rx @N_3	λ_1	5	0.97
	λ_2	7	0.97
L_1, L_2	λ_1	6	0.98
	λ_2	9	0.98
L_3, L_4, L_5	λ_1	$\notin \Lambda_{SWS}$: not avail. for S_1	
	λ_2	5	0.99

(b)

Fig. 1: Example WDM network: (a) physical topology, (b) S_1-specific network state representation.

In this example, the best path in terms of transmission quality appears to be Tx-N_1-L_1-N_2-L_2-N_3-Rx at λ_1, with overall transmission degradation Σd= 4+6+6+5=21 (<30). However, if the request on reliability for this service were $r(S_1)>0.90$ (90%), this path would show unsatisfactory performance: $\prod r=0.95\times0.98\times0.98\times0.97=0.885$, due to the "Tx($\lambda_1$)-bottleneck". On the other hand, by allocating λ_2 over the same route, an unacceptable degradation, i.e. $\Sigma d=31$, is obtained. Alternative path, e.g. Tx-N_1-L_5-N_5-L_4-N_4-L_3-N_3-Rx at λ_2, however, would satisfy both bounds (d and r), i.e. $\Sigma d=28$ (<30) and $\prod r=0.932$ (>0.90). Hence, optimising one single metric is necessary, but not sufficient to find a feasible path. Note that the NE quality attributes and the routing decision completely change, if a service of type S_2 has to be set-up.

4. DISTRIBUTED PATH DISCOVERY METHOD

To overcome the fundamental problems of multiple constrained path discovery, which are NP hard, we propose the following distributed method [2,4]. For simplicity, we will limit our consideration to three types of quality attributes: additive d (e.g. SNR, jitter, delay), multiplicative r (e.g. reliability), concave b (minimal value, e.g. residual capacity in wavelengths, inserted crosstalk minimisation), but any other attributes can be similarly treated.

Distributed discovery of wavelength paths (DWP)

Step 1:

Get a connection request between (src, dest) *for a service* S_r, *with particular QoS requirements [here: max. allowed transmission degradation* $d_{max}(S_r)$, *min. reliability* $r_{min}(S_r)$, *min. residual capacity* $b_{min}(S_r)$*].*
Forward this request and the message containing the initial values (e.g. [d, r, b]) from src *to all neighbouring NEs for every* $\lambda \in \Lambda_{SWS}[S_r, NE_{src}]$.

Step 2:

The neighbouring NEs forward the received requests and messages with actualised values of [d, r, b] to all their neighbours which have not been previously visited, until dest *is reached.*

The messages are actualised according to the following rule:
For all $\lambda \in \Lambda_{SWS}[S_r,$ *previous NE*$]$, *if* $\lambda \in \Lambda_{SWS}[S_r, NE_i]$, *i=NE-counter (e.g.* NE_i *is the* i^{th} *NE in the path), update the values of [d, r, b], so that d is summed along the precedent NEs, r is multiplied, while b of the considered path is the minimal value:*

$$d(p) = (\text{previous } d(p)) + d(NE_i),$$
$$r(p) = (\text{previous } r(p)) * r(NE_i),$$
$$b(p) = \min\{ (\text{previous } b(p)), b(NE_i) \}.$$

The wavelength paths and channels with insufficient quality in terms of [d(p), r(p), b(p)] are sorted out and not forwarded. Finally the NE-id (k) and wavelength (λ) *is added to the surviving path-sequences.*
At the destination node dest *we finally obtain:*

$$d(p) = \Sigma d(NE \in (path)) =$$
$$= \{..\{..\{\{d(srcNE)+d(1stNE)\}+d(2ndNE)\}+..+d(NE_i)\}+..+d(destNE)\},$$
$$r(p) = \Pi r(NE \in (path)) =$$
$$= \{..\{..\{\{r(srcNE) \times r(1stNE)\} \times r(2ndNE)\} \times .. \times r(NE_i)\} \times .. \times r(destNE)\},$$
$$b(p) = \min\{b(NE \in (path))\} =$$
$$= \min\{.\ \min\{..\min\{\min\{b(srcNE), b(1stNE)\}, b(2ndNE)\}.., b(NE_i)\}.., b(destNE)\},$$
$$path = [(k_{src}, \lambda), (k_1, \lambda), (k_2, \lambda), .., (k_i, \lambda),.. , (k_{dest}, \lambda)],$$

for each feasible path at every $\lambda \in \bigcap\{\Lambda_{SWS}[S_r, NE \in (path)]\}$.

Step 3:

From the obtained set of all feasible wavelength-paths $\mathcal{P}$ *for which* $r(\mathcal{P}) > r_{min}(S_r)$, $d(\mathcal{P}) < d_{max}(S_r)$, *and* $b(\mathcal{P}) > b_{min}(S_r)$, *i.e. each path from* $\mathcal{P}$ *satisfies the given requirements, the destination node* dest *can select the best path according to one or more criteria as follows: minimum hop, min. cost, minimum signal degradation, minimum number of traversed NEs of particular type, etc.* [*Obviously, a number of different strategies exists and might be chosen according to network properties, traffic conditions or types of services.*]

Step 4:

From dest, *send the acknowledgement message back along the selected path to* src, *for the purpose of resource reservation. At the time the acknowledgement message reaches the source node, the wavelength circuit is set up and ready for the optical network service provision.*

The distributed strategy proposed above refers to route and wavelength allocation, if applied to the whole network, while it returns wavelengths only, if applied on a statically routed, predefined path (e.g. N_1-L_1-N_2-L_2-N_3), where only those NEs are visited which belong to that path. As shown in the example from Fig. 1, routing over a single predefined path might lead to a poor blocking performance. On the other hand, by applying the method DWP to the whole network, the signalling effort can significantly increase, especially in wavelength-converting and densely connected networks.

To overcome these problems, one method might be to pre-route a set of candidate paths and apply DWP on this set. In order to define which paths are candidate paths, a single wavelength independent parameter, e.g. number of hops, can be chosen to break a tie. Refer to the example network shown in Fig. 1 once more. If for defining the candidate paths, the number of hops was limited to three, i.e. $n \leq 3$, we could find two candidate routes for a connection request between N_1 and N_3. At the same time, for a connection request N_1 and N_2, only a singe candidate path can be found (i.e. the connecting link).

Note that for a simple ring-topology, the reduction of the computational complexity is not as significant as for a mesh topology. For example, in a full-mesh topology with five nodes, a limitation in number of hops to two, e.g. $n \leq 2$, reduces the number of possible routes from 16 to 4, while the number of necessary updates for message-parameters (Step 2, DWP) is reduced from 49 to 7 per wavelength.

Finally, a possible implementation strategy might be to send *wavelength probes*, which run DWP at all idle wavelength channels independently, or to send *wavelength probes* within one separate service channel, updating the message-parameters for every surviving wavelength at each traversed node.

5. PERFORMANCE STUDY AND NUMERICAL RESULTS

For the simulation results, the connection requests arrive according to Poisson process with call holding time being negative exponentially distributed. Traffic distribution and service requests are uniform; all results are obtained with a confidence level of 95%. For simplicity, we assume the network elements with service independent properties along the network, which is not necessarily a realistic assumption. We also assume reliability to be constant over time, which is also not generally the case; availability might be a better measure, but this is out of the scope of this paper [3]. For distributed (multi-path) routing strategies, we adapted the Bellman-Ford shortest path algorithm for routing and connectivity information update at each node, for all loop-free paths [2]. In order to show the applicability of our method, we study three different network topologies: bi-directional ring, full-mesh and mesh-torus. The proposed algorithm (DWP) is compared with such a centralised one, where a single path is found according to a certain criteria (here: least load links) for which then, other quality attributes are checked (transmission degradation, reliability, residual capacity). Least load routing is used to overcome the problem of blocking due to the overloaded links and it provides an easy combination of resource and load optimisation.

In the first example, we study the bi-directional ring topology, the same one as shown in Fig. 1a. In order to demonstrate the handling of non-uniform NE-properties in multi-constrained routing four wavelengths are assumed, with which the wavelength-dependent properties related to the network elements are shown in Fig. 2a. The service-specific requirements on quality are as follows: $S_1[d<30\text{dB}, r>90\%, b\geq 1]$ and $S_2[d<90\text{dB}, r>85\%, b\geq 1]$.

As it can be seen from Fig. 2b, the distributed, multi-path strategy yields superior results for both services. This is because the distributed method yields several feasible paths, out of which the best can be selected (here: min. number of hops), in contrast to the non-DWP methods where a single path, optimised for one quality attribute (here: least load), is checked on remaining requirements (transmission degradation, reliability, residual capacity). While for service S_1 the results are as expected due to the fact that for some calls the shortest path found for a single constraint was not feasible with respect to other constraints, the results shown for service S_2 also improved. In Fig. 2b, it can be seen that in a lightly loaded network the multiple paths yielded by DWP reduce the blocking also for S_2, where all shortest paths (due to a single routing criteria) are feasible without constraints.

NE definitions		d[dB]	$r \times 100$ [%]
Tx	$\lambda_1, \lambda_2, \lambda_3, \lambda_4$	4, 4.66, 5.33, 6	0.95, 0.963, 0.977, 0.9
Rx	$\lambda_1, \lambda_2, \lambda_3, \lambda_4$	5, 5.67, 6.33, 7	0.97
L_1, L_2	$\lambda_1, \lambda_2, \lambda_3, \lambda_4$	6, 7, 8, 9	0.98
L_3, L_4, L_5	$\lambda_1, \lambda_2, \lambda_3, \lambda_4$	35, 25, 15, 5	0.99

(a)

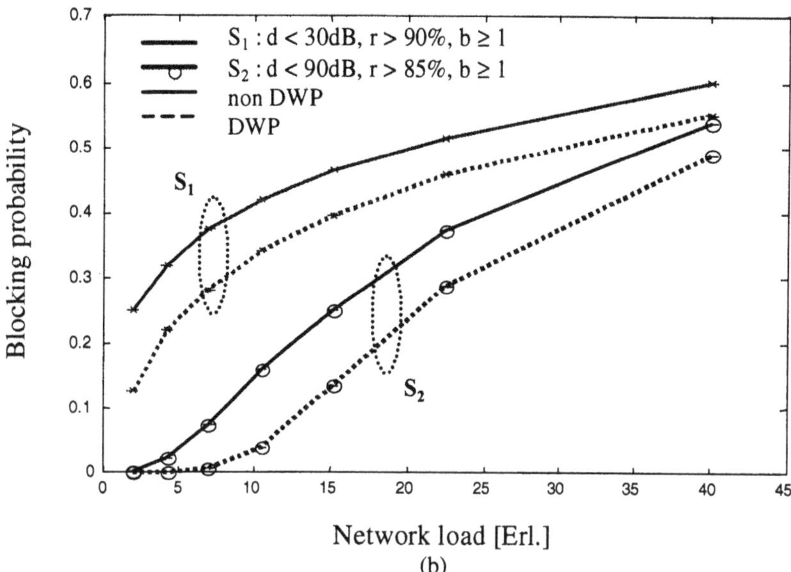

(b)

Fig. 2: Five nodes bi-directional ring: (a) NE-properties, (b) blocking probability per service.

We next study a five-node full-mesh network with the scope of proving the reduction of computational running time with the reduced set of the scanned, pre-routed paths, as previously discussed. To define a hop limit that automatically adapts to different calls and network loads, we use a limit relative to the number of hops of the shortest path found, according to a single constraint. For example, if the relative limit is defined as +1 hop, and if the shortest path according to a certain constraint is 2 hops, we consider only the paths up to 3 hops. The path length limitation as used here reduces the computation time, as shown in Section 4. Generally, the relative limit in the path length of the paths to consider can be set lower for a network with higher connectivity. Since we are using least load routing, we also use the accumulated load to limit the number of considered paths. An additional relative limit is set to +25% to the network load found by the 'shortest' path. (i.e. if the path with minimum accumulated load, the 'shortest' path , yields a total load in number of occupied wavelengths of 0.6, we pre-route paths only up to a maximum total load 0.75). Therefore the cost variable of the Bellman-Ford algorithm is set to link loads. Number of hops is an intrinsic variable of the algorithm and therefore always accessible [2].

For simplicity, we assumed only nodes and links as network elements, all with same properties per wavelength along the network (links[d=1dB, r=99%], nodes [d=3dB, r=99%]). Transmitters and receivers are modelled as ideal [d=0dB, r=100%]. The network uses 8 wavelengths for all NE's, which generally increases the traffic accommodation complexity, as all wavelengths are equal in terms of constraints. Due to this assumption and due to a higher connectivity, compared to 5-node ring network from the previous example, the network is simulated with higher loads. As in all examples, we assume reliability to be constant over time, and compare the proposed algorithm (DWP) with the centralised one. Here, two versions of DWP are used, one scans only a reduced set of paths, while the other considers all paths. The reduced set of paths is limited relative to the least loaded path (here: +1 hop, +25% of the accumulated link load). Since this exercise aims at measuring the simulation running time, only one single service is considered. The S_1-specific service constraints are as follows: [d<30dB, r>90%, b≥1].

Also here, the distributed, multi-path strategies yield superior results up to the limit where the load is too high and no idle (longer) paths can be found. Already below this limit the reduction in the number of scanned paths (relative as defined), compared to DWP presented directly to the network, where all paths are scanned, does not contribute to the increased blocking probability. As the longer paths in number of hops are more likely to be occupied, there are only a few feasible paths which are not included in the reduced set. However, the reduction in computation time is significant. In fact, the computational effort decreases for higher network loads, as many probes are rejected in course of the DWP process. Due to this effect, and, more important, due to the reduction of scanned paths, with the reduced set of prerouted paths, a faster solution for the multi-constrained routing can be achieved (Fig. 3b).

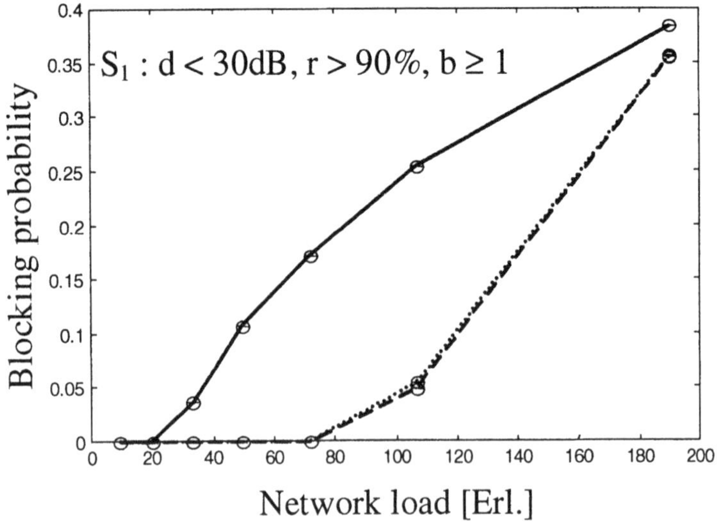

(a)

Single path (solid)	268 time units =100%
Reduced set (dashed)	672 time units =2.5x
All paths (small dashes)	2281 time units =8.5x

(b)

Fig. 3: Five-nodes full-mesh network: (a) blocking probability, (b) computation effort measured via total simulation time.

Finally, we study a mesh-torus network, a topology widely accepted for simulation and analytical results related to wavelength-routed networks. We study a mesh-torus network of size 5x5, with 8 wavelengths per link [5]. The properties related to the network elements are taken from the previous example (links[d=1dB, r=99%], nodes [d=3dB, r=99%], transmitters and receivers ideal). Here two different service classes: S_1 and S_2 are studied again, S_1 being the same as in the previous examples [d<30dB, r>90%, $b \geq 1$], i.e. requiring higher transmission quality and higher reliability than S_2 [d<60dB, r>75%, $b \geq 1$]. According to the previous example, the method DWP is applied to the reduced set of paths (here +1hop, +25% of accumulated load) and is again compared to the centralised method. Fig. 4 shows the comparison. For both services, the distributed, multi-path strategy again yields superior results.

6. CONCLUSION

In this paper, we proposed a distributed, QoS-based, methodology for the solution to the problem of wavelength routing tailored to specific clients of the upcoming optical transport networks. The distributed path-selection computation among the intermediate nodes between source and destination is likely to be the straightforward solution for appropriate handling of network state information with multiple resource and QoS attributes. In fact, the contradictory requirement for a scalable but accurate maintenance of the network state information is the main reason which makes the global (centralised) wavelength routing essentially impractical, particularly for multi-service networks with quality constraints.

While being strongly dependent on a particular service definition and networking scenario, the numerical examples show important directions for further study: (i) technological feasibility of service-specific characterisation of optical network elements, a necessary feature for co-ordination between higher-layer network elements (e.g. IP routers or data NEs) and optical NEs, and (ii) applicability of distributed protocols for QoS-delivery, resource management and restoration within the optical control layer.

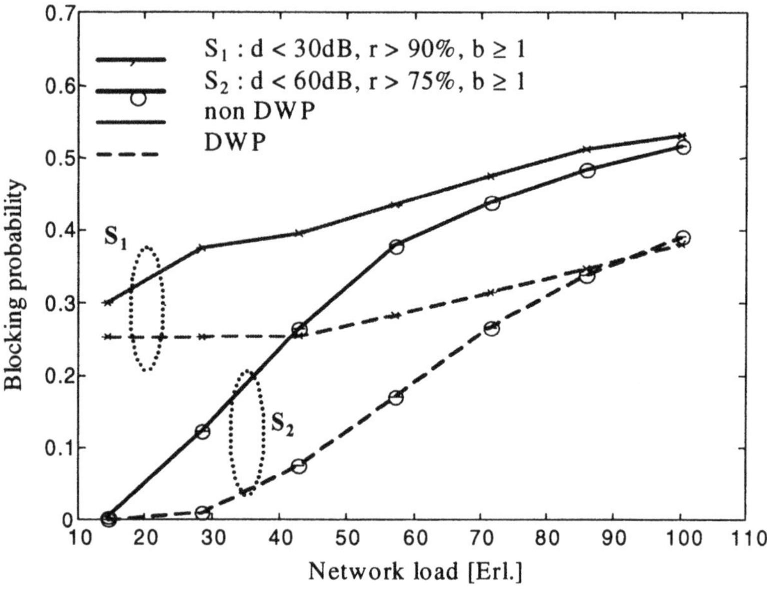

Fig. 4: 5x5 mesh-torus network: blocking statistic probability per service.

7. REFERENCES

[1] A. Jukan, H. R. van As: "Service-specific resource allocation in WDM networks with quality constraints", IEEE JSAC, 2000.

[2] D. Cavendish, M. Gerla: "Internet QoS-routing using the Bellman-Ford algorithm", Proc. HPN'98, 1998, pp. 627-647.

[3] R. Inkret, B. Mikac, F. Tillerot: "Availability Comparison of WDM Systems" Proc. ONDM'99, 1999, pp. 405-416.

[4] S. Chen, K. Nahrstedt: "Distributed QoS-routing in high speed networks", Proc. LCN '98, 1998.

[5] A. Jukan: "QoS-based wavelength routing in multi-service WDM networks", PhD thesis, Vienna University of Technology, August 2000.

AUTHOR INDEX

www.ingramcontent.com/pod-product-compliance
Ingram Content Group UK Ltd.
Pitfield, Milton Keynes, MK11 3LW, UK
UKHW020318200726
13857UKWH00001B/203

* 9 7 8 1 4 7 5 7 6 8 5 8 9 *